PESTS
OF THE
GARDEN
AND
SMALL FARM

A Grower's Guide to Using Less Pesticide

SECOND EDITION

Mary Louise Flint

Director, IPM Education and Publications
Statewide Integrated Pest Management Project
and
Extension Entomologist, Department of Entomology
University of California, Davis

Photographs by Jack Kelly Clark

UNIVERSITY OF CALIFORNIA
STATEWIDE INTEGRATED PEST MANAGEMENT PROJECT

UC DIVISION OF AGRICULTURE AND NATURAL RESOURCES
and the
UNIVERSITY OF CALIFORNIA PRESS
Berkeley   Los Angeles   London

WARNING ON THE USE OF CHEMICALS

Pesticides are poisonous. Always read and carefully follow all precautions and safety recommendations given on the container label. Store all chemicals in their original labeled containers in a locked cabinet or shed, away from food or feeds, and out of the reach of children, unauthorized persons, pets, and livestock.

Confine chemicals to the property being treated. Avoid drift onto neighboring properties, especially gardens containing fruits and/or vegetables ready to be picked.

Mix and apply only the amount of pesticide you will need to complete the application. Spray all the material according to label directions. Do not dispose of unused material by pouring down the drain or the toilet. Do not pour on ground: soil or groundwater water supplies may be contaminated. Follow label directions for disposing of containers. **Never burn pesticide containers.**

PHYTOTOXICITY: Certain chemicals may cause plant injury if used at the wrong stage of plant development or when temperatures are too high. Injury may also result from excessive amounts or the wrong formulation or from mixing incompatible materials. Inert ingredients, such as wetters, spreaders, emulsifiers, diluents, and solvents, can cause plant injury. Since formulations are often changed by manufacturers, it is possible that plant injury may occur, even though no injury was noted in previous seasons.

ISBN (UC DANR) 1-879906-40-6

ISBN (UC Press) 0-520-21810-8

Library of Congress Catalog Card No. 98-060419

© 1998 by the Regents of the University of California
Division of Agriculture and Natural Resources

Printed in Canada.

17.5m-rev 7/98-IPM/NS

Contents

Preface

Everybody is interested in reducing pesticide use these days. Citizens are concerned about pesticide residues on their food, in their homes, and in their environment. Workers are concerned about the long- and short-term effects of being exposed to pesticides in their workplace. Farmers are interested in reducing the risks, costs, and complications pesticides have brought to the agricultural industry.

Federal and state agencies and land grant universities have focused a substantial amount of time and funds over the last decade towards developing agricultural pest management methods that rely less on the use of pesticides. In California, the University of California's Integrated Pest Management Project has been a leader in researching, refining, and disseminating more ecologically sound pest management methods. However, most of this work has been directed at average-sized commercial growers who wish to reduce, but not necessarily eliminate, the use of conventional synthetic pesticide on their farms. There has been less practical information available from these sources for home gardeners, very small-scale farmers, and organic growers. This lack of information has been particularly unfortunate in the area of home and garden pest control, where consumers get most of their exposure to pesticides. For example, each year in California, home gardeners use about one pound of pesticide for every man, woman, and child in the state, yet they have few places to turn to find out about specific alternatives that could reduce this high pesticide consumption. I wrote this book to fill in some of these gaps.

For the last ten years I have been actively involved in developing integrated pest management (IPM) programs and publications for farmers through the University of California Statewide IPM Project. Many of these programs and publications have been widely adopted by mainstream agriculturalists. I have also spent a considerable amount of time talking to home gardeners, and organic and very small, diversified growers about their practices and problems. I have found that many of the IPM packages or programs that have been so successful for farmers growing 20 acres or more of a single crop have little practical value for those growing less than an acre. Yet the biological information and many of the techniques can be readily adapted for garden and small farm use. Additionally, although many of the IPM programs have been per-

ceived as unusable by organic growers because they involve some use of synthetic pesticides, most of them can be modified for use with organically acceptable methods. For all of these audiences, this book provides instructions on how to bring the IPM philosophy and IPM methods into their smaller scale and/or organic growing situations for use against specific problem pests.

A book of this depth and diversity could not have been written without the great resources of an institution such as the University of California. Many persons within the University spent a substantial amount of time sug-

gesting ideas, reviewing information, and providing me with information about what has worked in their experience. These people are named in the Acknowledgments. In compiling the book I have also relied heavily on many publications, especially those from the University of California Division of Agriculture and Natural Resources and the UC Statewide IPM Project, that have preceded it in the areas of IPM and home gardens. These publications are listed in the References.

—*Mary Louise Flint*
May 1990

Preface to the Second Edition

This Second Edition of *Pests of the Garden and Small Farm* has been completely reviewed and substantially revised since the publication of the first edition in 1990. To prepare for this revision, I solicited input from scientists, horticultural advisors, and Master Gardeners across California and beyond. New management practices have been added for dozens of pests with much more emphasis on biological control agents, use of environmentally benign pesticides such as insecticidal oils, and cultural practices such as solarization, disease suppressive composts, and mulches. Discussions and lists of

pests controlled through solarization, candidate insectary plants, commercially available biological control agents, and organically acceptable pesticides have been updated and augmented. The 30 Crop Tables in the back of the book have been substantially changed, with new pests or management practices added to almost every one. Over the last two years, there has been increasing concern about runoff of organophosphate pesticides into rivers and streams from gardens and farms. Additionally, the public has seen further regulation and warnings related to potential residues of

pesticides in food, especially that consumed by children. It is my hope that this revised version of *Pests of the Garden and Small Farm* will provide gardeners and growers with information that will allow them to reduce reliance on the most hazardous materials into the next century.

Visit the University of California Statewide IPM Project website at http://www.ipm.ucdavis.edu for a continually updated database of pest management information.

—*Mary Louise Flint*
November 1998

Acknowledgments

THIS BOOK WAS PRODUCED BY IPM Education and Publications, Mary Louise Flint, Director, under the auspices of the University of California Statewide IPM Project, Frank G. Zalom, Director, University of California, Davis.

COLLABORATORS

The following people provided substantial information and guidance in the development of the first drafts of chapters and substantial input during subsequent review.

Insects and Mites: Carlton Koehler, Entomology Extension, U.C. Berkeley
Diseases: Larry L. Strand, IPM Education and Publications, U.C. Davis
Nematodes: Philip A. Roberts, Nematology U.C. Riverside
Weeds: Robert Norris, Botany, U.C. Davis
Clyde Elmore, Botany Extension, U.C. Davis

PRODUCTION

Design and Production Coordination: Naomi Schiff
Photographs: Jack Kelly Clark
Drawings: David Kidd
Copy Editing: Janine Hannel, Margaret Brush
Manuscript Preparation: Christine Joshel

TECHNICAL COMMITTEE AND PRINCIPAL REVIEWERS

The following people provided ideas, information, and suggestions and reviewed many manuscript drafts. Some served on an advisory committee that was formed prior to writing the book; others were key resources and reviewers in specific subject matter areas.

William W. Barnett, Area IPM Advisor, U.C. Kearney Agricultural Center
Kate Burroughs, Harmony Farm Supply, Graton, CA

Leopoldo Caltagirone, Division of Biological Control, UC Berkeley

Clyde Elmore, Botany Extension, UC Davis

Donald Flaherty, Cooperative Extension, Tulare County

Doug Gubler, Plant Pathology Extension, UC Davis

Hunter Johnson, Botany Extension, UC Riverside

Carlton Koehler, Entomology Extension, UC Berkeley

Patrick Marer, IPM Education and Publications, UC Davis

James Marois, Plant Pathology, UC Davis

Arthur McCain, Plant Pathology Extension, UC Berkeley

Mike McKenry, Nematology, UC Kearney Agricultural Center

Robert Norris, Botany, UC Davis

William Olkowski, Bio-Integral Resource Center, Berkeley

Albert Paulus, Plant Pathology Extension, UC Riverside

Phil Phillips, Area IPM Advisor, Ventura County

Carolyn Pickel, Area IPM Advisor, Sutter-Yuba Counties

Dennis Pittenger, Botany Extension, UC Riverside

Paul A. Rude, Oakland, CA

Larry L. Strand, IPM Education and Publications, UC Davis

Charles Summers, Entomology, UC Kearney Agricultural Center

Paul Vossen, Cooperative Extension, Sonoma County

Special Thanks for the Second Edition

Special thanks to the following people who suggested revisions and improvements for the Second Edition.

Pam Bone, UC Cooperative Extension, Sacramento County

Mort Brigadier, Master Gardener, San Diego

Bob Bugg, UC SAREP, UC Davis

Ellen Cooper, Master Gardener, Santa Cruz

Steve Dreistadt, IPM Education and Publications, UC Davis

Clyde Elmore, Vegetable Crops, UC Davis

Sandy Ferguson, Master Gardener, Santa Cruz

Patricia Gouveia, IPM Education and Publications, UC Davis

Janine Hasey, UC Cooperative Extension, Sutter-Yuba Counties

Sharon Hite-Stoner, Master Gardener, Santa Cruz

C. Jay Hoyt, Master Gardener, San Diego

Vincent Lazaneo, UC Cooperative Extension, San Diego County

Julie Link, UC Cooperative Extension, Solano County

Peter Newberg, Master Gardener, Kern County

Phil Phillips, IPM Advisor, UC Cooperative Extension, Ventura County

Dennis Pittenger, UC Cooperative Extension, Riverside

Cheryl Weber Reynolds, IPM Education and Publications, UC Davis

Sharon Rossi, Master Gardner, Santa Cruz

Yvonne Savio, UC Cooperative Extension, Los Angeles County

Larry Strand, IPM Education and Publications, UC Davis

Nick Toscano, Extension Entomologist, UC Riverside

Sonya Varea-Hammond, UC Cooperative Extension, Santa Cruz County

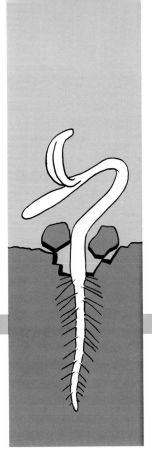

CHAPTER ONE

Introduction

THIS BOOK IS DESIGNED TO HELP gardeners and small farmers protect their vegetables and fruits from pests with minimum use of pesticides, especially the more toxic synthetic materials. While the focus of this book is on food-producing plants, most methods can also be adapted to manage pests on ornamentals. For more in-depth coverage of pests on woody ornamentals, consult University of California DANR Publication 3359, *Pests of Landscape Trees and Shrubs*. Many, but not all, of the methods will be suitable for use by organic farmers seeking to get their produce certified as "organic" according to California law. Management practices discussed here are primarily directed at gar-deners, very small commercial growers, or growers with small plantings of a number of crops. Larger growers should consult the series of *Integrated Pest Manage-ment* manuals on various major crops and other publications for more scale-appropriate recommen-dations. (See the References for suggested titles.) For instance, pest monitoring, sampling, and decision-making programs appro-priate mainly for larger acreages are not detailed here, although they may be mentioned.

Several different types of organisms cause crop losses. This book covers insects, mites, snails, slugs, nematodes, and other invertebrates; pathogens—includ-ing fungi, bacteria, and viruses—that cause disease in plants; and weeds. Rodents and birds, which are major pests in many gardens and farms, are not addressed here but are the subject of another University of California Agricul-ture and Natural Resources Publication, *Wildlife Pest Control around Gardens and Homes*, Publication 21385.

Many methods are available for managing pests of food crops. The key to their successful use is knowing when and how to apply them. Normally, the best approach is an integrated program that

1

relies on a spectrum of techniques both for prevention and control of problems. Alternatives to pesticides include biological control, resistant varieties, traps and barriers, crop rotations, solarization, tillage, and a host of other cultural practices—such as changes in planting times, irrigation methods, soil management, pruning techniques, and garden cleanup—that can reduce pest problems.

Pesticides are discussed in this book but other practical alternatives are emphasized when they are available. Because of the broad range of pests and crops covered in this book, it would be difficult to give a thorough treatment of all available pesticides. Organically acceptable pesticides such as soaps, oils, microbials, copper and sulfur sprays, and botanical insecticides may be mentioned by name in this book. Because of the large number of choices, other more "conventional" pesticides are not usually specified by name, although in some cases they are as safe or safer than organically acceptable materials—and many times they are more effective. Check with your farm advisor, county agricultural commissioner, or nurseryman for further information on available materials for specific crops. Always read labels before using any pesticide product. Most pesticides are packaged into many different products with varying toxicities, inert ingredients and recommended uses, and rates of application; only those crops listed on the label should be treated. Many products available for commercial use are not available to home gardeners; likewise, some products available to home gardeners are not packaged for use on commercial farms.

Every garden or farm will have a different complex of pests and problems. Pest problems vary with region, climate, crop variety, soil type, mixture of crops within the planting area, and farming or gardening practices. The great variety of growing conditions in California makes it impossible to give foolproof pest control recommendations for any garden or farm situation without careful inspection and knowledge of past practices and problems. Growers and gardeners must develop a solid understanding of the biology of crop plants as well as pests so they can tailor a pest management program that meets their specific needs. This book should give you the basic background for identifying pest problems and making pest management decisions. Additional books and articles are listed in the References.

The next chapter of this book is designed to give you a broad understanding of the concepts of integrated pest management (IPM). General principles of crop growth and developmental requirements are presented and general pest management practices are discussed. Subsequent chapters describe the biology and management of different groups of pests with individual treatment of almost 100 common pests; the pests, damage symptoms, and natural enemies are illustrated to help you identify many common problems. The last section of the book, the Crop Tables, lists common pest problems in 30 fruit, nut, and vegetable crops and directs the reader back to relevant chapters of the book. A Glossary, References and Index are provided at the back of the book.

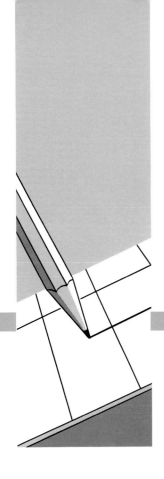

CHAPTER TWO

Designing a Pest Management Program

A WIDE ARRAY OF ORGANISMS occurs in any garden or farm situation. Only a small portion of these are pests. Many are beneficial —decomposing organic matter (such as dead plants), pollinating crops, killing pests, providing shelter or food for natural enemies of pests, or performing other useful functions. Still others are just incidental, having little or no impact on crop production. With many pest species, a few individuals or light damage can often be tolerated.

The first step in any pest management program, especially one that aims to minimize pesticide use, is to understand which pests can cause damage to your crop and under what conditions damage is likely to occur. After that you can begin to develop an integrated program that includes regular inspection of plants for pests or damage, use of crop management practices that prevent pest invasions, buildups, or damage, and control techniques that are applied only after pests begin to approach intolerable levels.

Crop Development in Relation to Pest Management

The real purpose of a pest management program is not to kill pests but to give the crop plant enough protection to allow you to harvest a food product in the quantity and of a quality that is acceptable to you. For instance, if you are harvesting the fruit, you can usually tolerate a little insect damage to the leaves; in the case of leafy vegetables, if the outer leaves are often removed at harvest, even these crops can tolerate some foliage damage with no loss to you. Also, most crop plants have certain stages of growth when they are particularly vulnerable to damage; at other times they are relatively resistant to damage and can tolerate higher densities of certain pests.

Other aspects of crop health may influence potential for pest damage. Vigorous, rapidly growing plants can tolerate competition from weeds and often partially overcome losses caused by some pests. Plants stressed by over or underwatering, lack of nutrients, too cold or too hot weather, or other environmental conditions are particularly prone to damage. Often these conditions produce symptoms that are confused with pest damage.

Growth requirements. All plant growth depends on the capture of solar energy through photosynthesis, which occurs in the plant's green, chlorophyll-containing parts. The sugars and starches produced in photosynthesis are either immediately used for growth, maintenance, or repair of plant parts damaged by pests or other stresses, or are stored for future use.

To carry out photosynthesis and grow, a plant needs water, light (as its source of energy), carbon dioxide, various nutrients, and sufficiently warm air temperatures. The supply of water can be partly controlled by irrigation and drainage and the amount of nutrients by application of fertilizers and other soil amendments. Temperature and the sun's energy vary according to weather, so they can be manipulated somewhat by planting in a suitable area at the proper time. Weed control also indirectly regulates sunlight, since it prevents weeds from shading crop plants. You can increase temperatures around your plants by using hot caps, row covers, mulches, or cold frames or by first planting seeds inside or in the greenhouse and transplanting them in the garden when conditions warm up. The amount of carbon dioxide in the outdoor air cannot be manipulated by the grower, but its levels can be increased in greenhouses. All the basic requirements for growth must be fully met for plants to have maximum resistance to diseases, compete with weeds, and tolerate damage by insects and other pests.

DEVELOPMENT OF VEGETABLE CROPS

Most vegetable crops are annual plants that normally go through three developmental stages: germination and seedling development, vegetative growth, and reproductive growth (development of flowers, fruits, and seeds). However, a few vegetables such as asparagus and artichokes are perennial plants that remain in the garden producing a crop year after year if properly cared for; these perennial vegetables follow a somewhat different cycle from that described below. (See the Crop Tables for references describing growth of these plants.)

Germination and seedling development. Vegetable crops are often most susceptible to pest damage and weed competition between the planting of seeds until plants have 6 to 8 true leaves. Early development of a tomato plant is sketched in Figure 2-1; most vegetables follow a similar developmental pattern. The root is the first part of the plant to emerge out of the seed. Several days to a week or more later, the seedling will poke through the soil surface. In most vegetables, the first leaves to appear are the seedling leaves, called cotyledons, which were contained inside the seed. However, there are variations in this process; for instance, the pea cotyledon remains beneath the soil surface. Over the next few days, much of the plant's energy is devoted to developing a strong root system; the first true leaves, which more closely resemble

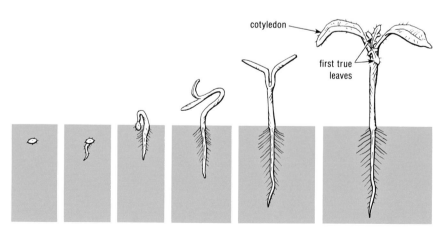

FIGURE 2-1. Development of a tomato seedling.

typical leaves of the plant, begin to develop, and after a time the cotyledons shrivel and drop off.

Young plants are susceptible to pest damage even before they push their first green sprouts through the soil surface. A variety of soil pathogens and invertebrates as well as vertebrates, such as birds, attack the newly planted or germinating seed or seedling root. It is often difficult to determine the actual cause of seed or seedling loss because the whole seed and root will be consumed. The seedling becomes vulnerable to an additional array of pests once it develops leaves, and with very few leaves and little energy stored in roots or other plant parts, even a small amount of damage can weaken the plant. Pest levels that do little harm to older plants can kill seedling plants. For this reason it is important to provide conditions that will get seedling plants growing as vigorously and rapidly as possible; it is especially important to choose varieties that grow well in your area and under your planting conditions. Plant when temperatures are warm enough, provide adequate moisture but avoid waterlogging, and monitor seedling growth closely, carrying out appropriate cultural practices as necessary. Where insects or snails and slugs frequently attack seedlings or where seedling growth is likely to be slow, the cost of various types of protective coverings can often be easily justified.

Vegetative growth. Once seedling plants have 4 to 6 true leaves, they are much more tolerant of pest damage and enter a

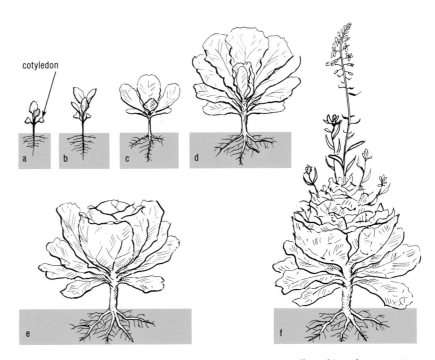

FIGURE 2-2. Stages of development of a cabbage plant: a) seedling, b) early vegetative growth, c) later vegetative growth, d) beginning of heading, e) mature head ready for harvest, f) bolting of flowers.

phase of vegetative growth when much of the plant's energy is focused on growth of new shoots and leaves, and the proportion of the plant that is above ground commonly increases compared to the size of the root system. Plants are usually thinned or transplanted at the start of this period. Transplanted plants need extra care during the first week or so following transplanting so they can reestablish a strong root system and acclimatize to the field or garden.

The intermediate stages of vegetative growth mark the period when you often have the greatest flexibility regarding pest management. In most crops, damage to foliage at this stage is not critical unless it is extensive. If weeds have been kept under good control during seedling development, many vegetable crops can now shade out or compete well with any weeds that germinate after this point.

Even crops such as lettuce and cabbage, with leaves that are harvested, have a period of four weeks or more after thinning when substantial leaf damage can be tolerated because outer leaves (which are being formed at this time) can easily be removed at harvest if judged too unsightly for consumption (Figure 2.2). However, once head lettuces or cabbages begin to form heads, pest management efforts must be stepped up—the curled leaves that make up the heads can rapidly become infested with aphids or other insects; these pests are protected from biological and chemical controls by their leafy coverings and are also difficult to remove with washing after harvest. Similarly, special care must be taken with any crop once the edible or marketable stage begins to develop; other crops that are harvested for their vegetative stage include spinach, chard, celery, and many herbs.

Although some localized leaf damage can be tolerated during the vegetative growth stage, it is important to remember that damage to roots or injury to the whole plant, such as overall stunting or injury to the nutrient or water conducting systems, will reduce yields and may kill the plant.

Reproductive growth. Growth patterns vary markedly from crop to crop; but generally, once the plant begins to produce flowers and set fruit, much of the plant's energies are redirected to fruit production and the portion directed to the production of leaves is reduced. Thus, if plants have been stressed early in their development and vegetative growth is less than normal, the plant may not later be able to support an adequate fruit production. Similarly, root pests or systemic disorders that affect water and nutrient flow within the plant will normally cause substantial loss.

In many vegetable crops, it is the flowers or fruit that are harvested or eaten. For most of these crops, a limited amount of leaf loss due to insect feeding or other pests later in the season will not reduce yield or quality, although heavy leaf loss can be serious if it exposes fruit to sunburn. On the other hand, pests that directly injure the flower buds, flowers, or fruit are normally of great concern. A single caterpillar in a pod or fruit can render it unmarketable or inedible, in the minds of some people. These kinds of crops require vigilant surveillance during fruit development. Many times pests are inside the fruit and hard to spot or control.

Underground tuber, bulb, or root growth. Some vegetables are grown for their fleshy underground roots, bulbs, or tubers. These plants, which include potatoes, carrots, beets, turnips, onions, and garlic, store substan-

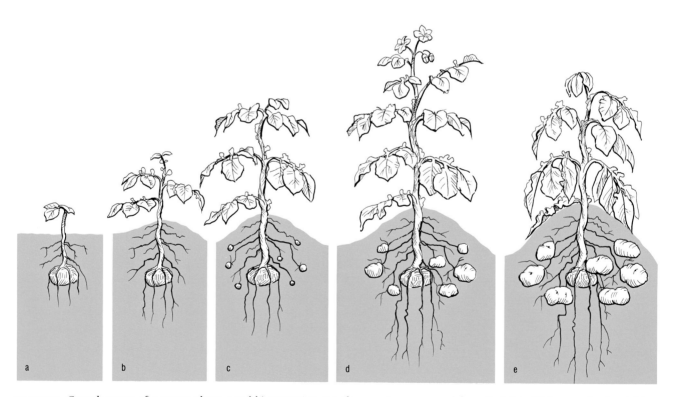

FIGURE 2-3. Growth stages of a potato plant: a and b) vegetative growth: sprouts emerge and form first 8 to 12 leaves; c) tuber initiation: tubers begin to form at tips of stolons; foliage continues to develop; d) tuber growth: most of plant production supports tuber enlargement; e) maturation: tuber periderm (skin) thickens, and dry matter content reaches a maximum; vines begin to senesce.

tial quantities of starch and sugar in these organs, providing an excellent food resource. Careful soil preparation, water management, and management of soil-dwelling pests and weeds are particularly important with these crops. Some foliage damage can usually be tolerated in these crops. Development of a potato plant is shown in Figure 2-3.

DEVELOPMENT OF FRUIT AND NUT TREES

Whereas the planting and harvesting of most vegetable crops occurs within a single season, tree, grapevine, and sometimes berry plants remain in the garden for many years. Therefore management must be directed toward long-term maintenance of overall plant health and vigor as well as toward producing a single season's harvest. The discussion below briefly describes development of fruit trees; see the Crop Tables for references describing growth of grapevines, strawberries, and caneberries.

The nonbearing years. During the first three to four years after planting a new fruit tree, pest management considerations are focused more on the establishment of the tree itself and less on protection of the fruit. Fruit production on most fruit trees is absent or light during this period. However, this is the time when major root growth occurs and the basic framework of the tree is being developed. These years are critical to the future well-being and productivity of the tree.

The first year that the tree is in the ground is the most important for root development. Stress caused by disease, nematodes, weed competition, or insufficient irrigation can hinder root development and hence top development, even in trees planted on resistant rootstocks. After the first few years, young trees become more tolerant to many of these stresses.

The second and third growing seasons are critical ones for developing a good framework for fruit production. Deciduous fruit and nut trees must be properly shaped and trained for structural strength while developing maximum fruiting area. (Citrus trees, on the other hand, do not usually require extensive pruning.) Pests, such as the shoot or twig boring insects, that cause distortion of early limb growth are most serious during these early years because their damage affects the ultimate shape of the tree. Young trees are often most susceptible to certain plant diseases such as bacterial canker (Figure 2-4). Suppression of growth due to weed competition can be very severe during the first three years of fruit tree development. Vertebrate pests are also most serious in newly planted orchards because trunks of young trees are more susceptible to their feeding and other limbs are easily accessible.

The seasonal cycle of bearing trees. The seasonal cycle of bearing trees begins in the spring with the enlargement of flower buds that will produce the season's crop (Figure 2-5). Even lemons, which may flower throughout the year, bloom most abundantly in the spring. Leaf and shoot growth is also generally most vigorous in the spring.

FIGURE 2-4. Young trees are often more susceptible to certain pests. This young almond tree was killed during the winter by the sour sap phase of bacterial canker. Because only the above ground portion of the tree is affected, roots put out suckers below the graft union in the spring.

Weather conditions or diseases that injure blossoms will have a great impact on the season's fruit production. Many deciduous fruit and nut trees must be pollinated by bees or other insects; pest control activities or other actions that disturb pollinators can also seriously reduce crop production. Most fruit tree crops have a natural thinning process that causes a portion of the tiny fruit to drop soon after their formation. Some commercial growers supplement this natural thinning by removing additional fruit so those that remain will be larger, better spaced, and located on parts of the tree that can withstand the weight. This practice is probably not worthwhile for home gardeners. Also, certain insect

dormant bud   budswelling   blossom

petal fall   fruit formation

FIGURE 2-5. Development of fruit and flower in almonds.

and pathogen pests can cause an undesirably high level of fruit loss at this stage.

During the bearing years of most tree fruit crops the most serious pests are those that injure the fruit or nut directly or those that interfere with the tree's water and nutrient conducting systems, reducing the long-term overall vigor of the tree. Pests, such as mites, aphids, or certain leafrollers, that limit their injury to leaf surfaces can often be tolerated at moderate levels because the tree normally produces more than enough leaf surface to provide adequate yields. Substantial yellowing or defoliation of the tree (especially early in the season) can reduce tree vigor and often have an impact on yields and fruit size in subsequent years or weaken a tree's resistance to disease.

Components of a Successful Pest Management Program

Effective, environmentally sound pest management requires considerable forethought and knowledge. If you wait to act until your crop is nearly dead or heavily infested with pests, often your only recourse will be to plow under or pull out the plants—or, at best, spray with a fast-acting pesticide, which may or may not correct the problem.

Plan for possible problems before you plant or, for deciduous tree or vine crops, during the dormant season. Review the crop tables in the back of this book for the crops you grow and familiarize yourself with the common pests and symptoms likely to occur. Keep records of problems in your garden or on your farm and talk to neighbors about pest histories on their property.

Whenever possible, take action to prevent problems before they occur by choosing pest resistant varieties, providing optimum conditions for plant growth, and taking preventive action against pests known to be special problems in your area. Once crops are growing, regularly check them for pests and damage symptoms. Some pests can be successfully hand-picked or rogued out in small plantings when populations are low. Learn to recognize when pest levels are approaching numbers that will require control. Select control methods that will be effective under your growing conditions and least likely to cause adverse effects on other aspects of the cropping system. Often more than one method can be employed to give the most reliable control. All these components—prevention, pest identification, regular surveying for pests, control action thresholds, and an integration of compatible, environmentally sound control methods—are part of a pest management strategy called *integrated pest management* (IPM).

Pest identification. Misidentification of pests accounts for many more pest control failures than most people realize. Some problems occur because the insect, pathogen, or other pest causing the problem is misidentified and the resulting control action is ineffective. Many pests look similar, especially to the untrained eye, and some can even be easily confused with beneficial or innocuous organisms. Frequently, people incorrectly associate damage symptoms with an insect or other organism that happens to be on the plant at the time the symptoms are observed when, in fact, that organism is not causing the problem: the pest that actually caused the damage may have left the site or may be hard to detect, such as a pathogen within roots or the plant's water conducting vessels. Damage symptoms can also be caused by factors other than pests such as over- or underwatering, over- or underfertilizing, toxins in the soil or water, air pollution, cold, heat, hail, wind, or genetic disorders.

The descriptions and photographs in this book will help you recognize many common pests found in gardens and small farms. However, because of the broad scope of this book, some of the pests you encounter will not be pictured or described here. Check other references at the back of this book for additional sources of identification information. Remember, some pest problems can only be diagnosed reliably by experienced professionals; do not hesitate to seek their help. Your University of California Cooperative Extension farm advisor, county agricultural commissioner, or garden or farm supplier may be able to assist in identification or direct you to professional diagnostic services.

Surveying for pests. One of the most important features of an integrated pest management program is going out to the field, orchard, or garden on a regular basis and systematically checking for pests and damage symptoms. For many major pests and crops, sophisticated sampling programs and monitoring techniques have been developed for use on farms with many acres of a single crop. Most of these are inappropriate for very small farms or gardens where only a few rows or a portion of an acre is grown in any one crop; in these situations, the grower or gardener can make a fairly intensive field check of most of the crop in a reasonable amount of time without resorting to complicated sampling schemes. Monitoring devices, such as pheromone traps, may provide less accurate information when only a few trees or plants are involved because they can draw in pests from outside the managed area. Pest-specific sampling programs that are useful in limited scale situations are described in the individual pest sections of this book. See the IPM manuals listed in the back of this book for methods appropriate for larger farms.

Check your vegetable plants or orchard trees at least once a week (twice a week in the peak of the growing season or more often when pest populations are building up) for signs of pests and pest damage. In orchards, pick a few leaves on each side of the tree to

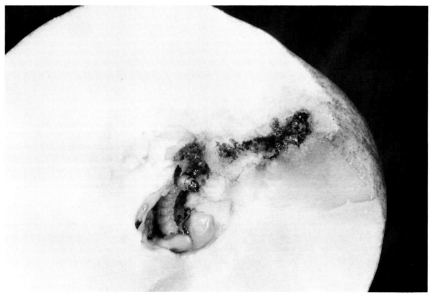

FIGURE 2-6. Often it is necessary to cut open fruit to confirm the presence of pests. This pear contains a codling moth larva.

FIGURE 2-7. When viewing with a hand lens, hold the lens close to your eye and move the object being viewed closer until it is in focus.

check for pests, pest damage, or disorders, check the trunk for injury, oozing sap or migrating insects. Observe fruit for scars or entry holes. If there are a lot of fruit on the ground, cut or crack them open to look for pests such as codling moth (Figure 2-6). In vegetable crops, walk through the furrow turning over leaves on every few plants to check for signs of insects or disease. If you observe wilted plants, pull them up and look for signs of nematodes, root-infesting insects, or pathogens. Keep written notes of your findings, recording date, time of day, stage of crop development, and any unusual weather conditions or association with cultural practices such as irrigation or pesticide application. If you do something to try to correct the problem or kill the pests, keep a record of what you did and the results. Following these procedures will allow you to develop a pest management program that works best for you.

Equipment for your monitoring program should include a notebook for keeping records, an 8- to 10-power hand lens (Figure 2-7) for viewing small arthropods and fruiting bodies of pathogens; and small vials and plastic bags for keeping samples to take back for identification. A beating sheet (Figure 2-8) can be useful for sampling various insects that are easily dislodged by shaking. Specialized trapping devices are useful for certain insects and are described in the individual pest sections. A minimum-maximum thermometer and other weather recording devices such as hygrothermographs and rain meters can help you learn to associate certain pest problems with weather conditions.

Tolerance levels or control action guidelines. Fundamental to integrated pest management is the concept that a certain number of pest individuals or a certain amount of pest damage can

FIGURE 2-8. A beating sheet or tray is useful for sampling insects that are easily dislodged by shaking.

be tolerated. The difficulty is in determining the point at which some action must be taken to control the pests to prevent unacceptable damage. For some pests in some crops, researchers have developed control action guidelines that indicate when management actions (usually a pesticide application) must be taken to avoid economic loss. Where these are applicable to small farm and backyard situations, they are mentioned in the individual pest sections and crop tables.

Your ability to tolerate certain levels of pest infestation on your crops depends on many variables. One is the ultimate destination of a crop. Food grown for home consumption can often be produced with higher levels of pests that cause blemishes on fruit or vegetables (sometimes called "cosmetic" damage) because it does not have to meet industry or grocers' standards; home gardeners are usually willing to cut out the blemishes. Another factor is your ability to control a pest effectively and rapidly. If you are limited to methods that take several days to provide control or only kill a fraction of the pests, you may have to allow for more lead time. As discussed earlier, crops are more vulnerable to pest damage at certain times in their development; for instance, seedling plants are particularly vulnerable to competition from weeds and damage from insects. These differences in susceptibility allow for variability in control action guidelines over the growing season. Other conditions or stresses affect a crop's ability to tolerate pest damage; for example, a plant weakened by water

stress, weed competition, root diseases, nematodes or other chronic pests, or physical damage must be more carefully protected.

Control action guidelines for insects and other invertebrate pests are often expressed as numerical thresholds indicating the population levels that will cause economic damage. Guidelines for weeds, diseases, nematodes, and vertebrates are usually based on the history of a field or region, the stage of crop development, weather, pest distribution, and other field observations. Control action guidelines are helpful only when used with accurate pest identification and careful field monitoring. Keep records of how you determined when to treat and results of the treatment. These records will help you develop guidelines that work best for your situation.

General Management Practices

Many different types of practices are useful for managing pest problems. Some—like the application of pesticides, hoeing for weeds, or release of natural enemies—are applied specifically for the control of a particular pest, usually after it has begun to cause a problem. However, the cheapest and often most reliable way to avoid many pest problems is to provide an environment that discourages pest activities or reduces the plant's susceptibility to damage. These types of methods often include adjustments in cultural practices such as planting time, soil or bed preparation,

water management, choice of crops or varieties, and management of areas adjacent to the garden, farm, or orchard. Some of these general types of management practices as well as more direct control methods are discussed briefly here. Specific pest control methods are presented in the individual pest discussions and the Crop Tables.

SOILS AND NUTRITION

Soil type and nutrient content affect crop growth. A vigorously growing crop can often tolerate more pest damage or better compete with weeds than a less healthy one. In some cases, the abundance of the pests themselves can be influenced by soil factors, particularly the presence of organic matter and the ability of the soil to drain or retain water. Table 2-1 gives some examples of pest problems sometimes associated with various soil conditions.

Although you cannot change your soil type in a backyard or very small farm situation, there are some practices that greatly improve growing conditions for plants. The most important is proper preplant soil preparation. Well-prepared seedbeds are easier to irrigate and hoe or cultivate. As a result, weed control is more effective and efficient and there is less chance of root diseases and other disorders associated with waterlogging of the soil. If furrow irrigation is to be used, grade the beds slightly so water will flow evenly to the tail end and not accumulate along the way. Break up compacted layers. Plant most vegetable crops on raised beds or hills between furrows. If you have

TABLE 2-1.

Some Pest Problems Commonly Associated with Soil and Nutrient Conditions.

SOIL AND NUTRIENT CONDITION	PEST
High organic content	symphylans, springtails, cutworms, wireworms, green fruit beetles
Poorly decomposed crop debris	tomato pinworms, squash bugs, damping-off, gray mold, white mold, many other fungi, plus all pests listed under high organic content
Poorly decomposed manure	root maggots, green fruit beetles, weeds
Dry, dusty conditions	mites
Sandy soils with poor water retention	root knot nematode
Acid soils	clubroot of cole crops
Poor drainage and overwatering	damping-off, big vein of lettuce, Phytophthora and other root rots, many other pathogenic fungi, weeds
Too high nitrogen	weeds, aphids, fungi favored by high humidity in the crop canopy

serious drainage problems, consider making beds up to 9 inches high.

Organic matter. Too much organic matter is a common problem, particularly in organically managed gardens and farms. High organic matter provides a good environment for pests such as springtails, symphylans, and wireworms that feed on decomposing plants as well as seedlings. Soil that contains only partially decomposed debris from previous crops can harbor many pests, especially fungi. Additionally, for nonorganic growers, soils with high levels of organic matter are difficult to fumigate successfully.

On the other hand, a moderate amount of organic matter in the soil is desirable. Organic matter added to clay soils will help improve water drainage and penetration and make them easier to work. In sandy soils, organic matter improves the soil's ability to hold water and nutrients. In all soils organic matter improves structure and aeration. It also provides a substrate for beneficial organisms such as earthworms and microbes in the soil. As these organisms break down the organic matter, they help maintain soil tilth and release nutrients from the organic matter into the soil. Some of these microorganisms are also important in the biological control of soil dwelling pests such as plant pathogens and nematodes.

Soil additives with the highest percentages of organic matter include commercial redwood composts, peat, and "planting mixes." Steer and chicken manure are higher in salts and lower in organic content. Although homemade and commercial mushroom composts can be inconsistent in content, they are also valuable sources of organic matter.

When adding organic matter to your soil, incorporate it a bit at a time. Large amounts of only partially decomposed organic matter will stimulate microbial activity, reducing the availability of nitrogen to crop plants. Except for manures, most additions of organic matter should be supplemented with additional nitrogen fertilizer to compensate for the increased microbial activity.

Apply animal manures several weeks or a month before planting. Mix them in the top few inches of the soil and irrigate to allow the manure to begin decomposition. Manures can cause severe injury or even kill seedlings if applied too heavily or too closely to planting time. Raw or poorly decomposed manure is also associated with root maggots and green fruit beetles, as well as many weed pests. Be sure manure is adequately aged or composted before applying it.

Fertilizers. Most California soils require regular addition of nitrogen and sometimes phosphorus (especially in cool seasons) for best growth of crops. Other minerals, including potassium, are not normally needed. Necessary nutrients can be obtained from many sources, including organic and inorganic materials. Table 2-2 shows typical percentages for various products. Remember that nutrient composi-

TABLE 2-2.

Nitrogen, Phosphorus, and Potassium Content of Some Common Fertilizer Materials.

(Many other fertilizers are available.)

	AVERAGE PERCENTAGE		
	Nitrogen (N)	Phosphorus (P$_2$O$_5$)	Potassium (K$_2$O)
ORGANIC MATERIALS[a]			
Poultry manure/droppings	1.6–4.0	1.3–3.2	0.9–1.9
Steer manure	2.0	0.5	1.9
Dairy manure	0.7	0.3	0.7
Goat manure	2.8	1.8	2.9
Rabbit manure	2.0	1.3	1.2
Horse manure	0.7	0.3	0.5
Hog manure	1.0	0.8	0.9
Dried blood	13.0	1.5	—
Fish meal	10.4	5.9	—
Fish emulsion	5.0	2.0	2.0
Bone meal	2.8	15.0	—
Alfalfa hay	2.5	0.5	2.1
Grain straw	0.6	0.2	1.1
Cotton gin trash	0.7	0.2	1.2
Compost[b]	0.1–1.0	0.1–1.0	0.1–1.0
INORGANIC MATERIALS			
Ammonium nitrate	33.5	0	0
Ammonium sulfate	21.0	0	0
Calcium nitrate	15.5	0	0
Urea	46.0	0	0
Single superphosphate	0	20.0	0
Triple superphosphate	0	45.0	0
Mono ammonium phosphate	11.0	48.0	0
Ammonium phosphate sulfate	16.0	20.0	0
Di-ammonium phosphate	21.0	53.0	0
Potassium chloride	0	0	62.0
Potassium sulfate	0	0	53.0
Potassium nitrate	13.0	0	44.0
MIXED FERTILIZERS[c]			
20–10–5	20.0	10.0	5.0
6–12–6	6.0	12.0	6.0
12–12–12	12.0	12.0	12.0
4–10–10	4.0	10.0	10.0

Source: *Western Fertilizer Handbook*, 6th ed., 1980, and Hunter Johnson, 1985. *Fertilizing Vegetable Gardens*. UCR, Mimeo.

[a]The percentage of plant nutrients in organic materials varies widely depending upon how the material is processed, how the animals were fed, and the moisture content.

[b]Compost is an excellent source of organic matter, but has little value as a fertilizer material.

[c]Mixed fertilizers are available in many ratios of nitrogen-phosphorus-potassium. The plant nutrients are generally derived from inorganic materials.

TABLE 2-3.

Suggested Application Rates for Various Fertilizers Applied Before Planting.

(Many other materials are available.)

| | AMOUNTS PER 1000 SQUARE FEET | | |
| | | Approximate Volume (dry processed) | |
	Pounds	Cubic Feet	Gallons
ORGANIC MATERIALS			
Poultry, goat, sheep, or rabbit manure	100–125	2½–3	18–20
Steer, dairy, hog, or horse manure	300–400	7½–10	60–75
Blood meal	20–25	½	3
Bone meal	20–30	⅓–½	2½–3
Fish meal	20–30		
INORGANIC MATERIALS		Pints	
Ammonium nitrate	4–5	5–6	
Ammonium sulfate	7–8	7–8	
Urea	3–4	4–5	
Single superphosphate	10–12	8–10	
Potassium sulfate	4–5	2½–3½	
Ammonium phosphate (16–20–0)	9–10	9–10	
12–12–12	12–15	11–14	
16–16–16	9–10	8–9	
19–9–0	7–8	7–8	

Source: Hunter Johnson, 1985. *Fertilizing Vegetable Gardens.* UC Riverside, Mimeo

tion of organic materials varies greatly. Compost, although a good source of organic matter, is not a good source of nutrients. For best results, apply all the phosphorus but only part of the nitrogen before planting. Table 2-3 shows suggested preplant rates for vegetable gardens. Nitrogen will be most effectively used by the plant if part is added later— when direct seeded plants have 4 to 6 true leaves or 4 to 5 weeks after transplants are set out. Inorganic fertilizers can be side-dressed as shown in Figure 2-9; manures are more difficult to use as a side-dressing and must be tilled into the soil.

Commercial growers should have their soil tested annually by a reputable commercial lab. Serious large-scale gardeners, especially those who suspect soil problems, should also consider annual soil testing. Soil tests can indicate nutrient deficiencies, excess salts, and pH problems. Regular testing by the same lab following the same sampling procedures can indicate if the soil amendments you are adding to your soil are doing their job. To be meaningful, soil tests must be taken in a consistent manner. (See UC ANR Publication 1879, *Soil and Plant Tissue Testing in California*, listed in the References for more information.)

Contrary to popular belief, vigorously growing crop plants are not less attractive to most pests; in fact, in some cases they are more attractive. For instance, lush vegetative growth is very attractive to aphids, and aphid problems are often increased by overfertilization with nitrogen. Lush leafy growth also increases humidity within the plant canopy, reduces ventilation, and can increase the incidence of pathogens such as gray mold or powdery mildew favored by high air moisture. Therefore, it is essential to keep nitrogen at recommended levels and be sure adequate levels of other nutrients are also supplied.

WATER MANAGEMENT

Mismanagement of water is a major contributing factor to many pest problems. Too little water can result in small plants, poor root systems, and slower growth, and exacerbates the injurious effects of pests such as root nematodes that injure roots. Excessive water can contribute to development of many diseases. Uneven distribution of water can encourage weeds and disease problems and prevent uniform maturing of the crop. Further-

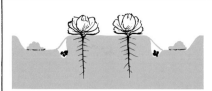

Make a small groove an inch or two deep on both shoulders of the bed, 4 to 6 inches from the plant row and band in the fertilizer.

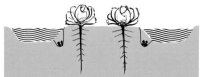

Replace the soil, and irrigate.

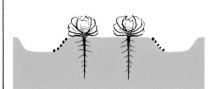

Fertilizers can also be scattered along the bed shoulders. This is less efficient than the banding method

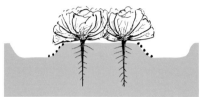

but will be more practical when the plants are so large that the bed shoulders are inaccessible for banding.

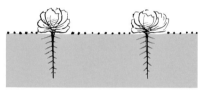

Where sprinklers are used, fertilizer may be scattered on the soil surface between rows before irrigating.

Where drip or trickle tubes are used, apply fertilizer on the soil surface near the drip tube.

FIGURE 2-9. Methods of side-dressing fertilizers.

more, damage caused by over- or underwatering is frequently misdiagnosed as pest damage.

To prevent problems, design your irrigation system and prepare soil to promote even water distribution and good drainage. In vegetable gardens and fields, break up compacted areas, reduce clod size, and firm the beds to increase water penetration. Irrigate to meet the needs of the particular growth stages and crops present. Commercial vegetable growers often sprinkle irrigate from planting until thinning and then switch to furrow irrigation as plants get larger. On older plants, sprinkling can splash soil onto leaves, promote disease problems if leaves remain wet after nightfall, and favor weed germination on beds. Drip irrigation is useful in some situations; it provides excellent control of water and is the most practical method for use under row covers and with certain types of mulches.

Orchards and trees can be irrigated with surface (flood or furrow), sprinkler, or drip methods. Poor water management in orchards can be just as serious as in vegetable crops although the problems are not so immediately obvious. Water stress early in the season slows shoot growth and can lower fruit or nut yields. Severe water stress can cause leaves to drop and nuts or fruit to remain on trees after harvest. Too much water, on the other hand, damages roots by depriving them of oxygen and creates conditions that favor infection by soilborne diseases, particularly Phytophthora root and crown rot, one of the most destructive diseases of tree fruit and nuts. Each type of irrigation system has its advantages and disadvantages. Flood or furrow irrigation may discourage ground squirrels and pocket gophers from digging burrows in an orchard. Sprinkler irrigation can increase disease problems if water is applied so leaves or fruit remain wet for extended periods; adjust sprinkler heads so this does not happen.

When to irrigate. More important than the irrigation method is determining when to apply water. In larger commercial farms, growers should determine when to apply water by developing a water budget, which estimates how much water the plants are using, and then irrigate to meet the plants' water needs. Irrigating on a calendar schedule can cause serious damage from over- or underwatering if weather, soil conditions, and stage of growth and crop species are not considered. Use of water budgets is described in the various IPM manuals for specific crops as well as in several UC ANR Publications listed in the References of this book.

In a garden or very small farm, it is probably most practical to examine your soil visually and with your hands to determine the

TABLE 2-4.

Judging the Depletion of Soil Water by Feel and Appearance.

	COARSE-TEXTURED SOILS	INCHES OF WATER NEEDED	MEDIUM-TEXTURED SOILS	INCHES OF WATER NEEDED	FINE-TEXTURED SOILS	INCHES OF WATER NEEDED
FIELD CAPACITY	Soil looks and feels moist, forms a cast or ball, and stains hand	0.0	Soil dark, feels smooth, and will form a ball; when squeezed, it ribbons out between fingers and leaves wet outline on hand	0.0	Soil dark, may feel sticky, stains hand; ribbons easily when squeezed and forms a good ball	0.0
TOO EARLY	Soil dark, stains hand slightly; forms a weak ball when squeezed	0.3	Soil dark, feels slick, stains hand; works easily and forms ball or cast	0.5	Soil dark, feels slick, stains hand; ribbons easily and forms a good ball	0.7
TIME TO IRRIGATE	Soil forms a fragile cast when squeezed	0.6	Soil crumbly but may form a weak cast when squeezed	1.0	Soil crumbly but pliable; forms cast or ball, will ribbon; stains hand slightly	1.4
TOO LATE	Soil dry, loose, crumbly	1.0	Soil crumbly, powdery; barely keeps shape when squeezed	1.5	Soil hard, firm, cracked; too stiff to work or ribbon	2.0

NOTE: Check the soil where plants are getting most of their moisture—e.g., surface few inches for seedlings, top foot for shallow rooted vegetables like lettuce, 2 feet for tomatoes at the fruiting stage. Once past the seedling stage, the top 2 inches of soil should be dry to the touch before irrigating again. The feel and appearance of soils change as they dry out. Starting at field capacity (top), the numbers in each column are the approximate amounts of water needed to restore one foot of soil depth to field capacity.

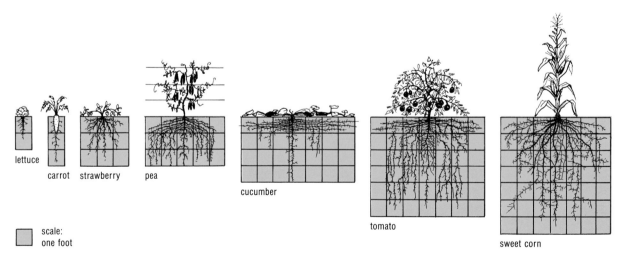

FIGURE 2-10. Comparative rooting depths of common garden vegetables.

need for additional water: Table 2-4 provides some guidelines. Remember that adequate water needs to be available to the full rooting depth of your plants and beyond (Figure 2-10). If you only allow water to seep through to the top few inches of your soil, root growth and overall plant growth will be stunted. Keep adequate moisture in the deeper soil layers even when plants are small. Deep moisture is harder to replace later in the growing season without overirrigating. Add enough water to fill the deepest part of the rooting zone, not just the surface; this will require that you irrigate for longer periods but will allow you to irrigate less often. When using sprinklers, keep the application rate low enough that water does not puddle or run off; if you cannot adjust the sprinklers low enough, turn them off for an hour or two and then reapply.

SANITATION

Where do pests survive when your crop plants are dormant or out of the garden? Although a few pests can migrate great distances, most pest populations come from sources within the garden itself or adjacent areas—surviving on alternate hosts, crop debris, or as dormant forms such as seeds, vegetative propagules, spores, eggs or pupae in the soil. Some pests are even inadvertently brought in on contaminated seed, transplants, soil, or equipment. Practices aimed at removing these sources of new pest populations are often called sanitation practices.

Most sanitation practices are just good common sense. Make

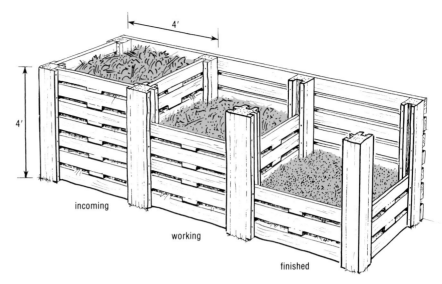

FIGURE 2-11. Compost bins. Although many people make compost in free-standing piles, a series of well constructed bins can make the process more efficient and tidier. Piles and bins should be a minimum of 3 feet in width and height. Open slats in this design allow good air circulation.

sure crop seed and organic soil amendments are free of weed seeds and pathogens; use certified seed or stock if available. Check transplants or other greenhouse stock for aphids, diseases, nematodes, and other pests. Make sure your supplier has taken precautions to prevent pest infestations. Clean equipment before moving it from infested areas. Do not transport soil you know to be infested with nematodes, propagules of perennial weeds such as bermudagrass stems, or nutsedge nutlets to other parts of the garden or farm. If you use surface irrigation water from canals, use screens to exclude large weed seeds and other plant parts.

Always be sure plant debris, especially residue from previous crops, is completely decayed or removed before planting a new crop. In orchard crops, be sure old fruit or nut culls remaining on trees or on the orchard floor are removed from the orchard or decomposed in the orchard soil

well before new growth begins in the spring. Also destroy prunings that might be infested with disease or borers. Sweep up and destroy or till June drop walnuts and apples as soon as they fall; they may contain codling moth.

Composting is an excellent way to destroy most crop and weed residues around the garden and small farm and also control pests that may be harbored in the residues (Figure 2-11). The compost may then be used as a mulch or incorporated into soil to add organic matter. However, composting must be done correctly to assure the destruction of many serious plant pathogens and weed seeds. The rapid composting method detailed in Table 2-5 will control all insect pests, nematodes, most pathogens with the exception of heat tolerant viruses such as tobacco mosaic virus, and most weeds and weed seeds with the exception of oxalis bulbs, burclover seeds, amaranth seeds and cheeseweed seeds.

TABLE 2-5.

Requirements for Assuring Adequate Decomposition in the Compost Pile.

1. Woody material should be chopped to ½ to 1½ inch bits. Soft succulent tissues need not be chopped.

2. Carbon-to-nitrogen ratio in a compost pile should be 30 to 1. A mixture of equal volumes of green plant material and dry plant material (dry leaves, woody prunings, straw, shredded paper) will normally achieve this. If the C/N ratio is less than 30 to 1, the compost will give off an ammonia odor. Add sawdust to counteract this.

3. Do not put soil, ashes from a stove or fireplace, or manure from meat-eating animals in the compost.

4. A compost pile should be at least 36"x36"x36" to assure adequate heating. Heat retention is better in bins than in open piles. Pile temperature should remain at 160° F to kill pest organisms. Get a thermometer to check the temperature. Do not let temperatures get more than a few degrees higher; decomposing microorganisms will be killed. Frequent turning will prevent overheating.

5. Turn the pile every 1 to 2 days to keep it from getting too hot and to aerate it. Piles turned every day should take about 2 weeks to compost; those turned every 2 days will take about 3 weeks.

6. Once the pile is started, do not add anything (unless your compost is not functioning properly and you add sawdust, water or nitrogen as described in numbers 2 or 7).

7. If temperatures do not get up to 160° F within 24 to 48 hours, the pile is too wet, too dry, or there is not enough nitrogen (green material). If too wet, spread out the pile to dry. If too dry, add water. If not enough nitrogen, add a material high in nitrogen (e.g., ammonium sulfate, grass clippings, fresh chicken manure).

8. A healthy compost will have a pleasant odor, give off heat as vapor when turned, have a white fungal growth on the decomposing material, will get smaller each day, and will gradually change color to dark brown. When no further heat is produced, the compost is ready to use.

Source: Raabe, R. 1981. *The Rapid Composting Method.* UC ANR Leaflet 21251.

TABLE 2-6.

Examples of Pests Often Associated With Weedy Adjacent Areas.

Armyworms	Slugs and snails
Crickets	Stink bugs
Cutworms	Thrips
Darkling beetles	Certain viruses vectored by aphids and leafhoppers
Earwigs	
Fleabeetles	Weeds—especially those spread by wind such as groundsel, sowthistle, dandelion
Grasshoppers	
Lygus bugs	

Properly prepared compost can also help suppress soil borne diseases. For the process to work, it is essential that the pile be of the proper size, be turned every one to three days, maintain a temperature of 160°F, and have nothing added to it once the composting process has begun. (See UC ANR Publications 21251 and 21514 listed in References for more details.)

A final sanitation method that merits mention is to keep weeds under control at all times and be sure to remove weedy areas around your garden or field well before the new crop emerges. Not only do weeds produce seeds that may find their way into your garden, they also are likely sources of a number of immigrating pests; a few examples are listed in Table 2-6. If you wait to remove or mow these weeds until your crop emerges, the pests will move into your crop. Waiting for the weeds to dry out naturally will also encourage pests. Certain weeds are often touted for their benefits around the edges of the garden, but they must be carefully managed to prevent potentially serious problems.

RESISTANT VARIETIES

Use of crop varieties (often called cultivars for cultivated varieties) that resist pest attack or damage is one of the most economically and ecologically sound forms of pest control. Some resistant varieties inhibit pest attack through toxic or repellent compounds or through physical factors; other resistant varieties are attacked by pests but suffer little or no damage. (This phenomenon is sometimes called tolerance.) Varieties of a number of garden crops are available that are resistant to various pathogens and root knot nematode. Many rootstocks are available for fruit and nut trees that resist attack by common pathogen and nematode pests. Few vegetable or fruit cultivars that resist insects are available, although some varieties are known to be less susceptible to damage by certain species.

Use pest-resistant varieties whenever possible, but take into account other agronomic factors that may be important. Varieties will differ in required planting times, resistance to frost and heat, time from planting to harvest, and other cultural requirements as well as flavor, color, ability to be stored, and consumer acceptance. Remember that resistance is not synonymous with immunity. Some resistant varieties or rootstocks are still somewhat susceptible to high levels of pest inoculum or when stressed by other factors. To get reliable information about resistant varieties suitable for growing in your area, contact your local nurseryman, seed supplier, or Cooperative Extension office.

CROP ROTATION

The practice of changing the type of crop growing in a field or portion of a garden is known as crop rotation. Crop rotation is an important agronomic practice on farms but has a more limited use in gardens because of space limits and the impracticality of growing many of the most effective rotation crops. However, it is always a good idea to avoid planting the same crop two years in a row in the same part of your garden. A more formalized rotation scheme may be worth the effort even in backyard gardens with serious

root knot nematode problems and certain root diseases. Common rotation crops in medium- to large-scale farms include legumes (especially alfalfa and dry beans), field corn, small grains, and certain resistant varieties (e.g., root knot nematode-resistant varieties of tomato and bean); each of these crops adds nutrients, improves soil texture and drainage, competes with weeds, and/or is an unsuitable host for certain pathogen and nematode pests of vegetable crops.

Crop rotation is a feasible management practice only for a limited number of pests. Pests that can be successfully controlled with crop rotation must meet specific biological requirements: (1) the source of the pest must be within the field (i.e., the soil); (2) the pest cannot be so mobile that it is likely to move in from adjacent fields or other areas; (3) the host range of the pest must not be so wide that practical alternate crops cannot be found; (4) the pest must not be able to survive in the soil for more than a year or two in the absence of living host plants. Good candidates for management by rotation include soil and root-dwelling nematodes and soil-borne pathogens that do not produce airborne spores. Rotation can be an important component in weed management if crops, such as onions, that do not compete well with weeds are followed by very weed competitive crops such as corn, potatoes, or cole crops. No major insect pests in California can be controlled by rotation. Further discussion of rotation plans for specific pests can be found in the weed, nematode, and disease chapters.

SOIL SOLARIZATION

Soil solarization provides a method for reducing or eliminating many soil-inhabiting pests, and it is a simple technique that can be done by hand in the home garden or by commercial machinery on larger farms. Solarization works by heating up the soil in the presence of moisture to levels lethal to many fungi, nematodes, weeds and weed seeds, and other pest organisms.

To solarize your soil, you must leave a clear plastic tarp on the soil surface for 4 to 6 weeks during the hottest part of the year. Use transparent plastic; black or colored tarps will not allow the soil to heat to the highest possible temperatures. Polyethylene plastic 1 mil thick is the most efficient and economical; however, it is easier to rip or puncture and is less able to withstand high winds than thicker plastic. Plastic 1.5 to 2 mils thick could be used in windy areas; avoid thicker (4 to 6 mils) transparent plastic because it is more reflective and the soil beneath does not heat up as much. Plastic containing ultraviolet (UV) inhibitors is slower to deteriorate and can be left in place as a mulch or reused; however, this type of plastic is difficult to purchase in small quantities for the home garden.

Plastic tarps can be laid by hand for small farms or gardens or by commercial machinery for larger acreages (Figure 2-12). Before laying the plastic down, level the soil, remove all weeds, plants, debris and large clods of soil, and prepare the soil for planting. Maximum soil heating occurs when the tarp is close to the soil; air pockets caused by clods, debris, or furrows decrease

FIGURE 2-12. Application of plastic for solarization. Smooth the soil surface to make it as flat as possible. Roll out tarps, smoothing out air pockets as you go along. Bury edges of tarp with soil.

the effectiveness of the treatment. Disc, rototill, or spade the soil, and rake it smooth. Not only do these procedures provide an even surface, they will also help heat penetration.

Roll tarps out over the soil surface, smoothing out air pockets as you go along. Anchor the tarp by burying the edges with soil. Where large areas are being solarized, adjacent plastic strips may be held in place by soil or glued or heat-fused together. In some situations, strips may be placed so only the tops of planting beds are covered and furrows are left untreated; this may be the easiest method for small areas. Solarization with strips works best if beds are 48 to 60 inches wide and oriented north and south rather than east and west.

For successful solarization, the soil under the tarp must be wet. The moisture causes organisms to be more sensitive to heat and also allows faster and deeper

penetration of heat into the soil. For small areas in gardens, wet the soil immediately before laying the tarp. For larger applications, it is more effective to wet the soil after putting the plastic in place; insert one or more hose or pipe outlets under one end of the tarp and turn on the water to soak the soil. Drip irrigation under the tarp on beds will also work. When solarizing only the bed tops, it is best to apply water under the tarp with a drip system. Otherwise, the uncovered furrows will become quite weedy and special care will be required to keep weed seeds and plant parts out of the treated areas during weed removal.

The degree of pest control achieved with soil solarization is related to the intensity, depth, and duration of the elevated soil temperatures. Although some pests may be killed within a few days, 4 to 6 weeks of treatment in full sun during the summer is

required to assure control of most pests. Highest soil temperatures will be obtained when days are long, air temperatures are high, the sky is clear, and there is no wind. The heat peak in most areas of California is around July 15; therefore the best time for solarization is in June and July. In some areas satisfactory results can also be achieved in May, August, or September. It may be difficult to obtain complete control with solarization in the cooler coastal areas of California where foggy conditions occur frequently.

Do not leave an untreated (i.e., not ultraviolet light protected) transparent polyethylene tarp on the soil suface for more than 6 to 7 weeks during summer heat. It will become brittle, tear, and be difficult to remove. If you wish to leave the tarp on longer to use it as a mulch, use plastic that contains a UV inhibitor.

Diseases and weeds known to be effectively controlled with solarization are listed in Tables 2-7 and 2-8. Many more species are probably also controlled. However, note that certain problem weeds such as bermudagrass, field bindweed, johnsongrass, and yellow nutsedge are suppressed but not completely controlled by the procedure. Nematode populations are also reduced, but control is not complete because the heat may not penetrate deep enough to destroy those below the top foot of soil. The treatment should give satisfactory control of nematodes for shallow-rooted crops and home gardens.

TABLE 2-7.

Pests Controlled by Soil Solarization.

FUNGI

SCIENTIFIC NAME	DISEASE CAUSED (CROP)
Didymella lycopersici	Didymella stem rot (tomato)
Fusarium oxysporum	Fusarium wilt (cucumber)
Fusarium oxysporum f. sp. *fragariae*	Fusarium wilt (strawberry)
Fusarium oxysporum f. sp. *lycopersici*	Fusarium wilt (tomato)
Fusarium oxysporum f. sp. *vasinfectum*	Fusarium wilt (cotton)
Phoma terrestris	pink root (onion)
Phytophthora cinnamomi	Phytophthora root rot
Plasmodiophora brassicae	club root (cruciferae)
Pyrenochaeta lycopersici	corky root (tomato)
Pythium ultimum, *Pythium* spp.	seed rot or seedling disease (many crops)
Pythium myrothecium	pod rot (peanut)
Rhizoctonia solani	seed rot or seedling disease (many crops)
Sclerotinia minor	drop (lettuce)
Sclerotium cepivorum	white rot (garlic and onions)
Sclerotium rolfsii	southern blight (many crops)
Thielaviopsis basicola	black root rot (many crops)
Verticillium dahliae	Verticillum wilt (many crops)

BACTERIA

SCIENTIFIC NAME	DISEASE CAUSED (CROP)
Agrobacterium tumefaciens	crown gall (many crops)
Clavibacter michiganensis	canker (tomato)
Streptomyces scabies	scab (potato)

NEMATODES

SCIENTIFIC NAME	COMMON NAME
Criconemella xenoplax	ring nematode
Ditylenchus dipsaci	stem and bulb nematode
Globodera rostochiensis	potato cyst nematode
Helicotylenchus digonicus	spiral nematode
Heterodera schachtii	sugarbeet cyst nematode
Meliodogyne hapla	northern root knot nematode
Meliodogyne javanica	Javanese root knot nematode
Paratylenchus hamatus	pin nematode
Pratylenchus penetrans	lesion nematode
Pratylenchus thornei	lesion nematode
Pratylenchus vulnus	lesion nematode
Tylenchulus semipenetrans	citrus nematode
Xiphinema spp.	dagger nematode

WEEDS

SCIENTIFIC NAME	COMMON NAME
Abutilon theophrasti	velvetleaf
Amaranthus albus	tumble pigweed
Amaranthus retroflexus	redroot pigweed
Amsinckia douglasiana	fiddleneck
Avena fatua	wild oat
Brassica nigra	black mustard
Capsella bursa-pastoris	shepherdspurse
Chenopodium album	lambsquarters
Claytonia perfoliata	minerslettuce
Convolvulus arvensis (seed)	field bindweed
Conyza canadensis	horseweed
Cynodon dactylon (seed)	bermudagrass
Digitaria sanguinalis	large crabgrass
Echinochloa crus-galli	barnyardgrass
Eleusine indica	goosegrass
Lamium amplexicaule	henbit
Malva parviflora	cheeseweed
Orobanche ramosa	branched broomrape
Oxalis pes-caprae	Bermuda buttercup
Poa annua	annual bluegrass
Portulaca oleracea	purslane
Senecio vulgaris	common groundsel
Sida spinosa	prickly sida
Solanum nigrum	black nightshade
Solanum sarrachoides	hairy nightshade
Sonchus oleraceus	sowthistle
Sorghum halepense (seed)	johnsongrass
Stellaria media	common chickweed
Trianthema portulacastrum	horse purslane
Xanthium strumarium	common cocklebur

Source: *Soil Solarization*, UC ANR Leaflet 21377

TABLE 2-8.

Pests Partially Controlled by Soil Solarization.

FUNGI

SCIENTIFIC NAME	DISEASE CAUSED (CROP)
Fusarium oxysporum, f. sp. *pini* *Macrophomina phaseolina*	Fusarium wilt (pines) charcoal rot (many crops)

BACTERIA

SCIENTIFIC NAME	DISEASE CAUSED (CROP)
Pseudomonas solanacearum	bacterial wilt (several crops)

NEMATODES

SCIENTIFIC NAME	COMMON NAME
Meloidogyne icognita	southern root knot nematode

WEEDS

SCIENTIFIC NAME	COMMON NAME
Convolvulus arvenis (plant)	field bindweed (plant)
Cynodon dactylon (plant)	bermudagrass (plant)
Cyperus esculentus	yellow nutsedge
Cyperus rotundus	purple nutsedge
Eragrostis sp.	lovegrass
Malva niceansis	bull mallow
Melilotus alba	white sweetclover
Sorghum halepense (plant)	johnsongrass (plant)

Source: *Soil Solarization*, UC ANR Leaflet 21377

Solarization has also been used experimentally to help in establishing fruit and nut trees. The solarization temporarily reduces pathogens, nematodes, and weeds in the upper foot of soil and may stimulate beneficial soil microorganisms. There is no evidence that tree roots are damaged if procedures are followed carefully. Plant trees on raised beds to prevent waterlogging. Smooth soil and remove debris to assure good film-soil contact and preirrigate. Carefully fit a 5-foot by 5-foot square of transparent plastic with a small slit in the center over the tree. Bury the edges of the plastic with soil. Although tree damage has never been observed with nursery planting stock, avoid using the method on the most tender seedlings and applying the plastic on the hottest days. However, as with open areas, solarization must be done during the hottest months of summer. Remove film after 8 to 10 weeks to avoid possible oxygen depletion of the soil.

Once solarization is completed, minimize soil movement. Turning the soil from beneath the solarized zone can bring up viable weed seeds and pathogens. Most soil movement can be avoided by preparing the seed beds for planting prior to solarizing.

In addition to controlling certain pest organisms, solarization seems to stimulate plant growth beyond what might be expected due to the reduction of weed competition and soil pathogen activity. The reasons for this increased productivity are not completely known; however, it is believed that the treatment may increase the availability of some soluble nutrients and also enhance the activity of beneficial microorganisms in the soil.

REFLECTIVE MULCHES

Aluminum foil or silver colored mulches placed on the soil surface will repel aphids and whiteflies, especially when plants are small. These types of mulches have been shown to delay and reduce aphid transmission of virus diseases to vegetables. Aluminum-coated construction paper is available or you can spray paint clear plastic with silver paint. A combination of solarization and insect repellancy can be used for fall-planted vegetables. Place clear plastic on preshaped vegetable beds in July or August. When ready to plant in late August or September, paint the clear plastic silver and cut holes for seeds or transplants. Wait two or three days after cutting holes to allow the soil to cool before planting seeds or transplants. This is particularly important for transplants, which will be killed by the hot soil. See UC ANR Leaflet 21377 for more information on soil solarization.

PLANT CAGES, ROW COVERS, AND OTHER PEST BARRIERS

Sometimes the best way to prevent pest damage is to keep the pests away from the crop with physical barriers. The most common barriers are row covers, hot caps, and other types of plant cages, which may be placed over young plants to keep pests out. Many of these products were developed primarily to speed growth during early spring by raising air temperatures beneath the cover, thereby obtaining an earlier than normal crop. However, all have the secondary effect of keeping out migrating pests—especially maggots, cucumber beetles, and fleabeetles as well as whiteflies and aphids that can vector virus diseases.

Any type of covering that excludes insects can be used. For years, gardeners have made their own plant cages with wooden or wire frames covered with muslin or nylon (Figure 2-13). Once the cage was made, it could be used for several years. Recently, various synthetic materials have become available, including vented poly-ethylene, spunbonded polyester, point-bonded polypropylene, and woven plastics. Among the easiest to use are the spunbonded or floating row covers that can often simply be placed on top of the bed with no frames or hoops—the crop itself will lift the fabric up as it grows. These floating row covers are best used on sturdy crops that do not grow too tall, such as cucurbits. Sensitive growing tips on plants that grow upright can be damaged by floating row covers, so covers should be held up by hoops (often called plastic tunnels, Figure 2-14).

Normally plant covers or cages are kept on just during the sensitive seedling and young plant stages. Once the young plant gets big enough to tolerate some damage, the cover may be removed. By mid-May or June temperatures beneath plastic tunnels or other tunnels that trap heat may get too hot in many parts of California and damage plants, so removal of these types of covers is essential. Floating row covers of a spun-bonded or nonwoven nature that allow for air movement through many small pores do not usually reach temperatures lethal or damaging to warm season crops. Row covers can be used in the winter and early spring to protect melons and squash from aphids and whiteflies that spread serious viral diseases, but the covers are kept on only until the first flowers appear. Covers are then removed to allow bees access to pollinate the crops.

When row covers are kept on for an extended period of time, weed control can be difficult. Commercial growers control weeds with preplant fumigation or black plastic mulch under the cover. In gardens, the plastic can be carefully lifted off hoops to facilitate handweeding or hoeing. Irrigating under the row cover can be tricky as well. A drip or furrow system must be used.

Other pest barriers. Other barriers for specific pests are discussed in individual pest sections. These include sticky barriers around tree trunks to exclude migrating ants, and copper sheeting and other types of barriers to keep snails and crawling insects out of trees and vegetable gardens.

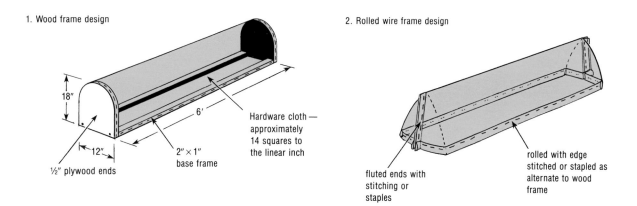

1. Wood frame design

18"

6'

12"

½" plywood ends

2" × 1" base frame

Hardware cloth — approximately 14 squares to the linear inch

2. Rolled wire frame design

fluted ends with stitching or staples

rolled with edge stitched or stapled as alternate to wood frame

FIGURE 2-13. Plant cages for keeping insects out of establishing plants. Cover frames with muslin, nylon, wire screen or spunbonded polypropylene.

FIGURE 2-14. Polypropylene fabrics may be placed over hoops for protecting tender young plants.

INTERCROPPING, COMPANION PLANTING, AND COVER CROPS

The concept of interplanting more than one crop in a field at the same time to reduce pest problems has received considerable attention in recent years. The idea is that large expanses of a single crop grown in monoculture provide an ideal environment for epidemic outbreaks of the pathogens and insect pests that thrive on that crop. Planting two or more crops in alternating portions of the field might slow the spread of the pests, provide alternate habitat for natural enemies, and also allow the grower to have at least one crop to harvest if the other is severely attacked by pests.

However, despite its appeal and benefits in some situations, the generalization that a diversity of plant species in the field or garden plot decreases pest problems does not always hold true. While pests that attack only one crop may be reduced, more generalist feeders may be increased; therefore, each combination and situation needs to be evaluated individually.

On large-scale farms, there has been little research showing that the added labor and expense of having to adapt production practices to manage two different crops in the same field is justified by enhanced pest control; thus few California growers operating larger farms have been convinced to try intercropping as a means of pest control. On the other hand, gardeners and growers operating smaller farms with multiple crops grow a diversity of crops in close proximity out of necessity. These growers often wonder if they can take advantage of this diversity by planting specific combinations of crops or crops and noncrop plants that would actually decrease pest damage.

Companion planting. Many gardening books and magazines have published lists of plants that are said to protect neighboring plants by repelling pests, a concept sometimes called companion planting. Research in this area has been limited but it has consistently shown no overall benefits under controlled conditions. For instance, University of California research showed that while a cabbage plant surrounded by four companion plants of various herbs had fewer imported cabbage worm eggs laid on it than a control plant not surrounded by herbs, yield reduction was severe due to the close proximity of the herb plants and competition for resources. When the number of companion plants per cabbage was reduced to two, the herbs gave no significant pest protection and yields were still substantially lower (see Table 2-9).

Plants that improve biological control. Certain plants within the garden, around field borders, or on the orchard floor appear to have benefits for enhancing the activities of natural enemies of insect pests and thus indirectly reduce some pest problems. Most of the tiny wasps that parasitize certain insect pests survive longer and thus lay more eggs if they are able to feed on nectar. Nectar-producing flowering plants within or near the garden or orchard can attract, support, and increase the activities of these beneficials. Many flowering plant species can serve this purpose. Species that flower throughout the season and those with small flowers with nectar and pollen readily accessible to the tiny parasitic wasps will probably provide the most benefits. Choose species that are not difficult to control as weeds when their seeds become

TABLE 2-9.

University of California Study on the Effect of Companion Plants on Cabbageworm Injury to Cabbage and Whiteflies on Beans.

COMPANION PLANT	WORM INJURY RATING TO CABBAGE	AV. WEIGHT CABBAGE HEAD (GRAMS)	NO. WHITEFLY NYMPHS ON BEANS	
			July 26	*Aug. 9*
Nasturtium	2.2	128	82	301
Marigold	2.9	56	40	172
Catnip	2.4	24	105	194
Summer Savory	2.2	12	75	194
Basil	2.5	90	100	214
Control (None)	2.9	301	32	123

Source: Koehler, Barclay and Kretchum, 1983, *California Agric.* 37 (9+10): 14–15.

food for egg production; in alfalfa, researchers found that 8 aphids per stem were necessary to attract, retain and stimulate high egg production by green lacewings. Sometimes, high populations of aphids and other insects that prey on neighboring weeds or crops will support populations of these general predators when pest populations are low on a crop.

As a source of general predators for field and vegetable crops, few crops can compete with alfalfa. Because alfalfa stays in the field for 4 or more years and has a long growing season, from 7 to 10 months in most of California, it provides a good habitat for

distributed in the garden. Also avoid plants that support insects, nematodes, or pathogens that can become problems on your crop plants. Examples of flowering plants that can provide pollen and nectar are listed in Table 2-9a.

Noncrop plants or alternate crops can also provide food for predatory insects and mites, allowing them to build up to high enough levels to successfully control pests on the crop plant. Syrphid flies (Figure 2-15), for instance, must feed on pollen before they can reproduce. A few predators, notably the predaceous mite, *Amblyseius hibisci,* can survive on a diet of pollen when pest populations are low. However, most predators require a fairly high population of the pest to support reproduction. For example, if adult convergent lady beetles do not find sufficient numbers of aphids in an area, they will fly elsewhere or fail to reproduce. Although they do not feed directly on insects, adults of green lacewings rely on aphid honeydew as

TABLE 2-9A.

Examples of Flowering Plants That Can Provide Nectar or Pollen for Predators and Parasites of Insect and Mite Pests.

Carrot Family
  caraway — *Carum carvi*
  coriander (cilantro) — *Coriander sativum*
  dill — *Anethum graveolens*
  fennel — *Foeniculum vulgare*
  flowering ammi or Bishop's flower — *Ammi majus*
  Queen Anne's Lace (wild carrot) — *Daucus carota*
  toothpick ammi — *Ammi visnaga*
  wild parsnip — *Pastinaca sativa*

Sunflower Family
  blanket flower — *Gaillardia* spp.
  coneflower — *Echinacea* spp.
  coreopsis — *Coreopsis* spp.
  cosmos — *Cosmos* spp.
  goldenrod — *Solidago* spp.
  sunflower — *Helianthus* spp.
  tansy — *Tanacetum vulgare*
  yarrow — *Achillea* spp.

Mustard Family
  Basket-of-Gold alyssum — *Aurinium saxatilis*
  hoary alyssum — *Berteroa incana*
  mustards — *Brassica* spp.
  sweet alyssum — *Lobularia maritima*
  yellow rocket — *Barbarea vulgaris*
  wild mustard — *Brassica kaber*

Other plant families
  buckwheat — *Fagopyrum sagittatum*
  cinquefoil — *Potentilla* spp.
  milkweeds — *Asclepias* spp.
  phacelia — *Phacelia* spp.

FIGURE 2-15. Syrphid fly feeding on pollen in a flower. Although syrphid fly adults do not feed on insects, their larvae are important predators. Syrphid fly adults must feed on pollen before they can reproduce.

beneficial insects when other crops are out of the field. For years, the University of California has suggested that cotton growers consider growing strips of alfalfa in cotton fields to help improve biological control. Alfalfa has its drawbacks, however. To maintain its populations of predators, it must be kept healthy and vigorous, and requires different cultural practices from the crops it might be grown between. Also, alfalfa can be a major source of pests—especially lygus bugs—which move out of the alfalfa when it is mowed or becomes weakened, often causing serious problems, especially in strawberries, tomatoes, or beans.

Cover crops. Cover crops are secondary plantings grown not for harvest, but to enhance production of the primary crop. Orchard crops frequently are intercropped with cover crops or ground covers and, especially when orchards are young, commercial crops. Orchard floor vegetation can have both positive and negative effects on pest problems. Orchards with cover crops have often been observed to have higher populations of certain natural enemies, largely due to increased habitat and alternate food sources for beneficial insects or mites; they also may have fewer problems with mites because of reduced dust.

However, if not properly managed, a cover crop can be a source of pests as well as natural enemies. Rodent, crown rot, ant, stink bug, and lygus bug problems have been associated with ground covers that get too overgrown. Legume ground covers—often popular because of their addition of nitrogen to the soil—can increase nematode problems and water use.

Cover crops may also be grown in areas designated for annual plantings and plowed into the soil a few weeks before planting a new crop of vegetables. Cover crops commonly planted include clovers, vetch, fava beans, phacelia, oilseed radish, and various grains. Cover crops can provide many benefits, such as addition of organic matter to the soil, improvement of soil structure and water penetration, and contribution of nutrients. Some cover crops like fava beans have extrafloral nectaries that attract many natural enemies; others may enhance beneficial microorganism populations in the soil. However, cover crops can harbor and increase populations of pests such as soil pathogens and nematodes and must be applied cautiously. For instance, recent research has shown that cover crops of phacelia, lana woollypod vetch, and Australian winter pea cover crops increased incidence of lettuce drop; other cover crops—oilseed radish, barley, and fava bean—did not.

BIOLOGICAL CONTROL

Biological control can be broadly defined as any activity of one species that reduces the adverse effect of another. Predators and parasites of insects (Figures 2-16 and 2-17) are probably the most well known biological control agents; however, pathogenic, predatory, competitive, and antagonistic organisms can also play an important role in the biological control of many pests. For instance, naturally occurring epidemics of viral, fungal, and bacterial diseases frequently keep pest insects, mites, nematodes, and weeds below damaging levels. Competitors and antagonists are also often important in the management of pests, especially weeds and pathogens. For example, cover crops in an orchard or living mulches in a vegetable garden are chosen and managed with the idea that they will be able to outcompete weeds that might later establish themselves, yet not be severely competitive with crop plants. Antagonists are organisms that release toxins or otherwise change conditions so that activity or growth of the pest organism is reduced; the inoculation of fruit tree rootstock with a strain of the bacterium *Agrobacterium radiobacter*, for example, can prevent infection by damaging strains of *Agrobacterium tumefaciens*, the pathogen that causes crown gall. Some plants release substances that are toxic to certain plants, a phenomenon called allelopathy; although these relationships are poorly understood, allelopathy is likely to sometimes have an impact on the abundance of certain weeds species in certain crops—such as walnuts.

FIGURE 2-16. The parasitic wasp, *Hyposoter exiguae*, lays its egg in a young beet armyworm. After hatching, the wasp larva will complete its development within the armyworm, killing it before it emerges. Many species of small parasitic wasps are important in controlling insect pests in gardens and farms.

FIGURE 2-17. Predators consume numerous individuals during their development. Many species—such as the bigeyed nymph shown here sucking out the contents of a caterpillar egg—feed on many different species and life stages.

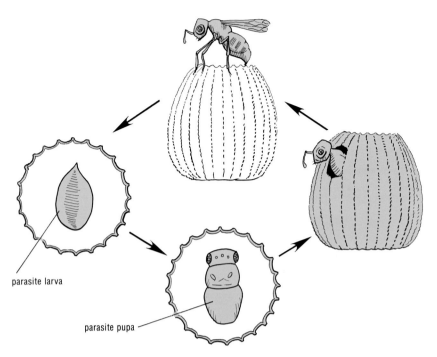

parasite larva

parasite pupa

FIGURE 2-18. Life cycle of a *Trichogramma* wasp. The tiny *Trichogramma* wasp attacks eggs of caterpillars. Larval and pupal stages take place entirely within the egg of the host, and the life cycle may take only a week. Each female wasp attacks many host eggs.

Taking advantage of naturally occurring biological control.

Biological control agents are constantly at work in gardens and farms. Most biological control occurs naturally without assistance from the grower or gardener. Often its importance is not appreciated until a broad spectrum pesticide, which kills certain natural enemies as well as targeted pests, is applied and a new pest—suddenly released from biological control—becomes a serious problem. This type of phenomenon, known as secondary pest outbreak, occasionally occurs in gardens. One example might be the sudden outbreak of aphid, scale, mite, or whitefly populations throughout a garden soon after a large tree has been sprayed with a broad spectrum insecticide such as carbaryl; not only are the pest insects in the tree destroyed, but also the insect parasites and other

natural enemies in, beneath, and adjacent to the tree canopy.

There are several things you can do to encourage the activities of biological control agents already present in your garden or farm. The most important is to avoid the use of broad spectrum pesticides whenever possible. Although the impact of pesticides on natural enemies is most well documented for insecticides, there is also evidence that fungicides and soil fumigants inhibit activities of certain biological control organisms. Find out about less disruptive ways to manage pests; this book contains many suggestions.

Sometimes you can provide a habitat that is more favorable for biological control agents. The previous section discussed the concept of choosing adjacent plants that can supply nectar, pollen, alternate hosts, and habitat for natural enemies. There is

evidence that providing adequate organic content in soil may be important in ensuring a good habitat for soil microorganisms that can control a variety of pests, particularly soil pathogens and nematodes. Another example, discussed in more detail in the insect chapter, is managing the honeydew-loving ants that protect aphids, scales, and mealybugs from their natural enemies.

Commercially available biological control agents. You can buy biological control agents to release for controlling pest species (Table 2-10); however, mass release of natural enemies is really a minor part of biological control. Most commercially available biocontrol agents are directed against insect or mite pests. Mass release of predatory mites has been successfully used in many crops. Most other agents have yet to have their value convincingly demonstrated in many situations. Entomopathogenic nematodes, which parasitize insects, have only been recently introduced but show great promise for use against certain boring insects, soil dwelling insects, or insects in other types of moist, confined locations. Release of *Trichogramma* wasps (Figure 2-18) for control of numerous caterpillars and lacewings for aphids and other small insects have potential, but use of these two agents has had mixed results because of variation in the quality of agents available and lack of reliable release procedures. Finally, release of natural enemies has been successful against a wide number of citrus pests, including scales, mealybugs, and the brown garden snail. Other biological control

TABLE 2-10.

Some Biological Control Agents Commercially Available in California.

BIOLOGICAL CONTROL AGENT	DESCRIPTION
APHID PREDATOR *Aphidoletes aphidimyza*	A predatory midge that has been used to control aphids in greenhouses; also occurs naturally in the field; attacks many different aphid species.
BLACK SCALE PARASITE *Metaphycus helvolus*	A parasitic wasp that attacks black scale (*Saissetia oleae*) in citrus, olives, and ornamentals; may occur naturally but is often disrupted by insecticides, ant infestations, or cold weather. Supplemental releases are best made in late summer or early fall.
CALIFORNIA RED SCALE PARASITE *Aphytis melinus*	A parasitic wasp that attacks California red scale (*Aonidiella aurantii*) in citrus; occurs naturally, but supplemental releases may be made as necessary. Attacks primarily mature virgin females so releases are most effective just before emergence of male scales. Extreme temperatures in Central and Imperial valleys limit ability of this parasite to survive through summer and winter.
CALIFORNIA RED SCALE PARASITE *Comperiella bifasciata*	Another parasitic wasp attacking California red scale in citrus; occurs naturally in citrus groves throughout the state; another strain of the same species controls yellow scale. Normally not as effective as *Aphytis*, but supplements control. Survives better in Central Valley than *Aphytis*.
CALIFORNIA RED SCALE PREDATOR *Chilocorus nigritus*	A predaceous lady beetle that attacks California red scale, usually released as eggs. Related species occur naturally.
CHINESE PRAYING MANTID *Tenodera aridifolia sinensis*	A general predator, interesting to watch but does not provide reliable control; mantids eat each other and other beneficial and predaceous insects as well as pests.
DECOLLATE SNAIL *Ruminia decollata*	A predaceous snail used to control brown garden snail, primarily in citrus and perennial ornamentals (may damage seedling plants). In California, decollate snails can be legally released only in Southern California and certain San Joaquin Valley counties, check with your county agricultural commissioner.
FLY PARASITES *Muscidifurax zaraptor, Nasonia vitripennis, Pachycrepoideus vindemiae, Spalangia endius, S. cameroni, Sphegigaster* spp., *Tachinaephagus zealandicus*	Various parasitic wasps used to control manure-breeding flies, mostly *Musca domestica* (the housefly) and *Fannia* (the little housefly) species, in feedlots, chicken houses, and dairy farms, etc.
GREEN LACEWING *Chrysoperla carnea* and other species	Larval stage is a general predator that feeds on many small insects, including aphids, thrips, pear psylla, insect eggs, and mites; used in greenhouses, orchards, and fields.
CONVERGENT LADY BEETLE *Hippodamia convergens*	A general predator that feeds on small insects, especially aphids. Most released lady beetles will rapidly leave site. High numbers may reduce aphids on small plants in concentrated areas.
MEALYBUG DESTROYER *Cryptolaemus montrouzieri*	A lady beetle species that feeds on mealybugs (and some small aphids and scale stages), has been most widely used in citrus and in greenhouse situations; does not survive cold winters well.
NAVEL ORANGEWORM PARASITE *Goniozus legneri*	A recently imported parasitic wasp that attacks the navel orangeworm.
PARASITIC NEMATODES *Steinernema* and *Heterorhabditis* species primarily	These tiny parasitic worms are used against many soil dwelling and burrowing insects including weevils, scarabs, carpenterworm, clear winged moths, and artichoke plume moth.
PREDATORY MITES *Galendromus (Metaseiulus) occidentalis, Phytoseiulus* and *Amblyseius* species.	These predaceous mite species are used against many mite species; effectiveness is related to pest species, species released, and field conditions. *Continued*

TABLE 2-10.

Some Biological Control Agents Commercially Available in California, continued.

BIOLOGICAL CONTROL AGENT	DESCRIPTION
TRICHOGRAMMA SPECIES	Parasitic wasps that attack the eggs of many caterpillar species including tomato fruitworms (corn earworms), loopers, hornworms, and codling moth.
WHITEFLY PARASITES *Encarsia* spp.	Parasitic wasps used against greenhouse whiteflies in greenhouses.

NOTE: Registered pesticides such as *Bacillus thuringiensis* are not included. For commercial sources, check the California Department of Pesticide Regulation Web Page: http://www.cdpr.ca.gov/

agents available for sale are useful in only a very limited number of situations, and a substantial number (e.g., release of praying mantids) are of questionable value. Microorganisms for the control of insects and plant pathogens are registered and sold as pesticides and will be discussed in the next section on pesticides.

Although naturally occurring biological control can be important in the control of all types of pests, purchase and release of natural enemies and habitat management practices to encourage biological control are currently limited almost exclusively for insect, mite, and mollusc control. Insects and pathogens have been successfully released against weeds on an areawide basis, mostly for rangeland and aquatic weeds, but this approach is not generally a suitable strategy in gardens and small farms—although geese, goats, and other vertebrates are sometimes used. There are no current practices recommended to improve naturally occurring biological control of plant pathogens and nematodes, although some soil management practices may be shown to be useful in the future. The use of biological control is discussed in more detail in specific pest chapters, especially the insect chapter.

PESTICIDES

Any substance applied to control a pest is called a pesticide. Over the last 40 years, growers and gardeners have become more and more reliant on pesticides to manage pest problems. Unfortunately, some gardeners equate pest control with pesticide application and have little knowledge of the other types of management methods discussed previously in this chapter. Although it can be argued that pesticides have been overused in recent decades, they remain an important pest management tool. While it is not difficult to eliminate all or most pesticide use in home gardens, many crops cannot be grown commercially at certain times of the year or in certain regions without occasional pesticide use. Examples include stonefruits, apples, pears, cole crops, grapes, and strawberries. These crops frequently require some use of pesticides either because produce cannot be grown in the quality or quantity necessary for an economically viable operation or because overall tree health would decline severely from chronic pest infestations. Weed control in commercially grown vegetable crops would be prohibitively expensive to achieve with-

out some use of herbicides because of the high cost of hand labor. A great variety of materials are registered as pesticides in the United States. They vary dramatically in toxicity, spectrum of pests controlled, hazard to health and environment, and mode of action. The most well known are synthetically produced chemicals; examples of materials in this category commonly used in gardens and very small farms include the insecticides malathion, carbaryl (Sevin), and diazinon; the herbicides glyphosate (Roundup) and 2,4-D; and the fungicide captan. Because these chemicals contain carbon and hydrogen in their basic structure, they are known as "synthetic organic" chemicals. Some naturally occurring substances are registered as pesticides as well. These include ones derived from plant or animal sources (e.g., botanical insecticides and antibiotics) and pesticides taken from naturally occurring rock sources (e.g., sulfur, copper sulfate, lime sulfur). A final category of naturally occurring substances are the microbial agents, such as *Bacillus thuringiensis*, that cause diseases in certain pests.

TABLE 2-11.

Pesticides Often Accepted for Use On or Around Some Types of Organically Grown Produce.[a]

COMPOUND	TYPE	USE
INSECTICIDES *Bacillus thuringiensis*	microbial	Controls many species of lepidopteran larvae, mosquito larvae or beetle larvae (depending on the variety of the *B. thuringiensis* used). Registered for many crops.
Cryolite (mined forms only)	inorganic	Controls moth larvae, beetles, and other insects. Toxic to fish. Registered for commercial growers of grapes, citrus, and some vegetables during seedling growth.
Diatomaceous earth	inorganic	Diatomaceous earth is a sorptive dust derived from the skeletons of microscopic marine organisms. As a desiccant, controls household pests such as cockroaches and ants. Also said to control some plant pests; some forms can be hazardous if breathed in.
Granulosis virus	microbial	Controls codling moth.
Insecticidal oils	hydrocarbon	Control aphids, psylla, scale insects, mites, aphid and mite eggs. May provide some control of other overwintering insects.
Lime sulfur	inorganic	Controls mites and psylla, registered for most fruit trees.
Neem (azadirachtin)	plant derivative	Controls a broad range of insects. Low human toxicity.
Pheromones	attractants	Used mainly for monitoring to time other control measures. Sometimes used to confuse insects in localized area to disrupt mating. Occasionally used to catch large numbers of specific insects to reduce future generations.
Pyrethrum	plant derivative	Broad spectrum of pests are controlled, including mosquitoes, flies, aphids, beetles, moth larvae, thrips, and mealybugs. Provides rapid knockdown to flying pests.
Rotenone	plant derivative	Contact and stomach poison. Controls beetles, weevils, slugs, loopers, mosquitoes, thrips, and flies. Also used for control of unwanted fish. Acts as a repellent and acaricide. Rotenone is slow-acting and has a short residual. It is nontoxic to honey bees. Most products, especially those labeled for fruit trees, also contain pyrethrins. Not labeled for home use.
Ryania	plant derivative	Controls codling moth, thrips, and the European corn borer. Products containing ryania are not currently registered.
Sabadilla	plant derivative	Has contact and stomach poison action against several species of bugs: leafhoppers, caterpillars, and citrus thrips. Not currently registered in California. Is toxic to honey bees. Not highly toxic to mammals.
Soaps	soap	Control mites, aphids, and other plant-sucking arthropods. Can be phytotoxic to plants under certain conditions. Soap must be specifically labeled for use as an insecticide. Home use products allow use on a wide range of fruits and vegetables.
Sulfur dust	inorganic	Controls mites. Available for use in a wide range of crops.
Vegetable oils	plant derivative	As a contact spray, control scale insects, aphids, and mites.

Continued

[a] Registrations change frequently and the above information may not reflect current registrations. Use all pesticides only in accordance with current federal and state labels.

TABLE 2-11.

Pesticides Often Accepted for Use On or Around Some Types of Organically Grown Produce, continued.

COMPOUND	TYPE	USE
FUNGICIDES Basic copper sulfate	inorganic	Controls early and late blight, scab, blotch, bitter rot, fire blight, downy mildew, black rot, leaf spot, melanose, greasy spot, brown rot, anthracnose, angular leaf spot, and others. Registered for use on most vegetable and fruit tree crops.
Bordeaux mix	inorganic	Bordeaux is a slurry made of hydrated lime and copper sulfate. Controls brown rot and shot hole diseases in tree fruit. Controls some grape diseases. Also controls apple scab, blotch, apple black rot, melanose, anthracnose, early and late blight of potatoes and tomatoes, downy mildew, fire blight, leaf spot, peach leaf curl, and many other fungal diseases.
Copper ammonium carbonate	inorganic	Controls angular leaf spot, alternaria leaf spot, cercospora leaf spot, early and late blight, bacterial blight, common blight, anthracnose, powdery mildew, downy mildew, and others. Registered for citrus, stone fruit, almonds, and walnuts.
Copper hydroxide	inorganic	Controls cercospora leaf spot, bacterial blight, septoria, leaf blotch, anthracnose, halo blight, helminthosporium, downy mildew, leaf curl, early and late blight, angular leaf spot, scab, walnut blight, and others. Registered for use on many fruits and vegetables.
Copper oxychloride sulfate	inorganic	Controls peach blight, peach leaf curl, damp-off, anthracnose, fire blight, shot hole, pear blight, bacterial spot, walnut blight, brown rot, celery blight, downy mildew, early and late blight of vegetables, cherry leaf spot, septoria leaf spot, powdery mildew, scab, and others.
Copper sulfate	inorganic	Suppresses development of fungal and bacterial organisms such as fire blight, cercospora leaf spot, early and late blight, bacterial blight, and others. Registered for use on most vegetable and fruit crops.
Fungicidal soap	soap/sulfur	Controls same species as sulfur. Currently available as a home garden product.
Lime sulfur	inorganic	Controls powdery mildew, anthracnose, apple scab, brown rot, peach leaf curl, and others. Registered for use on most fruit trees.
Sulfur	inorganic	Controls brown rot, peach scab, apple scab, powdery mildew, downy mildew, and others. Registered for use on most fruits and vegetables.
Terramycin and Streptomycin	antibiotics derived from fungi	Controls certain bacterial diseases.

Naturally occurring or "organic" pesticides.

Growers and gardeners seeking to grow food that can be certified in California as organically grown must limit their use of pesticides to those that occur naturally. The term "organic" is somewhat confusing because synthetic pesticides that contain organic chemicals cannot be used. Pesticides allowed for use include microorganisms, microbiological products, and materials derived or extracted from plant, animal, or mineral-bearing rock substances. Bordeaux mixes, trace elements, soluble aquatic plant products, botanicals, lime-sulphur, naturally mined gypsum, dormant oils, summer oils, fish emulsion, and insecticidal soaps are permitted. Table 2-11 lists the materials most commonly used as organic pesticides. In California, the term organic is defined in the California Health and Safety

Code, Section 26569.11. Additional standards are established by the California Certified Organic Farmers (CCOF) Statewide Certification Standards Committee, a private, nonprofit, membership organization established to promote member produce and to control the quality of food produced as organic by its members.

Generally there are less environmental and health hazards with pesticides allowed for organically certified produce than for synthetic pesticides. However, "organic" pesticides are not without hazard for users—acute toxicities of some of them are higher than for synthetic alternatives (Table 2-12). What makes organic pesticides safer environmentally is their tendency to break down more rapidly into nontoxic substances than synthetic materials; as a result, they are much less likely to leave toxic residues in soil, groundwater or marketed produce. However, their lack of persistence in the environment makes organic materials more difficult to use effectively. For instance, botanical insecticides lose their effectiveness within one to two days after application. Any eggs hatching or insects immigrating in after this time will not be controlled and will require a second application. In contrast, typical synthetic insecticides such as diazinon will remain effective for up to 5 to 10 days after application. Microbial insecticides are also effective only for a short time after application, one to two days in the case of *Bacillus thuringiensis* and less than 24 hours for the codling moth granulosis virus.

TABLE 2-12

Oral LD$_{50}$ Values for Some Pesticides Used in Small Farms and Gardens.

CHEMICAL	COMMON TRADE NAMES	ORAL LD$_{50}$[a]	TYPE OF PESTICIDE
Nicotine	Black Leaf 40	55	insecticide
Rotenone*		132	insecticide
Bordeaux*		300	fungicide
Diazinon		300	insecticide
2,4-D		375	herbicide
Carbaryl	Sevin	500	insecticide
Acephate	Orthene	866	insecticide
Copper hydroxide*	Kocide	1000	fungicide
Copper oxychloride sulfate*	C-O-C-S	1000	fungicide
Ryania*		1200	insecticide
Malathion		1375	insecticide
Pyrethrum*		1500	insecticide
Propargite	Omite	2200	acaricide
Sabadilla*		4000	insecticide
Glyphosate	Round-up	4300	herbicide
Cryolite*	Kryocide	10,000	insecticide
Benomyl	Benlate	>10,000	fungicide
Bacillus thuringiensis	Dipel	15,000	insecticide

NOTE: Some materials on this list may not be currently registered as pesticides or their use may be restricted.

*asterisk indicates chemical is acceptable for organically grown produce.

[a] LD$_{50}$ indicates the amount of pesticide that will kill half of a group of test animals. The smaller the LD$_{50}$, the more toxic the pesticide is considered to be. LD$_{50}$ values indicated here are for milligrams of pesticide per kilogram of an animal's body weight. LD$_{50}$ figures do not provide an indication of chronic health hazards or persistence in the environment.

Two exceptions to the rule about persistence include the inorganic copper fungicides, which under certain circumstances may remain in soils of treated orchards for years, and the inorganic insecticide cryolite, which should not be sprayed on edible parts of food crops, because it may remain for several weeks.

Using pesticides safely. If you decide to use pesticides, you must take special precautions to assure the safety of people who may come in contact with the spray and to prevent environmental contamination. Many people are less cautious when using "organic" pesticides, but all have potential to harm if improperly used; for instance, many dusts—a common organic pesticide formulation—have potential to irritate skin or the respiratory tract. Always read the pesticide label. Read it before you buy the pesticide to be sure it is legally allowed for your situation. Then read it again before opening it to be sure you properly mix and apply the material and are prepared to handle emergencies.

Wear the proper clothing. At a minimum, keep skin covered during spraying. Long-sleeved shirts and long pants are essential. Unlined rubber gloves are frequently advisable or required. Wash your clothes carefully after spraying and keep them separate from other laundry. Always wash your face and hands after spraying. Many of the more toxic materials used in commercial agriculture require rubber boots, facemasks, plastic hats, or waterproof pants and aprons. Minimum requirements will be stated on the label.

Choose a sprayer that is appropriate for your operation and the pesticide you are applying. For spot treating in home vegetable gardens, a plastic trigger pump sprayer may be all that is needed; watch for leakage around the pump, though. Larger jobs can be completed with hose end sprayers, although these are not strongly recommended because they are the hardest to adjust for proper application rate and as a result have the poorest record for control success. Hand pump or compressed air sprayers are more useful for larger backyard areas including small fruit trees. Pump sprayers can also be mounted as backpack or knapsack sprayers. Larger fruit trees and commercial operations require powered equipment. Power equipment ranges from power backpack sprayers to tractor mounted equipment. UC ANR Publication 3324, *The Safe and Effective Use of Pesticides*, explains the use of this equipment in detail.

Choose a pesticide material that is appropriate for managing your pest problem. Always be sure that the crop plant you are treating is on the pesticide label; it is illegal as well as unwise to use pesticides on nonlisted plants —for instance, never use a material formulated solely for use on ornamentals on vegetable crops; it may leave unhealthy residues or damage the plant. It is also a good idea to make sure the pest is listed on the label as well; this indicates the manufacturer believes the product will control the species in question. As a general rule, choose the safest material that you know will be effective. Select materials that are least toxic to honey bees and other beneficials in your garden. Table 2-13 lists relative toxicities of common garden pesticides to bees. Consult your farm advisor, other University of California publications, and your nurseryman for more information.

TABLE 2-13.

Relative Toxicity to Honey Bees of Some Pesticides Commonly Used in Gardens and Small Farms.

GROUP I—HIGHLY TOXIC
Severe losses may be expected if used when bees are present at treatment or within a day thereafter
Acephate (Orthene)
Azinphosmethyl (Guthion)
Carbaryl (Sevin)
Diazinon
Dimethoate (Cygon, Defend)
Malathion
Methomyl (Lannate, Nudrin)
Parathion
Phosmet (Imidan)
Synthetic pyrethroids [includes permethrin (Pounce) and fenvalerate (Pydrin)]

GROUP III—RELATIVELY NONTOXIC
Can be used around bees with minimum injury.
Bacillus thuringiensis (Dipel, Thuricide, etc.)
Cryolite (Kryocide)
Dicofol (Kelthane)
Nicotine
Propargite (Comite, Omite)
Pyrethrum (Natural)
Rotenone
Sabadilla

All fungicides
All herbicides

Finally, remember that careless pesticide applications not only endanger your health and the environment but are also less likely to control the target pest properly.

CHAPTER THREE

Common Insects, Mites, Other Arthropods, Snails and Slugs

Many insects, mites, and other invertebrates are commonly found living on and around crop plants and trees in gardens and farms. Some are pests that injure crops, but many are beneficial. Beneficials include natural enemies such as parasites or predators that destroy other pests, scavengers that help break down dead organic matter such as mulches and composts, and pollinators such as honey bees. This chapter concentrates on pest species, but it is important to remember that most species of invertebrates are beneficial or harmless and should not be destroyed. Actions you take to control pest invertebrates may also damage these beneficial species as well.

Damage

The pests covered in this chapter are a very diverse group. Most damage plants through their feeding activities, but few other generalizations can be made about them. The type of damage is determined by many factors including crop, stage of development, site of attack, and size and stage of pest. However, a principal factor is the type of mouthparts the pest has. Insects and related pests generally have chewing mouthparts or tubular sucking-rasping mouthparts (Figures 3-1 and 3-2). Pests with chewing mouthparts chew identifiable holes in leaves, twigs, stems or fruit or cut stems right off; pests in this group include grasshoppers, sowbugs, earwigs, sowbugs, crickets, most beetles and caterpillars. Pests with sucking or rasping mouthparts, such

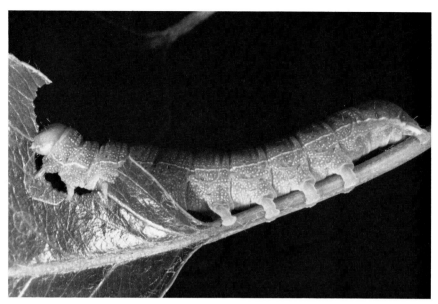

FIGURE 3-2. Pests with sucking mouth-parts, such as the stink bug shown here, insert their tubular mouthparts into leaves, stems, buds, or fruit and suck plant juices out.

FIGURE 3-1. Insects with chewing mouthparts, such as the green fruitworm caterpillar shown here, chew distinct holes in leaves, fruits or stems.

as aphids, scales, true bugs, leaf-hoppers, thrips, and mites, cause buds, leaves or fruit to discolor, distort, or drop, but never cut away pieces of leaves or fruit. Most sucking insects insert their mouthparts directly into vascular tissues and take up the sap of plants; a number of sucking pests can transmit plant diseases, especially viruses, during this feeding process.

Life Cycles

Most of the pests discussed in this chapter (many species of aphids are the main exception) begin life as an egg which hatches into an immature form called either a nymph or a larva. Insect and mite eggs are often quite tiny and difficult to spot; however, they can be a good indicator of future outbreaks, so farmers and gardeners should be on the lookout for them and learn some of the more distinctive shapes (Figure 3-3).

Insects, mites, and other arthropods grow by shedding their outer skin or exoskeleton, a process called molting. In addition to the change in size, there are usually modifications in the animal's shape with each successive molt. The whole process of change is known as metamorphosis. Some species (Figure 3-4) grow

in size while undergoing very little change in form from immatures to adult. In other cases, such as butterflies, flies, and beetles, there are major changes in form (Figure 3-5) between the immature and adult stages, with the transformation occurring within a nonfeeding, resting stage known as the pupa. Metamorphosis in these types of species is called "complete metamorphosis" and immatures are generally referred to as larvae. Those insects that go through more gradual metamorphosis, such as true bugs (Figure 3-6), may develop wings and change body proportions during metamorphosis but have no pupal stage. Immature individuals of these species are known as nymphs. Development of mites (Figure 3-7) is similar to gradual metamorphosis among insects; however, the stage that hatches out of the egg has six legs and is normally called a larva and later immatures, which, like adults, have 8 legs, are called nymphs.

A

B

C

D

E

F

G

H

I

FIGURE 3-3. Examples of insect eggs. Pest species include: a) corn earworm egg, b) cabbage looper egg, c) imported cabbageworm egg, d) the barrel-shaped eggs of the harlequin bug, e) the overlapping fish-scale like eggs of the omnivorous leafroller, f) beet armyworm egg mass with some of the cottony covering removed. Natural enemy species include g) praying mantid egg case, h) two tachinid fly eggs on the head of an amorbia caterpillar, i) a syrphid fly egg, j) convergent lady beetle eggs. See also photos of general predator eggs on the next pages.

J

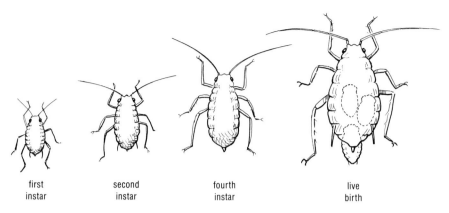

first
instar

second
instar

fourth
instar

live
birth

FIGURE 3-4. Development of aphids—simple metamorphosis. Many species of aphids can give birth to live aphid nymphs rather than laying eggs.

Species that undergo complete metamorphosis commonly have different feeding habits during their larval and adult stages. In many cases, such as among butterflies, moths, and flies, only the larval stage causes damage whereas the adult stage does not feed or takes in only nectar. A number of beetle species, however, such as the spotted cucumber beetle and flea beetles, cause damage as adults but do little harm to crop plants during their root-feeding larval stages.

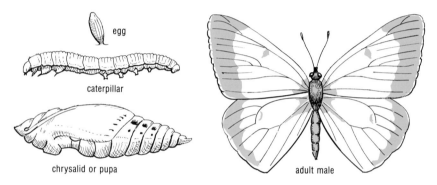

egg

caterpillar

chrysalid or pupa

adult male

FIGURE 3-5. Stages of development of the alfalfa butterfly showing complete metamorphosis. Most butterflies go through 4 or 5 molts during the caterpillar stages.

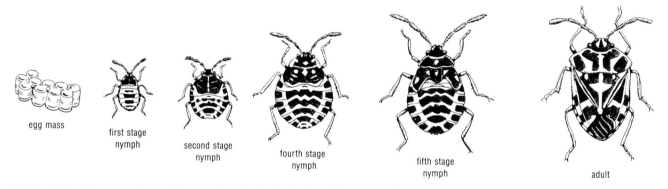

egg mass

first stage
nymph

second stage
nymph

fourth stage
nymph

fifth stage
nymph

adult

FIGURE 3-6. Development of a stink bug species, the harlequin bug, showing gradual metamorphosis with wing development.

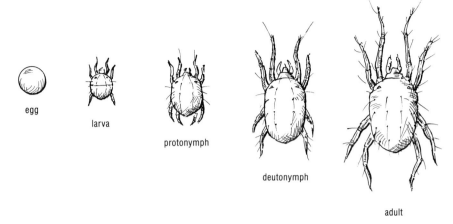

egg

larva

protonymph

deutonymph

adult

FIGURE 3-7. Development of a typical plant-feeding spider mite. The stage that hatches out of the egg is called a larva. Remaining immatures are nymphs.

Monitoring and Diagnosing Problems

Insects and mites often migrate and build up rapidly in gardens and farms, so regular checks for these pests are essential. Start surveying weeds and surrounding crops and vegetation for pests even before you plant annual crops, removing infestations several weeks before planting if possible. Likewise, orchard trees should be checked in the dormant season for overwintering stages of pests and at least weekly during the active growing season.

It is a good idea to purchase a small 8- to 10-power hand lens to check leaves for small insects, mites and eggs. Other than that, the home gardener needs few other tools beyond a notebook to keep a record of pests and control measures. Growers and those with larger gardens or orchards may wish to make or buy a beating sheet or tray (see Figure 2-8 in Chapter 2) to help in checking for pests. Simply place the beating sheet beneath the plant and give stems or branches a few good shakes. Pests and beneficials will dislodge and can be examined or counted on the sheet. Specialized equipment and techniques for monitoring, including traps, are discussed in the individual pest sections of this chapter.

Proper identification of pests and natural enemies is essential for successful pest management. Pest and beneficial insects and mites often look quite similar. Use the photographs in this book and the references listed in the back to help identify common pests and beneficials. Take pests you cannot identify to your county agricultural commissioner's office or county Cooperative Extension office for help.

Management Methods

General pest management methods have been reviewed in Chapter 2. Many of these methods can contribute to the control of insects, mites and other invertebrates. The key element in an integrated pest management program is preventing outbreaks before they begin. Important preventative practices include maintaining a healthy crop through the use of good cultural practices (picking off, pruning out, or otherwise removing small infestations before they spread) and choosing crops that are less prone to problems. Biological and chemical control are discussed below. Specialized traps, barriers, and other techniques are available for some species. Consult the individual pest sections for more detailed information.

BIOLOGICAL CONTROL

Biological control, as discussed in the earlier chapters, is very important in the management of insects and mites. Some of the most common general predators are listed and pictured below. Other more specialized predators and parasites, as well as pathogens that cause disease in insects, are discussed and pictured in other parts of this chapter. When you check your garden or orchard, watch as closely for signs of these natural enemies as for pests. Often they will control pests before serious damage occurs. Whenever you can, choose practices such as cultural and mechanical methods, water or soap sprays and microbial pesti-

Green lacewing larva.

cides that interfere as little as possible with the activities of natural enemies.

General predators. A predator is an animal that attacks and feeds on other animals. Some predators and many parasites are quite specialized and confine their feeding to only one or a few closely related species, whereas a general predator feeds on a wide variety of prey. General predators are very important in gardens and small farms with a diversity of crops and potential pests. Because they subsist on a variety of insects and mites, general predators will often be present in the garden even before serious pests arrive and will be able to provide immediate assistance in their control. More specialized predators and parasites, in contrast, may not provide much help until pest populations rise to near damaging levels because of their dependence on specific prey species. Learn to recognize general predators and encourage their activi-

ties. Some of the most common general predators are pictured here. Other common but more specialized predators are discussed and illustrated along with their primary prey in other sections of this chapter. For instance, syrphid flies and the convergent lady beetle feed primarily on aphids (although they may occasionally feed on other small insects) and are covered in the aphid section.

Green Lacewings. Chrysopa *spp.* Green lacewing larvae are tapered at the tail and flattened, with distinct legs and long curved mandibles for grasping their prey. Coloring may vary somewhat. Sometimes called aphidlions for their habit of feeding on aphids, they also feed on mites and many other types of small insects and insect eggs. Green lacewing eggs are easy to recognize because they are laid on tall stalks. Lacewings attach their spherical pupae with silk to tree trunks, undersides of leaves and other

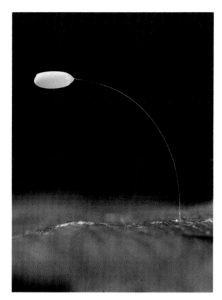

Stalked egg of green lacewing.

Green lacewing pupa.

Adult of green lacewing.

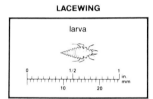

LACEWING

larva

0   1/2   1 in.
|+++++++|+++++++| mm
    10   20

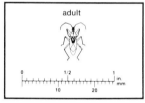

DAMSEL BUG

adult

0   1/2   1 in.
|+++++++|+++++++| mm
    10   20

Adult (top) and nymph of damsel bug.

protected places. Adults are slender insects with large lacey wings and copper colored eyes; adults do not feed on insects.

Damsel Bugs. **Nabis *spp.*** Damsel bugs feed on aphids, leafhoppers, plant bugs, and small caterpillars. Both adults and nymphs feed on insects. They are usually dull brown or gray and superficially resemble other plant bugs that feed on crops; however, their heads are longer and narrower than most plant feeding species.

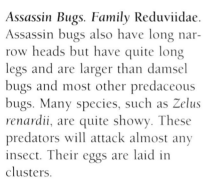

Assassin bug eggs.

Adult of one showy species of assassin bug, *Zelus renardii*, attacking a lygus bug.

Assassin Bugs. Family Reduviidae.
Assassin bugs also have long narrow heads but have quite long legs and are larger than damsel bugs and most other predaceous bugs. Many species, such as *Zelus renardii*, are quite showy. These predators will attack almost any insect. Their eggs are laid in clusters.

Bigeyed Bugs. Geocoris spp.
Bigeyed bugs are small bugs that feed on many small insects, insect eggs, and mites in a wide range of crops in both the adult and nymph stages. Eggs of bigeyed bugs develop a distinctive red spot soon after they are laid. Bigeyed bugs also feed on nectar from flowers, which helps sustain them when prey is scarce.

Nymph of bigeyed bug, sucking out contents of corn earworm egg.

Adult bigeyed bug.

Bigeyed bug egg showing red eye spot.

Adult of the minute pirate bug.

Minute pirate bug nymph.

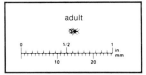

MINUTE PIRATE BUG

adult

Minute Pirate Bugs. **Orius** *spp.*
The tiny minute pirate bug is an important predator that often goes unnoticed because of its small size. The adult is black with distinctive white wing patches and the nymph is often yellow or orange. Although they feed on almost any tiny insect or mite, they are especially attracted to colonies of thrips, which often appear early in the spring.

Soldier Beetles. Family **Cantharidae.** Soldier beetles are common predators of aphids and other insects in the garden and orchard. They are long, narrow beetles usually red or orange with black, gray, or brown wing covers. Larvae are soil dwellers that are also predaceous.

Predaceous Ground Beetles. Family **Carabidae.** Predaceous ground beetles are very common garden insects that feed on many soil-inhabiting pests such as cutworms and root maggots. One species even feeds on the brown garden snail. Shapes and colors vary widely but carabids are

Adult soldier beetle.

Predaceous ground beetles are usually shiny and their antennal segments are all the same size. This is a species in the *Amara* genus.

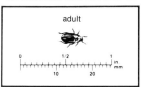

SOLDIER BEETLE

adult

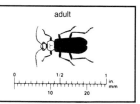

PREDACEOUS GROUND BEETLE

adult

usually shiny. Some species are black, but other species may be marked with bright colors. Be sure to distinguish them from darkling beetles (covered in a later section), which are common pests of seedlings. Darkling beetles are dull and the segments on the tips of their antennae are slightly larger than those at the base; carabid antennae rarely have antennal segments enlarged.

Spiders. Spiders are important predators of a great variety of insects and mites in almost every habitat. They come in many sizes and shapes, but virtually every spider is a predator.

A crab spider from the Family Thomisdae.

PESTICIDES

Pesticides may sometimes be necessary to control pests in commercial operations, and home gardeners may also choose to use them under some circumstances. Select the least toxic materials that will do a good job. It is not usually necessary to kill off 100% of the pests in your crops; often it is sufficient to bring pest populations down to a level where natural enemies can keep them adequately under control. Sometimes water sprays can be used to knock off aphids, mites, or whiteflies without the addition of a pesticide chemical. Some of the more common and least toxic materials are briefly described here. Always follow rates and precautions listed on the label. See Chapter 2 for more information on pesticides.

Horticultural oils. Horticultural oils (sometimes called mineral oils) are highly refined petroleum oils that are manufactured specifically to control pests on plants. Horticultural oils are paraffinic, degrade rapidly through evaporation, and have very low toxicity or almost no toxicity to humans or wildlife at the rates used to control pests. Properly used, they also have low toxicity to most plants. This distinguishes them from the more familiar aromatic and napthenic oils that are used as motor fuels and solvents and which can be toxic to wildlife—especially birds—and plants.

Horticultural oils kill mites and insects primarily by suffocating them, although there may be additional toxic reactions in some species. Oils are most effective against eggs, immature forms and soft-bodied adults. Scales, mealybugs, aphids, leafhoppers, whiteflies, mites, and eggs of all types of species are among the pests controlled.

Dormant treatments with oils. Yearly applications of oil sprays for control of various insect and mite pests of fruit trees during the dormant or delayed dormant period has been an important component of IPM programs for many years. In California, deciduous fruit trees generally are dormant from December to January and the delayed dormant period begins in February as buds begin to swell and continues until the green tip bud developmental stage. Precise timing varies according to tree species and variety and location. Consult your local Cooperative Extension office for proper times for your situation. Dormant oil sprays kill eggs of western tussock moth, green fruitworm and give some control of aphid and leafroller eggs on trees. They also control European red mites and brown mites, San Jose scale and other scales, mealybugs and pear psylla. Very severe infestations of some pests such as San Jose scale may require addition of organophosphates to the oil spray to provide adequate control. Narrow range oils, such as supreme or superior oils should be used. Applying oils or oil and organophosphate treatments during the dormant season allows for excellent coverage of the tree and also dramatically reduces the negative impact of applications on natural enemies. Oils should not be applied to walnut trees during the dormant season because of possible damage to trees.

Inseason treatments with oils. Many gardeners are unaware of the benefits of using narrow range oils during the growing season to control a wide range of softbodied insects (e.g., aphids, mite, scales, whiteflies, psyllids, mealybugs, leafhoppers) and mites.

Summer treatments can be used safely on most fruit trees, including deciduous species and citrus, but check the label to be sure. Several general precautions should be followed to avoid damaging trees or new growth. Never spray oils when trees are water stressed; water trees a few days beforehand to be sure. Never spray oils when it is very foggy or when relative humidity is above 90% for 48 hours. Never spray oils when temperatures are expected to rise above 90°F. Use supreme or superior-type oils with a minimum unsulfonated residue (UR) of 92 and a minimum percent paraffin (% Cp) of 60%. These types of oils, also called narrow range oils, are also recommended for dormant treatments. Follow label instructions carefully.

Little research has been done on using horticultural oils against pests on annual crops. One exception is sweet corn where applications of horticultural oil are made in silks to prevent damage by the corn earworm. Oils have also been used to control whiteflies on cucurbits, cabbage and a few other vegetables. Future research may show more applications; however, the succulent tissues of annual plants are more likely to be damaged by oils than trees.

Soap sprays. Insecticidal soaps are now readily available in garden centers for controlling various insect and mite pests. They are most effective against mites and aphids but can also reduce populations of whiteflies, immature scale insects, thrips and leafhoppers. Soap sprays have little negative impact on the environment; they are similar in toxicity to a solution of soapy dishwater. They can be toxic to some plants, especially those with a dull surface (and thus little protective wax) or those with many hairs. Seedlings or very small plants may also be quite sensitive. Some gardeners avoid burning plants by hosing them off with fresh water within a few hours after application. Again, check labels for application rates, precautions and hazard to plants, but most important, avoid using soaps on plants stressed for water or under conditions of high heat and high humidity. Soap sprays are discussed further in the aphid and mite sections.

Sorptive dusts. Various dusts (e.g., boric acid powder and silica gel) have been used successfully to control insects in indoor environments but little research has been carried out in gardens and farms. Sorptive dusts work by abrasing or removing the protective wax on the insects' outer cuticle. Most would be inappropriate for garden use because they lose their effectiveness when they get wet. There has been some interest in diatomaceous earth, but no recommendations can be made. Although most inert dusts are low in toxicity, avoid inhaling them because they can cause serious lung irritation.

Sulfur dust. Sulfur dust is most commonly used to control fungal pathogens that cause plant disease; however, it can also be effective in controlling mites. It can be toxic to some plants, so check the label. Never apply sulfur when temperatures rise above 90°F or after a heavy dew or fog. Sulfur dust is a skin irritant and can be harmful if inhaled. Wear appropriate protective clothing and a face mask when applying it.

Microbial insecticides. Microbial insecticides are commercially available formulations of microorganisms that can cause disease in specific groups of insects but have no effect on other organisms. They are almost ideal insecticides with little impact on the environment or human health. Presently, the only microbial insecticide available for use in California vegetables and fruits is *Bacillus thuringiensis*. Currently, the primary use is to kill larvae of butterflies and moths; however, products specific for killing the leaf-feeding larvae of some beetles, e.g. Colorado potato beetle and elm leaf beetle, are also available. *Bacillus thuringiensis* is discussed in more detail in the leaf- and fruit-feeding caterpillar section.

Botanical insecticides. Botanical pesticides are those which are derived from plants. Several botanicals are registered in California. Because botanical pesticides breakdown to harmless substances faster than most synthetically compounded pesticides, they are probably safer for the environment and human consumers. However, their short life also makes them harder to use effectively; they must be applied precisely when and where susceptible insects are located. Any pests immigrating in after the application will probably not be controlled. Available botanicals and pests they control are listed in Table 2-11 in Chapter 2.

The most widely available products are combinations of the botanical pyrethrum products. These products are registered for a vast array of food crops and their labels claim control of most garden pests. In reality, their effectiveness varies among pests; among those not reliably controlled are mealybugs, mites, scales, grasshoppers, and most insects that feed within fruit or burrows. An Asian botanical insecticide, neem, has recently been registered for ornamentals and food crops. Neem is apparently quite nontoxic for humans and controls a wide range of insects and mites.

Synthetic pesticides. The insecticides and miticides most commonly used on gardens and farms are synthetically compounded materials. While not regarded as acceptable for use on food that is to be sold as "certified organic," some of these materials are as safe as organically acceptable materials and often easier to use because they remain effective for a longer time than botanicals. Table 2-12 in Chapter 2 compares relative acute toxicity of various pesticides.

Malathion is probably the safest of the widely used synthetic insecticides, somewhat less toxic to humans than rotenone and similar in toxicity to ryania and pyrethrum. Malathion will remain effective on plant surfaces for one to three days. It will control aphids, scale insects,

mealybugs, leaflhoppers, whiteflies, cucumber beetles, and numerous other insects that feed on exposed areas of plants. It is especially useful on vegetables and, because of its short life, is less toxic to beneficials than other synthetic materials.

Diazinon can control insects for up to a week after application and thus is more effective against insects that tend to hide or migrate in than malathion. It is also more toxic to beneficials and bees, but not as toxic as carbaryl (Sevin). Diazinon controls a wide range of insects in orchards as well as in vegetable gardens. Diazinon is a special concern where drainage into waterways is a possibility because of its hazard to aquatic organisms.

Carbaryl (Sevin) is not very toxic to humans and quite effective against many insects; however, it is more toxic to bees and beneficials than malathion and diazinon, so its use should be avoided whenever possible. It should never be used near flowering plants that are visited by bees. Bait forms for cutworms and other seedling pests are not as likely to cause harm to bees and natural enemies. Carbaryl is especially toxic to the natural enemies of spider mites and also seems to enhance reproduction by spider mites. Many spider mite problems can be traced back to the use of carbaryl. Carbaryl is quite persistent in the environment and remains on leaf or bark surfaces longer than most other insecticides available to homeowners. For that reason, it can be useful for controlling extremely difficult pests such as codling moth.

LEAF- AND FRUIT-FEEDING CATERPILLARS

Every gardener is familiar with various caterpillars that feed on fruit and vegetables from time to time. Leaf-eating species such as the tomato hornworm, armyworms, and loopers chew large, irregular holes in leaves; the largest caterpillars may devour entire leaves along with stems and, in some cases, flowers or fruit as well. This damage is obvious but the injury often looks more serious than it is, especially on well established plantings such as fruit trees. Many species of smaller caterpillars fold or roll leaves together with silk to form shelters. Others feed on leaves beneath a canopy of silk, sometimes creating dense "nests" in foliage.

Codling moth, corn earworm, and other caterpillars that feed directly on fruit or nuts are apt to be much more damaging than leaf-feeding species. Often they burrow deep into the fruit, making it difficult for natural enemies to attack them or for gardeners and growers to control them.

BIOLOGY

All the caterpillars discussed here are the larval stages of moths or butterflies in the order Lepidoptera. After mating, the female butterfly or moth lays her eggs on the host plant. Eggs may be laid singly, such as those of the cabbage looper or corn earworm, or in distinctive egg masses like those of most leafrollers and armyworms. After a few days or weeks (or the whole winter in the case of species that spend the winter in the egg stage), the eggs hatch and larvae (or caterpillars) migrate singly or in groups to feeding sites on the plant. Many fruit-feeding pests, such as navel orangeworm and codling moth, begin to enter protected areas in fruit or nuts right after hatching. Leafrolling species often spend their first larval instar feeding exposed on leaves and do not begin making leafrolls until after they molt to the second instar.

Most caterpillars eat voraciously and grow rapidly, shedding their skins (molting) three to five times before entering a nonactive pupal stage. Some species spin cocoons to pupate in; others do not. The pupal stage may be passed anywhere, but most species pupate in a characteristic location, such as in the soil, on leaves, in overwintering nuts, or on the trunk of trees. Moths may emerge from the pupal case after several days to several months, depending on the species and season. Some common caterpillar pests, such as the fruittree leafroller and the western tussock moth, only have one generation a year, but most pest species have three or more under California's mild climate, so growers and gardeners must be on the lookout for these pests throughout the growing season.

DAMAGE

Damage caused by leaf-eating caterpillars can be unsightly, but the importance of the damage depends on the age of the plant and the purpose for which the plants are grown. Seedlings and very small plants are especially vulnerable, since good-sized caterpillars such as cutworms can quickly destroy an entire plant or an entire row of seedlings. In food crops grown for the leaves, such as lettuce or cabbage, a relatively small number of caterpillars may have a noticeable impact; also, caterpillars or droppings present at harvest may make the crop unacceptable for commercial sale. However, in most crops grown for a fruit, root, or tuber, caterpillars may damage a substantial part of the foliage before reducing the yield.

In contrast, caterpillars that attack fruit cause more serious damage; worms or holes in the fruit cannot be ignored. Actual damage varies from species to species and crop to crop. Some caterpillars feed primarily on the fruit surface whereas others bore deeply into the center. However, whenever caterpillars bore holes through the skin of the fruit, they open it up to infection by rot-producing fungi and bacteria that may produce even more serious damage.

Although holes in fruit attacked by large caterpillars such as the tomato fruitworm may be obvious, damage by smaller species may be harder to spot. For instance, the tomato pinworm bores into tomatoes under the green, leaflike calyx at the stem end of the fruit and often is not detected until the fruit is cut open to reveal the pest's extensive tunneling inside. Similarly, the oriental fruit moth generally enters fruit such as peaches or nectarines at the stem end and usually is not discovered until the fruit is eaten. However, in both these cases, a close examination of the stem end of the fruit often reveals a small accumulation of brown granular frass or excrement. Many pests of nuts, including the codling moth and navel orangeworm, remain undetected until harvest unless the grower periodically cracks a few samples open searching for worms.

MONITORING

The first step toward preventing damage is to learn how to recognize infestations early. Examine plants carefully every few days to look for eggs and small larvae; a hand lens will make identification easier. Close inspection is especially important when plants are seedlings. Many caterpillar species, including loopers, corn earworms, and hornworms, have eggs that are large enough to see if you inspect leaves closely. Others, such as armyworms, leafrollers, the saltmarsh caterpillar, and some cutworms, lay eggs in clusters that are easy to spot. Once caterpillars are large, they are often easier to find because their damage or other evidence of their presence such as frass, leafrolls, or webbed leaves may be readily apparent; however, by this time, caterpillars are more difficult to control and may have caused significant damage already. When you find caterpillars or eggs, remove and destroy them; handpicking can provide good control in a small garden if done on a frequent basis.

In vegetable gardens, the best way to look is to get on your hands and knees, turning over leaves and looking in crowns of seedlings for eggs and small caterpillars. If you find hatched eggs but no small caterpillars or damage nearby, chances are natural enemies are active in your garden. If damage is evident, caterpillars may have moved on to other feeding areas or are night feeders that hide during the day. In fruit trees, it is a good idea to get on a ladder and look deeper into the canopy. Most caterpillars are best monitored by simple leaf or fruit examination, but a few devices are available for monitoring caterpillars on larger acreages; a tray or framed cloth sheet placed beneath a branch or plant while it is tapped or shaken is the most common. This shake-and-beating tray method of sampling is not effective for caterpillars that web themselves in leaves, are in fruit, or are difficult to dislodge for other reasons.

Pheromone traps, which use sex attractants to lure adults of one sex (usually the male), are available for many caterpillar species (see Table 3-1); these can be valuable for determining when and if certain pests are present. Integrated pest management programs using pheromone traps to determine the best time to treat are available for codling moth, peach twig borer, and oriental fruit moth in commercial fruit and nut orchards. However, treatment guidelines based on pheromone trap catches are generally not available for caterpillar pests in vegetable crops (an exception is the tomato pinworm) and are not appropriate for small gardens and individual backyard fruit or nut trees.

Pheromones are also some times used in the control of caterpillar pests. See the sections on codling moth and oriental fruit moth.

TABLE 3-1.

Common Vegetable and Orchard Caterpillars for which Pheromones are Commercially Available.

CATERPILLAR	CROP
Black cutworm, *Agrotis ipsilon*	vegetables
Codling moth, *Cydia pomonella*	apples, pears, walnuts
Omnivorous leafroller, *Platynota stultana*	grapes, orchard crops
Orange tortrix, *Argyrotaenia citrana*	grapes, citrus, strawberries, apples
Oriental fruit moth, *Grapholita molesta*	orchard crops
Peach twig borer, *Anarsia lineatella*	almonds, stone fruit
Redbanded leafroller, *Argyrotaenia velutinana*	orchard crops
Tomato pinworm, *Keiferia lycopersicella*	tomatoes
Beet armyworm, *Spodoptera exigua*	vegetables
Cabbage looper, *Trichoplusia ni*	vegetables
Diamondback moth, *Plutella xylostella*	cole crops
Fruittree leafroller, *Archips argyrospilus*	orchards
Tentiform leafminer, *Lithocolletis blancardella*	apples
Apple pandemis, *Pandemis pyrsuana* (same pheromone as three-lined leafroller, *Pandemis limitata*)	apples
Variegated cutworm, *Peridroma saucia*	vegetables
Obliquebanded leafroller, *Choristoneura rosaceana*	orchards
Variegated leafroller, *Platynota flavedana*	orchards
Artichoke plume moth, *Platyptilia carduidactyla*	artichokes
Sunflower moth, *Homoeosoma electellum*	sunflowers
Carpenterworm, *Prionoxystus robiniae*	fruit trees
Potato tuberworm, *Phthorimaea operculella*	potatoes, tomatoes
Peachtree borer, *Synanthedon exitiosa*	stone fruit

BIOLOGICAL CONTROL

Many caterpillar species are kept under control by natural enemies and natural outbreaks of disease. General insect predators such as spiders, bigeyed bugs, pirate bugs, lacewing larvae, ground beetles, damsel bugs, and assassin bugs consume eggs, small caterpillars, or sometimes larger caterpillars. Eggs are also destroyed by tiny parasites such as *Trichogramma*. Larger caterpillars may be killed by large stinging wasps or eaten by birds. Most caterpillars have one or more important small parasitic wasps or flies that attack them. Some species of parasites, such as *Hyposoter* wasps and tachinid flies, parasitize a number of different pest species.

You get the greatest benefit from natural enemies if you avoid the use of insecticides that destroy them. Very selective insecticides, such as formulations of *Bacillus thuringiensis*, leave most natural enemies unharmed. You may also be able to increase the activity of natural enemies by maintaining plants in the garden or farm which supply nectar, shelter, or alternate insect hosts. Such planting regimes are discussed in Chapter 2.

Caterpillars are often killed by disease in the garden. Viruses, fungi, and bacteria can all affect caterpillars; but of those that occur naturally, diseases caused by viruses are the most common. Caterpillars killed by viruses and bacteria often turn dark and their bodies become soft and limp, eventually degenerating into a sack of liquified contents, which, when broken, releases new viral particles or bacterial spores that infect other caterpillers. Naturally occurring virus outbreaks can reduce populations rapidly under favorable conditions, although such outbreaks are difficult to predict and often do not occur until after substantial damage has been inflicted.

When checking your garden or farm for pests, also look closely for evidence of biological control. Presence of predators or disease-killed caterpillars, dead caterpillars, pupae or eggs with holes from which parasites emerged, or hatched caterpillar eggs with no evidence of caterpillars or damage can all indicate effective biological control. If you have an increasing number of pest caterpillars but also many natural enemies, wait a few days before using insecticides to see if the pests will be controlled naturally. In situations, such as seedling beds, where the population must be controlled immediately, use a material (*Bacillus thuringiensis* is effective against many caterpillars) or method that will have minimal impact on natural enemies.

Mass release of *Trichogramma* wasps. *Trichogramma* species are a group of tiny wasps that attack the eggs of many moths and butterflies. Several species are available from commercial insectaries (see References for publications listing current vendors). It is important to get the species of *Trichogramma* that is best adapted to your situation. Most insectaries claim to sell one or more of three species: *T. pretiosum*, *T. platneri*, and *T. minutum*. Generally, *T. pretiosum* is used to control caterpillar pests in vegetable crops throughout California. *T. platneri* is effective against caterpillars in fruit trees and grapevines in southern California. *T. minutum* is sold for release in orchards and vines in northern and central California, but the species of the wasps sold under this name have not been confirmed.

Commercial sellers of *Trichogramma* normally ship the parasite in the form of parasitized caterpillar eggs glued to a piece of cardboard. The wasps, which complete the immature stages of their life within the caterpillar eggs, should emerge as adults soon after the shipment is received by the purchaser. The wasps will be most effective if they are allowed to feed on a solution of honey and water for 24 hours after they emerge and then are immediately released. The species of caterpillar egg the parasite is reared on has an impact on the success of the release. Most commercial insectaries rear *Trichogramma* on a species (the Angoumois grain moth) that has substantially smaller eggs than the cabbage looper eggs that have been used successfully by University of California researchers. The smaller eggs result in smaller wasps and a lower reproductive ability of the commercially reared *Trichogramma*. As a result, larger numbers and more frequent applications of these smaller *Trichogramma* wasps will probably be necessary to achieve equivalent control.

Trichogramma must be released in large numbers at peak egg laying of the pest for maximum effectiveness. To find out when a pest is laying a new generation of eggs, check crops for fresh eggs; the pest and descriptions that follow tell you where to look for most major pest species. Using pheromone traps to determine when adults are prevalent will help you determine when searching for eggs will be most productive. Degree-day calculations (see Appendix) for predicting when a new generation of eggs is likely to be laid are available for some orchard pests.

Generally 100,000 to 300,000 wasps must be released per acre to have an effect on pest populations. In commercial tomato fields in southern California, releases of *T. pretiosum* were successfully used to control fruitworms, loopers, and hornworms. Releases of about 15,000 wasps per acre were made twice weekly over a period of 10 weeks in the summer. In commercial avocados, once-a-week releases of 50,000 wasps in each of four randomly spaced trees per acre over a period of three weeks controlled amorbia and omnivorous looper.

In other crops, release of *Trichogramma* alone may not provide adequate control of economically damaging populations. However, when properly timed, it can be a useful component in pest management programs against many caterpillars that lay exposed eggs; such pests include corn earworm, loopers, amorbia, codling moth, and omnivorous leafroller. *Trichogramma* is not very effective against caterpillars, such as beet armyworm, that lay their eggs under a protective covering of scales. Consider the scale of your operation and the seriousness of infestation before choosing to release natural enemies. For instance, in backyard situations, where only a few rows of vegetables are grown, *Trichogramma* release is probably not an economically viable control method.

HANDPICKING

In small gardens, one of the most effective ways to control large caterpillars is to pick them off plants and destroy them or snip them in half with pruning shears. Because they are so large, hornworms are an obvious candidate for this method. Sometimes the caterpillars are hidden among the leaves and the best way to find them is to look for fresh fecal droppings or new holes in leaves. Caterpillars that roll or web leaves can be picked off with the whole leaf—but be sure to destroy

the caterpillar. Fruit with entry holes indicating presence of caterpillars should also be picked and destroyed to prevent further generations. Egg masses of pests such as beet armyworm and tussock moth are also easily spotted and destroyed; but study the photographs in this book and be sure to distinguish pest eggs from egg masses of beneficial insects that may also be in your garden.

BARRIERS

Caterpillars, such as the saltmarsh caterpillar and armyworms, that migrate in large groups into gardens and fields can be stopped with various kinds of physical barriers placed along the edge of the garden or field. The simplest barrier is a dry ditch, plowed or dug so the steepest side is next to the garden edge to prevent the caterpillars from crawling up; such a ditch can also be filled with water; however, both of these barriers are extremely difficult to maintain. A 6-inch-high strip of heavy aluminum foil set on edge with the top part bent away from the garden or copper screening as described for snail control are easier to maintain. Be sure to bury the bottom edge about 1½ or 2 inches into the soil so caterpillars can't get under it.

To successfully use barriers, it is essential to put them up before very many of the caterpillars have entered the garden. Such insect migrations are not common but usually occur after a nearby field or weed patch dries up or is cut down or harvested; check these areas for migrating pests as part of your regular monitoring routine.

For caterpillars attacking seed-

lings, protective coverings, such as screens, collars, row covers, protective cheesecloth, or nonwoven fabric or caps, can be very effective in reducing or eliminating damage. See Chapter 2 for more discussion of protective coverings.

CULTURAL CONTROLS

Keeping weed growth under control in areas surrounding the garden and keeping the garden patch relatively weed-free during periods when you are not growing a crop can help reduce pests that might invade the newly planted garden. Certain leaf-feeding caterpillars, including cutworms, armyworms, and saltmarsh caterpillars, are particularly associated with weeds because they feed on weeds in the spring before vegetable plantings are available. Likewise, rapid removal and destruction of crop debris after harvest is also important to reduce overwintering populations of some pests. For instance, areawide removal of mummy nuts remaining after harvest is the most important technique in keeping navel orangeworm populations low in walnut and almond orchards.

Another factor to consider in backyard situations is outdoor lights. Lights on, in, or near the garden attract night-flying moths, luring them to the garden to lay eggs, and may lead to greater damage by caterpillars of these species.

PHEROMONE TRAPPING AND MATING CONFUSION TECHNIQUES

Pheromones are chemicals that insects and other animals release to communicate with other individuals of the same species. Sex pheromones, which are given off by one sex (usually—but not always—the female) to attract the other sex for mating, have been identified for many insect species. Table 3-1 lists caterpillar pests for which pheromones are commercially available. These pheromones are used widely on commercial farms to monitor for the presence of pests; for some pests, researchers have developed guidelines that allow growers to use pheromone traps to determine when to treat and/or if a treatment is necessary. These programs are not time-efficient for gardens and very small farms, but publications describing these programs for larger operations are listed in the References.

Pheromones are also occasionally used to control caterpillar pests. Two approaches are used. The first approach (sometimes called mass trapping) is to try to catch all the male moths in traps so the females will not be able to mate and lay eggs. A second strategy is to apply such large concentrations of the pheromone to an area that males become confused and too disoriented to find females to mate with.

The mass trapping technique has been used successfully against the codling moth in backyard and small orchard situations. (See the codling moth section for more details.) However, because of the expense and difficulty involved in mass trapping, many other applica-

FIGURE 3-8. Pheromone dispensers, such as these Isomate dispensers wrapped around a peach tree twig for managing the oriental fruit moth, are being used against a number of different caterpillar pests. Dispensers vary in type and shape according to pest and manufacturer.

tions of this method are not likely. On the other hand, a new method of dispensing pheromones out of flexible polyethylene tubes has made future expansion of the use of pheromones as confusants quite likely. These tubes (Figure 3-8) can be wrapped around branches in orchards or stakes in gardens. The tubes are commercially available for oriental fruit moth under the trade name Isomate-M. (See oriental fruit moth section for more details.) The method has been used successfully in commercial orchards; however, it has not been tested for backyard use. Currently the method is also being used for tomato pinworm, artichoke plume moth, peach

twig borer, codling moth, and a few other insects.

MICROBIAL INSECTICIDES

Microbial insecticides are commercially available formulations of microorganisms that cause disease in insects. They are almost ideal insecticides from the environmental and health hazard point of view. They are generally very selective, killing only the target pest and leaving natural enemies, honey bees and other beneficials largely unharmed. They are not toxic to humans and other vertebrates. Whenever possible, microbial pesticides should be the first choice for pesticide treatment in the home garden.

Bacillus thuringiensis (Dipel, Thuricide, Attack) is effective against many leaf-feeding caterpillars (see Table 3-2). Bt, as *Bacillus thuringiensis* is often called, breaks down and loses its effectiveness faster than most other more toxic insecticides, so its application must be carefully timed and often must be repeated. For best results Bt sprays should be applied as soon as caterpillars hatch; look for fresh eggs on crop leaves and spray as soon as you find eggs hatching. Caterpillars that have been infected by the disease stop eating and their internal tissues disintegrate and liquify. It may take several days for caterpillars to die, but they usually stop doing damage soon after infection. To the untrained eye it may at first appear that the application has been ineffective; however, wait 72 hours before drawing such conclusions. Good coverage is essential; be sure all leaf surfaces are well coated. Species that lay eggs over a period of time may require multiple applications. Because Bt loses its effectiveness after 24 to 48 hours, caterpillars hatching more than a day or two after the original application will not be killed and will require a second application for control. Follow label directions for mixing and application. *Bacillus thuringiensis* is available in formulations that must be diluted with water and sprayed on and in ready-to-apply dust formulations in hand-shaker-type packages.

Although a microbial pesticide (a virus) for control of codling moth is under development, no other microbials are currently available for caterpillar control on food crops. Other materials may become available in the future. Some organic gardening sources suggest making your own microbial sprays by macerating apparently healthy caterpillars in water, straining them, and applying them to plants. However, University of California trials of this technique showed no control of cabbage looper or imported cabbageworm. More research is required to determine effectiveness for other species.

TABLE 3-2.

Some of the More Important Caterpillars Attacking Fruits, Nuts, and Vegetables in California Killed by *Bacillus thuringiensis*.

CATERPILLAR	MAJOR CROPS
Alfalfa looper, *Autographa californica*	cole crops
Anise swallowtail, *Papilio zelicaon* (California orange dog)	citrus
Artichoke plume moth, *Platyptilia carduidactyla*	artichokes
Cabbage looper, *Trichoplusia ni*	many vegetables, lettuce, cole crops
Celery looper, *Anagrapha falcifera*	celery
Citrus cutworm, *Xylomyges curialis*	citrus
Corn earworm *Helicoverpa (Heliothis) zea* (Tomato fruitworm)	lettuce, corn, beans, tomatoes
Diamondback moths, *Plutella* spp.	cole crops
Fall cankerworm, *Alsophila pometaria*	orchards
Fall webworm, *Hyphantria cunea*	trees
Fruittree leafroller, *Archips argyrospilus*	orchards
Grape leaffolder, *Desmia funeralis*	grapes
Green fruitworms, *Orthosia hibisci, Amphipyra pyramidoides*	pomefruit, stone fruits
Hornworms, *Manduca* spp.	tomatoes, grapes
Imported cabbageworm, *Artogeia rapae*	cole crops
Omnivorous leafroller, *Platynota stultana*	orchards and vines
Peach twig borer, *Anarsia lineatella*	orchards
Redhumped caterpillar, *Schizura concinna*	orchards
Tobacco budworm, *Heliothis virescens*	lettuce
Western grapeleaf skeletonizer, *Harrisina brillians*	grapes
Western tussock moth, *Orgyia vetusta*	orchards

NOTE: Check pesticide product labels to be sure crop and pest are listed before using.

PARASITIC NEMATODES

Commercially available formulations of parasitic nematodes, the most widely known being *Steinernema* and *Heterorhabditus* species, have been used to control caterpillars of certain boring moths such as the artichoke plum moth, the carpenter worm, and peachtree borer as well as soil dwelling grubs, cutworms and fungus gnats.

OIL SPRAYS

Spraying fruit trees with specially refined horticultural oils, preferably supreme or superior-type oils can kill overwintering eggs or pupae of leafrollers, tussock moths, and other caterpillars on trunks, thereby lowering summer populations of these pests substantially. The primary reason for applying such oil treatments during the dormant season is to control scales or as a delayed dormant treatment for aphids or mites; caterpillar populations alone do not usually justify such treatments. See the scale section for more information on dormant season oil treatments.

OTHER PESTICIDES

If more toxic pesticides than the ones discussed above are needed. use spot treatments when possible and time treatments to coincide with the pest's most vulnerable stage—usually the newly hatched caterpillars. Even among more conventional materials, some are less toxic than others or are sold in less disruptive formulations. For instance, insecticidal baits attract and kill cutworms and other seedling pests

and minimize impact of the pesticide on natural enemies and other nontarget species.

A number of materials designated as acceptable for use on crops to be labeled "organically grown" are effective against certain caterpillar species. In addition to the microbials and pheromones already mentioned, these include certain botanicals and cryolite, an inorganic formulation of sodium fluoaluminate, which is effective against orange tortrix and other caterpillars in grapes if there is no fog, rain, or irrigation water to wash it off. (See table 2-11 on page 31.) Check labels to determine registration status, crops these materials should be used on, and pests likely to be controlled.

CATERPILLARS AFFECTING FRUIT AND NUT TREES

Codling Moth
Cydia pomonella

Crops attacked: Apples, pears, walnuts, plums.

Damage. Codling moth is the most serious caterpillar pest of apples, pears, and walnuts; it is a less important pest on plums and other stonefruit. On apples and pears, larvae penetrate fruit and bore into the core or cause more

shallow blemishes on the fruit surface. In plums, and occasionally other stonefruit, codling moth bores into the fruit all the way to the pit. Codling moth larvae feed within walnuts on the kernel. Early in the season, damaged nuts drop off trees soon after damage occurs; later in the season damaged nuts remain on trees but kernels are inedible.

Identification of caterpillar. Caterpillars are not very distinctive; they are white to pink with a mottled brown head and shield on the first segment behind the head. However, few other caterpillars are found inside walnut, pear or apple fruit. Navel orangeworms may also be found in walnuts but have crescent-shaped markings on the second segment behind the head; also, unlike the codling moth, the navel orangeworm leaves a lot of webbing in the nut.

Where eggs are laid. Eggs are laid singly on fruit or nuts or on nearby leaves. Eggs are disk shaped, opaque white, and not easy to spot. Later they develop a red ring, and just before hatching the black head of the caterpillar inside becomes visible. Eggs can be found in apple and pear orchards, but are almost impossible to find in walnuts.

Other monitoring. Codling moths usually occur in fairly distinct generations, laying eggs and hatching in broods over a period of time. Emergence of moths of each generation can be monitored with pheromone traps placed in orchards; distinguish codling moths from other moths

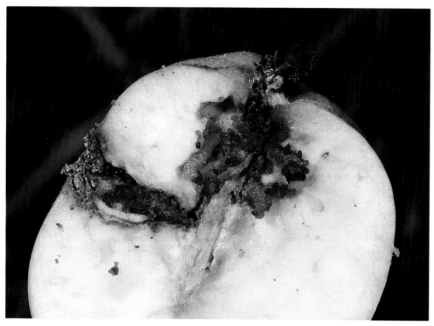

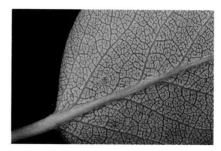

The small, disk-shaped object above the midrib of this leaf is a codling moth egg; eggs are extremely difficult to find in the field.

CODLING MOTH

egg	last instar larva	pupa
·		
		adult

0   1/2   1 in.
10   20   mm

Codling moth larvae bore into the center of fruit; they are cream colored with a brown head.

accidently caught in pheromone traps by the coppery band at the tip of the codling moth's wings. The hatching of eggs laid by these moths can be predicted using degree-day calculations. (See Appendix at the end of the book for more information on how to use degree-days.) Predicting egg hatch is important for timing treatments with pesticides or release of parasites.

Biological control. Naturally occurring natural enemies are not very effective against the codling moth. Some commercial growers report some control with mass releases of the egg parasite *Trichogramma* applied when peak egg laying by codling moth is occurring. To be effective parasite releases must be supplemented with other control methods, especially pheromone mating disruptants.

Sanitation. Various sanitation practices may be helpful in limiting codling moth numbers, especially in isolated orchards or backyard trees. A portion of the codling moth population will overwinter as pupae in protected areas on tree trunks, props, and in rubbish around the base of the tree; clean these areas in winter and remove loose bark that may harbor pupae. Another major source of overwintering codling moths is unsold apples stored in bins through the winter; be sure to clean these out well before the overwintering generation starts to emerge. In spring, clean up fallen fruit and nuts as soon as possible. Many of the fruit attacked by the first generation of caterpillars in May or June drop from trees with the developing larvae still in them. If these tiny fruit are disked under or collected and destroyed within a few days after they fall, there is some evidence that popu-

lations of the next generation can be significantly reduced. Daily removal is recommended in backyard situations for maximum effectiveness.

Trunk banding. A traditional nonchemical method for controlling codling moth is to trap mature larvae in a trunk band of tanglefoot, burlap bags, or corrugated cardboard as they move down the tree to pupate under debris. Trapped caterpillars (or pupae) must be killed before they can emerge as adult moths that can lay a second generation of eggs. Place bands on trunks after bloom just before the caterpillars begin to move down the tree. If you keep degree-day records, this would be about 485 degree-days (°F) after the eggs were laid. Less precise timing can be achieved by putting the bands up in early May in the Central Valley and by the end of May along the coast.

The corrugated cardboard method is preferred. Cut a 2 inch-wide strip of large-core corrugated cardboard (size Flute A 18-inch rolls from Oakland Paper Company, 1311 63rd Street, Emeryville CA 94608) and wrap it around the trunk of the tree so that the corrugation tubes are vertical. Staple bands to trees about 18 inches or more from the ground; pick the smoothest part of the trunk. The codling moth caterpillars will crawl into the corrugations and pupate. Remove the cardboard bands before moths begin to appear (725 degree-days after egg laying or the end of May in the Central Valley, and the last week of June in the coastal areas) and burn them to destroy all caterpillars and pupae. Be sure to mechanically destroy any pupae remaining on the trunk after you remove the band. To help suppress the overwintering generation, put new trunk bands up in August and remove and burn them between November and January. Tanglefoot can also be used, but you cannot be sure the sticky band will kill all the codling moths; also, tanglefoot tends to accumulate dust and debris and must be renewed frequently.

Banding works best on smooth-barked apple varieties such as Red Delicious that do not provide good alternative pupation sites; scaly varieties like Newtown Pippin have so many crevices that many caterpillars will pupate before they get to the banded area. Even in the best situations, banding will only control a small percentage of the codling moths because many pupate elsewhere in the tree or drop to the ground, bypassing the trunk.

Any of the banding techniques should be used in combination with other methods for effective control because they trap only a portion of the caterpillars.

Bagging fruit. Enclosing fruit that are on the tree in bags so that they are not accessible to codling moth larvae can give excellent control of this pest. Bagging can be done over the entire tree, or on as many fruit as desired. To bag the fruit, thin fruit to one per cluster. The best time to do this is when the fruit is from ½ to 1 inch (1.2 to 2.5 cm) in diameter. Use No. 2 paper bags (standard lunch size) that measure 7¼ inches (18.5 cm) by 4 inches (10 cm). Cut a 2-inch (5 cm) slit in the bottom of the bag and slip this opening over the fruit to form a seal around the stem. Staple the open end shut. While this technique does not affect the maturity or quality of the fruit, it may have an impact on the color development of red varieties. On the plus side, bagging the fruit protects it from both codling moth and sunburn. In addition, thinning the fruit to one per cluster, results in larger fruit at harvest.

Mass trapping. Mass trapping of codling moth males can provide some control under certain circumstances. The idea behind mass trapping is to attract all the male moths in the area into a sticky trap. Traps are baited with a synthetic sex attractant (pheromone) that mimics the chemical female moths use to lure males for mating. Once trapped in the sticky trap, males are no longer available for mating and unmated females cannot produce viable eggs. For this method to be effec-

tive, however, you must catch all the males in the area before they can mate. Thus mass trapping works best where trees are isolated by at least a mile from other trees harboring codling moth (especially apple, pear, and walnut trees) and where populations are low.

To carry out a mass trapping program, start in mid-March by placing two to four traps in each large tree and one to two in each small tree, away from the trunk about 6 feet above the ground. Check the traps every week or two and remove dead moths. Put new pheromone lure caps in every 2 weeks, and change the sticky bottoms every 2 weeks or sooner if they become dirty or more than 200 moths have been caught. Also scratch or score the sticky surface to maintain its adhesive quality. Codling moth traps, caps, and sticky bottoms are available from many commercial sources.

Pheromones are also used in commercial orchards in controlled release form in plastic dispensers or polyethylene tubes. Dispensers or tubes are tied around twigs in the same way as those for oriental fruit moth are applied (see Figure 3-8). The high dosages of the pheromone confuse the male, and males and females are unable to find each other for mating; thus, few viable eggs are laid. They are most useful in orchards with light to moderate populations. As is the case with mass trapping with pheromones, trees would have to be well isolated from other sources of codling moth for this method to be successful.

CHEMICAL CONTROL

In backyard trees grown for home use, insecticide applications can be difficult to time accurately and effectively, so consider using a combination of the cultural methods described earlier and accepting the presence of some wormy fruit. Reducing damage to very low levels usually requires application of broad spectrum insecticides.

Proper timing of insecticide applications is critical for them to be effective against codling moth. They must be applied just as eggs are hatching. Once the caterpillar has gone into the fruit or nut, it is protected from pesticides. The most effective way to time insecticide applications involves the calculation of heat units, known as degree-days. This method is recommended for commercial orchards but is probably too labor intensive and difficult for home gardeners. In order to calculate degree-days, you need daily maximum and minimum temperatures and a pheromone trap. Details on timing treatments using the degree-day method are given in *Integrated Pest Management for Apples and Pears* listed in the References. Calculation of degree-days is also discussed in the Appendix of this book. First egg hatch will occur when 125 degree-days (using a lower developmental threshold of 50°F and an upper threshold of 88°F) have accumulated from the first 62°F sunset after moths are caught in traps. Sprays for the first generation should begin at about 250–300 degree-days.

While timing sprays is best done with the use of degree-day calculations, if you are unable to

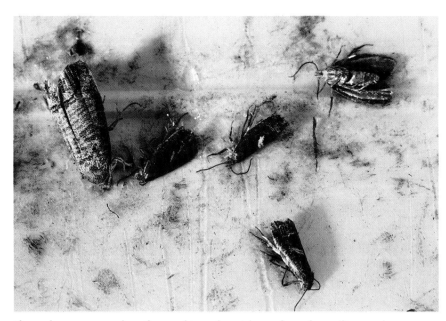

Shown here are several moths caught on the sticky surface of a codling moth pheromone trap. The larger moth with the coppery wing tips is the codling moth. The smaller ones are *Grapholita* species.

calculate degree-days and do not live in an apple-and pear-growing county where this information is available from your local Cooperative Extension office, you can monitor your trees to detect the beginning of egg hatch. Check a sample of fruit at least twice a week for entry holes or "stings" that indicate larvae have begun boring into fruit. Be sure to examine the fruit where it touches another fruit; this is the most likely place to find an entry hole. Spray as soon as you see the first sting. Unless insecticide applications are properly timed, they are not worth applying; the alternative control methods listed earlier will probably provide more reliable control. If a treatment is planned, most home orchards only require treatment of the first spring generation. However, where populations have been very heavy or tolerance

for damage is very low, the second and perhaps third generation may also need to be treated.

A number of insecticides are available to homeowners for managing codling moth. Broad-spectrum materials such as diazinon and carbaryl are most effective and remain effective for the longest time. Among these, diazinon has the advantage of being the least disruptive to natural enemies of insects and mites. However, carbaryl maintains its effectiveness against codling moth longer (28 days versus 14 days for diazinon), and thus may be easier for homeowners to use effectively without repeat applications. Egg hatch extends over about 28 days, depending on weather, and it is during this period that the crop must be protected.

Summer oils, which work by suffocating the eggs, have no

residual activity but are less toxic than the broad-spectrum materials. While more effective at controlling codling moth than the other organically acceptable materials, they may injure leaves or fruit if applied within 30 days of a sulfur, copper, or zinc application. Because they have no residual, 3 to 5 applications of oil are usually required to treat the egg hatch period. Other materials such as *Bacillus thuringiensis* (B.t.), ryania, cryolite, and pyrethrin/rotenone combinations have not been found to be effective at controlling codling moth.

Navel Orangeworm
Amyelois transitella

Crops attacked: Walnuts, almonds, pistachios.

Damage. Caterpillars enter nuts after the hulls or husks split, feed directly on nutmeats, and contaminate nuts with large quantities of excrement and webbing.

Identification of caterpillar.
If there is a large amount of excrement and webbing in the nut, the caterpillar is almost always a navel orangeworm. The caterpillar is milky white to pink with a dark reddish brown head. A pair of crescent-shaped marks on the second segment behind the head distinguishes it from codling moth, oriental fruit moth, and most other larvae found in nuts.

Where eggs laid. In spring, first and some second generation eggs are laid singly on damaged or mummy nuts that remained on trees after harvest and through the winter. Later generations lay eggs on hulls of newly forming or ripening nuts.

Other monitoring. Check trees in winter for unharvested mummy nuts. All mummies should be removed and destroyed by February 1. Crack open mummy nuts to find navel orangeworms and gauge the extent of infestation.

In commercial orchards, place egg traps baited with almond press cake by the first week in April to determine when egg laying is beginning. Observe eggs on traps to determine when hatching begins and when to time insecticide sprays if they are needed. (See *Integrated Pest Management for Almonds* and *Integrated Pest*

NAVEL ORANGEWORM

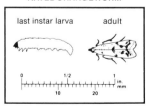

last instar larva   adult

Navel orangeworm leaves copious webbing and frass in the nuts it invades. The caterpillar is ivory with a brown head; the brown shiny pupa is partly exposed at left.

Management for Walnuts in the References for further information on this program.)

Control. Remove all old nuts after harvest from trees by hand poling before February 1. Infestation is unlikely in the subsequent year if cleaned trees are ¼ mile or more from other infested trees (including loquats, figs, pomegranate, and other deciduous fruit). Harvest nuts as soon as hullsplit occurs in almonds and as early as possible in walnuts, and pick them up to be stored in a sealed container as soon as they are dry. Insecticide sprays should be unnecessary if these two conditions are met.

Two parasitic wasps, *Copidosommopsis (Pentalitomastix) plethorica* and *Goniozus legneri*, have been introduced to control navel orangeworm, but populations of these beneficials do not remain high in properly cleaned orchards. It may be possible to maintain populations of parasites by keeping infested, parasitized nuts in screened cages throughout the orchard. Screen mesh must be small enough to be sure emerging navel orangeworm moths cannot escape into the orchard, but large enough that the parasites, which are smaller than the navel orangeworm moth, can.

Sprays of the parasitic nematode *Neoaplectana carpocapsae* have been used experimentally with some success to control navel orangeworm in almonds. Sprays must be applied just after hullsplit so the newly hatched navel orangeworms will become infected as they enter the nut. Nematodes remain effective for

about one week, so a second spray of nematodes should be applied a week after the first.

Orange Tortrix
Argyrotaenia citrana

Crops attacked: Grapes, citrus, stonefruits, strawberries, apples; mostly in coastal areas and interior valleys of coastal areas.

Damage. Caterpillars feed on leaves, often webbing them together; they may also feed on fruit, creating shallow scars that are unsightly and can be invaded by decay organisms. Grape berries may be webbed together and individual grapes or clusters killed; varieties with tight bunches suffer the most damage. In apples, varieties with short stems are damaged more severely by orange tortrix.

Identification of caterpillars. Caterpillars are greenish to bright yellow or pale straw colored with a golden head and shield on the segment behind their head. Like other leafrollers, they feed inside webbed nests and when disturbed, wiggle and drop from a silken thread.

Where eggs are laid. Eggs are laid in groups, overlapping like fish scales. They are cream colored and usually laid on smooth surfaces such as stems, fruit, and upper surfaces of leaves.

Other monitoring. Look for webbed leaves, larvae, and eggs, especially in the south and east quadrants of trees. Orange tortrix

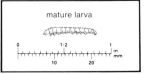

Two orange tortrix caterpillars and their webbing can be seen among this bunch of grapes.

prefers the top half of trees and twice as much damage occurs on the top as on the bottom half of trees. Look in grape clusters for webbing, but do not confuse spider webs with orange tortrix webbing. Confirm diagnosis by checking for other evidence of orange tortrix such as caterpillars or feeding damage. In dormant periods, check weeds and debris for caterpillars, eggs, and pupae. Pheromone traps are available but not very useful because orange tortrix tends to occur in overlapping generations after spring activity begins.

Control. Clean up orchards and vineyards in the dormant period. Destroy host weeds (e.g., curly dock, filaree, lupine, mallow, California poppy, mustard, oats and barley), mummy fruit, and trash at least a month before buds begin to break. Hand removal of webbed leaves and larvae can partially control orange tortrix in the backyard situation. Early harvest in grapes can eliminate some damage since populations increase in late summer and fall. For commercial grapes grown for the "organically grown" market, the ground mineral cryolite is an effective insecticide against the orange tortrix. In apples, thinning the crop to one or two fruit per cluster will reduce damage.

The most important parasite is a species of wasp in the genus *Exochus* that attacks larvae; it can control orange tortrix in some situations. Other parasites include *Apanteles aristolilae* and a tachinid fly, *Nemorilla pyste*. General predators also feed on eggs and larvae.

The fruittree leafroller has been taken out of its protective leafroll here; note the shiny black head.

Fruittree Leafroller
Archips argyrospilus

Crops attacked: Citrus, apples, almonds, pears, stonefruit.

Damage. Caterpillars feed on young leaves, buds, and developing fruit, often rolling and webbing leaves and sometimes fruit together. Damage to fruit may open it up to decay organisms. Most damage occurs in the spring and early summer; there is only one generation a year; after midsummer the pest remains inactive in the egg stage until the following spring.

Identification of caterpillar. The caterpillar is green with a shiny black head. It feeds inside rolled leaves or blossoms and wriggles and drops on a spun thread when disturbed.

Where eggs laid. Irregularly shaped egg masses of the fruittree leafroller are covered with a white coating and laid mostly on twigs and small branches in the upper half of the tree. The pest overwinters in this stage from midsummer until mid-March.

Search out and destroy egg masses or, if you intend to use insecticides, watch eggs in the spring to determine when hatching starts. Hatching begins first in warm, elevated areas. Treatments are most effective at 25 to 50% egg hatch.

Other monitoring. Fruittree leafroller is an early spring pest; the caterpillars are the small green worms with black heads that appear when buds of deciduous fruit trees are just opening up. Later in the spring look for webbed leaves and fruit on the outer edges of the tree canopy. Destroy them as you spot them.

The fruittree leafroller, like other leafrollers, wriggles and drops on a spun thread when disturbed.

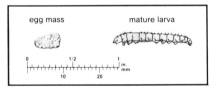

FRUITTREE LEAFROLLER

egg mass   mature larva

After midsummer, egg masses on twigs and branches will be the only stage of this leafroller present.

Control. Fruittree leafroller is rarely a serious pest in most unsprayed orchards, despite a lack of known natural enemies. *Trichogramma* may parasitize eggs, and general predators attack larvae and eggs. Handpicking and destroying rolled leaves or webbed plant parts will help keep numbers low.

Oil sprays applied in January or February before buds begin to open will kill egg masses on twigs of apples, almonds, and stone fruit. Spraying must be thorough to kill eggs on smaller twigs. *Bacillus thuringiensis* is effective if applied during warm or dry weather at 25 to 50% egg hatch.

Fruittree leafroller egg masses are laid on twigs and small branches. The lower egg mass shows the exit holes left by emerged larvae.

Omnivorous Leafroller
Platynota stultana

Crops attacked. Most fruit trees, including citrus, grapes, kiwi, some row and vegetable crops, especially peppers

Damage. Caterpillars feed on leaves, flowers, and fruit, webbing leaves together and onto fruit—often causing unsightly fruit. In grapes injury to fruit creates entry sites for rot-producing organisms. Omnivorous leafroller has many generations a year, and damage can occur throughout the growing season.

Identification of caterpillar. The smallest larvae are cream colored with brown heads and shields on the segment behind the head. Older larvae have lighter brown heads and are creamy white to green, but are so translucent that the main blood vessel is visible running down the back. Each abdominal segment has about six chalky white bumps with bristles growing out of them; similar bumps on related species, such as the orange tortrix, are the same color as the rest of the body. After hatching, larvae are solitary and web nests in leaves and other plant parts. When disturbed, they retreat to nests or wriggle and drop suspended from a silken thread.

Where eggs laid. Eggs are laid in groups overlapping like fish scales, on upper surfaces of leaves, and fruit. Eggs turn from green to greenish brown as larvae develop within.

Other monitoring. Look for this pest in the sunny south and east quadrants of the tree. Look especially for webbed nests in leaves and fruit and for egg masses on upper surfaces of leaves and fruit. In citrus, check for small larvae under the sepals of the fruit where it attaches to the stem. Destroy larvae and eggs when you see them. Pheromone traps are available and are useful for tracking populations, although there are no specific guidelines for using them to time treatments.

Control. Naturally occurring parasites often effectively control omnivorous leafroller. Important parasites include a tachinid fly, *Erynnia tortricis*, and a eulopid wasp, *Elachertus proteoteratis*. Look for tachinid fly eggs on the heads of the larvae. Larvae attacked by *E. proteoteratis* have many small parasite larvae or pupae along their backs. *Apanteles* wasps and *Trichogramma* may also attack omnivorous leafroller. Some growers have used mass releases of *Trichogramma* against this pest.

High populations of omnivorous leafroller can be tolerated during periods when fruit is not susceptible. Usually it takes 2 to 3 years for a damaging population to build up after a treatment has been made. Light to moderate infestations can be controlled with *Bacillus thuringiensis*. In grapes, cryolite is used effectively on commercial produce for the "organically grown" market.

Sanitation is important. During the spring and summer, remove webbed leaves and fruit and egg masses. Remove trash, mummies, and remaining grape bunches in

The omnivorous leafroller larvae are so translucent you can see the main blood vessel running down their backs. Note the chalky white bumps on each segment and the brownish head that distinguish this species from other common leafrollers.

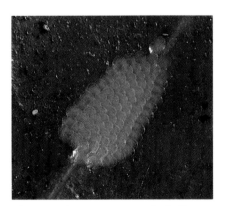

The eggs of the omnivorous leafroller are disk shaped and laid in overlapping clusters.

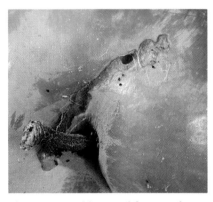

The gouges, webbing and frass on this plum were caused by omnivorous leafroller attaching a leaf to the fruit.

OMNIVOROUS LEAFROLLER

mature larva

0   1/2   1
| | | in.
                              mm
          10   20

the dormant season. Control host weeds. In commercial vineyards french plowing is recommended to bury infested trash, but drivers must be skilled to avoid damaging vines and stakes.

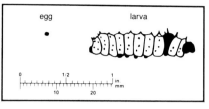

Mature redhumped caterpillars are very distinctive with red heads and humps and yellow, black and white stripes.

These redhumped caterpillars have just hatched from the adjacent egg shells.

This fluffy mass is a group of cocoons from an *Apanteles* parasite of the red-humped caterpillar.

Redhumped Caterpillar
Schizura concinna

Crops attacked: Deciduous fruit trees, especially plums, prunes, and walnuts.

Damage. This caterpillar is an occasional pest; it skeletonizes leaves and can defoliate trees but does not damage fruit. Often a single branch near the top of the tree will be defoliated. Serious damage is normally limited to young trees since older trees can tolerate substantial leaf loss with little economic loss.

Identification of caterpillar. Redhumped caterpillars feed in groups. Young larvae are yellow with black heads and black stripes and warts along their backs. Older larvae are yellow with black and white stripes, a distinc-tive red hump on the back, and a red head.

Where eggs are laid. Round, creamy eggs are laid in clusters of 25 to 200 on the undersides of leaves.

Other monitoring. In May, start looking for eggs and young larvae feeding in groups on the undersides of leaves. Once cater-pillars are large, you can often spot infestations by checking for piles of frass on the ground. The pest has three generations a year with the second brood beginning to hatch in July.

Control. The redhumped cater-pillar has many natural enemies including the parasitic wasps *Hyposoter fugitivus* and *Apanteles* species. Both types are larval parasites whose pupae can be found attached to skins of killed pest caterpillars. General preda-tors also feed on redhumped caterpillar.

Physically removing clusters of redhumped caterpillars from young trees is easy and costs less than spraying. Clip infested leaves and drop them into a can of soapy water. *Bacillus thurin-giensis* will control redhumped caterpillar. Spot treatments of affected trees are all that are generally necessary, even on young trees.

Western Tussock Moth
Orgyia vetusta

Crops attacked: Apple, cherry, plum, prune, walnut, avocado, citrus.

Damage. Western tussock moth caterpillars can defoliate deciduous trees in spring or destroy the spring growth flush in citrus. They may also take small bites out of newly set or young fruit; these damaged areas later scab over. The pest rarely attacks older fruit.

Identification of caterpillar.
The long hairs on tussock moth caterpillars make them easy to identify. Young larvae are black with long bristles; older larvae have numerous red and yellow spots, four white tufts of hair, and many more groups of bristles down their backs.

Where to find eggs.
Tussock moth eggs are laid in midsummer and do not hatch until the following spring; there is only one generation a year. Eggs are commonly deposited on branches and twigs in an egg sac, usually on an empty tussock moth pupal case. The pupal cases, which are easier to find than the eggs themselves, are light brown, quite hairy, and about a half inch long. Look for them in the dormant season on leafless trees.

Other monitoring.
Although the most effective way to monitor is to check for eggs in the winter before they hatch, you can also check for larvae in the spring.

The long hairs and colorful tufts of the older tussock moth caterpillars make them easy to identify.

Most damage occurs between mid-March and mid-May.

Control. Removing egg masses and newly hatched caterpillars should provide sufficient control in backyard trees. However, an oil spray in the winter will also kill the egg masses on deciduous trees. In commercial citrus orchards, one egg mass per tree is considered an economic treatment level; treat after 90% of the eggs have hatched. *Bacillus thuringiensis* is available for use against the western tussock moth in most crops.

The tiny dermestid beetle, *Trogoderma sternale*, feeds on eggs and is an important natural control in southern California citrus groves.

These tussock moth larvae have just hatched out of an egg mass laid on an empty tussock moth pupal case.

WESTERN TUSSOCK MOTH

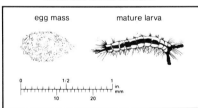

egg mass · mature larva

Green Fruitworm

Crops attacked: Apples, cherries, plums, apricots, strawberries, pears, prunes.

Damage. Early in the season, green fruitworms feed on leaves, occasionally tying them loosely together with silk; however, economic damage begins after fruit formation. They begin attacking fruit at petal fall; fruit may be almost entirely devoured or they may only have a few bites taken out of them. Badly injured fruit drop, but superficially damaged fruit remains on the tree, becoming misshapened and scarred. Feeding gouges on fruit are deeper than those caused by leafrollers. There is one generation a year and damage occurs only in spring on very young fruit.

Identification of caterpillars.
At least 10 species of relatively large green caterpillars can be included in the green fruitworm group; the most important in California are the speckled green fruitworm, *Orthosia hibisci,* and the humped green fruitworm, *Amphipyra pyramidoides.* Youngest larvae feed during the day whereas older larvae feed at night and hide during the day. Small larvae are difficult to distinguish from other small green caterpillars. Mature larvae of the speckled green fruitworm have cream-colored lines down the back and sides of their bodies. Humped green fruitworms have a prominent hump on the tail end of their body and bright yellow and white stripes down their backs.

This speckled green fruitworm is shown on an apple blossom.

Where eggs laid. Eggs are laid on twigs and begin to hatch when fruit buds begin to reach the cluster stage. Most species lay their eggs in the spring; an exception is the humped green fruitworm, which lays its eggs in fall. All species have only one generation a year.

Other monitoring. Check for caterpillars and damage by carefully examining fruit as soon as it begins to form. Beating trays can also be used, especially after petal fall when larvae are large and readily dislodged. Populations are often spotty within the orchard, so check a representative sample of trees and limbs.

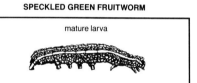

Control. Natural control factors probably hold green fruitworms below damaging levels in most seasons. One larva or damaged cluster per 100 clusters examined has been tentatively established as a treatment level in commercial pear orchards. Many insecticides are available for control, including *Bacillus thuringiensis.*

Peach Twig Borer
Anarsia lineatella

Crops attacked: Peaches, nectarines, almonds, apricots, plums, prunes.

Damage. Peach twig borer caterpillars bore into the growing shoots of twigs and ripening fruit or nuts. Damage to twigs is most serious to young trees, because death of new shoots can interfere with normal shoot growth and cause problems with tree shaping and pruning. Shoots and leaves wilt (flagging) and die back one to several inches from the growing tips of twigs. Slice open twigs and look inside for caterpillars. Damage to almond kernels is usually superficial scarring with no webbing. In fleshy fruit, such as peaches and nectarines, peach twig borers infest ripening fruit, especially around the stem end or along seams of apricots. Usually feeding injury is superficial, less than ⅜ inch deep on fruit. Examine caterpillars to distinguish the damage from oriental fruit moth injury in peaches, nectarines, and almonds.

Identification of caterpillars.
When they first hatch, twig borer caterpillars are dirty white or light brown with a black head. The head remains black, but the body turns chocolate brown as the caterpillar grows and the white portions between each body segment give the appearance of bands. Mature caterpillars do not grow much larger than ½ inch.

Peach twig borer caterpillars bore into growing shoots of twigs in spring; this one has been sliced back to expose the pest within.

Older peach twig borer larvae have a banded appearance. The pupa is shown on the almond husk above.

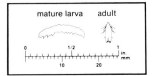

PEACH TWIG BORER

mature larva   adult

0   1/2   1 in.
|‖‖‖‖‖‖‖‖‖‖‖‖‖‖‖‖‖‖‖‖| mm
       10   20

Where eggs are laid. Tiny eggs are laid individually on twigs and fruit in the spring and summer and on young branches in the fall. Eggs are also laid on the undersides of leaves next to veins or the midrib.

Other monitoring. In the fall, winter, and very early spring, look for the caterpillar's overwintering cells (called hibernacula) in limb crotches. The overwintering cells are visible on the bark surface as minute, chimneylike, reddish piles of excrement and sawdust on 1- to 4-year-old wood. During the growing season, check

Peach twig borers overwinter in cells called hibernacula covered with chimney-like piles of frass and sawdust. Look for them in tree crotches in fall or early spring.

trees for wilting or wormy shoot tips and wormy fruit. Pheromone traps are used to monitor the adult moths.

Control. Although the peach twig borer has many natural enemies, none provides reliable control every year, even in many unsprayed orchards. The most important are the parasitic wasps, *Paralitomastix varicornis* and *Hyperteles lividus,* and the grain or itch mite, *Pyemotes ventricosus.*

Low levels of peach twig borer can be tolerated in backyard trees. If control is needed, the best control options are sprays of the microbial insecticide *Bacillus thuringiensis* at bloom or a spray combining oil and an organophosphate in the dormant season. Sprays of oil alone will not control peach twig borer. The dormant oil and organophosphate spray has the advantage of also killing a number of other insect pests such as scales and is easy for gardeners to time treatments during the December through January dormant period. If coverage is good, populations will be reduced over 95%. *Bacillus thuringiensis* sprays have the advantage of posing little risk to human health, the environment, or natural enemies of other pests. For Bt sprays to work, you must make 2 applications during bloom: the first between popcorn and the beginning of bloom, and the second 7–10 days later, but no later than petal fall. These sprays are aimed at killing the overwintering caterpillars just as they emerge from hibernacula.

Once fruit and twigs are infested in the spring, control is much more difficult. Sprays must be applied to control hatching larvae before they enter twigs or fruit. *Bacillus thuringiensis* is not effective at this time.

Oriental Fruit Moth
Grapholita molesta

Crops attacked: Peaches, nectarines, almonds, apricots, prunes.

Damage. Damage to twigs is similar to that caused by peach twig borer; young shoots wilt and die back one to several inches from the tip. Search for caterpillars to distinguish damage from the two pests; control options are different. On green and ripening fruit, oriental fruit moth caterpillars commonly bore right into the center to feed around the pit, distinguishing them from the more superficially feeding peach twig borers. Entry holes may be hard to find and plugged with hardened gum.

Oriental fruit moth, like the peach twig borer, bores into twigs and causes wilting. Note the frass at the tip of the twig that indicates the presence of a caterpillar. Split the twig open to find the caterpillar.

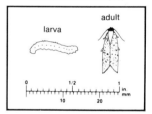

ORIENTAL FRUIT MOTH

larva   adult

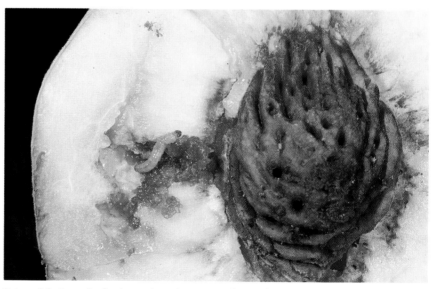

Oriental fruit moth also bores into the center of stonefruit. The caterpillar is white or pink with a brown head.

Identification of caterpillar.

The oriental fruit moth caterpillar is white or pink with a brown head; it grows up to ⅝ inch long. It does not produce webbing, distinguishing it from the navel orangeworm in almonds.

Where eggs are laid. Tiny disk shaped eggs are laid individually on leaves, twigs, and fruit.

Other monitoring. Look for caterpillars in wilted shoots and fruit. Sample fruit from the tops of trees, where the pest prefers to feed. Pheromone traps are available for monitoring.

Control. Naturally occurring biological control does not normally control oriental fruit moth. Programs using a mass release of the parasite *Macrocentrus* have been used with limited success in Colorado. Dormant oil sprays *do not* control this pest.

A new nontoxic control technique, which uses pheromones to disrupt mating and reproduction of the pest, became available in 1987 under the trade name Isomate-M. It has given good control in experiments in commercial orchards, although it has not been tested for use in backyards. Pheromones are dispensed from small polyethylene tubes tied on twigs (see Figure 2-8). Tie 3 or 4 tubes on to each tree in February before the first moths emerge. Replace tubes in 90 days to give control through the rest of the season.

If insecticides are used in commercial orchards, a special program using degree-days is available for timing application.

COMMON CATERPILLAR PESTS IN VEGETABLE GARDENS

Armyworms
Spodoptera species

Crops attacked: Lettuce, cole crops, tomatoes, beans, corn.

Damage. Armyworm caterpillars skeletonize leaves. In tomato they make shallow (and occasionally deeper) gouges in fruit.

Identification of caterpillar.

The tiniest armyworms feed in groups, distinguishing them from other common vegetable pests like corn earworms and loopers. Although markings on newly hatched armyworms are usually hard to distinguish, older larvae have distinct lengthwise stripes. Beet armyworm is green with light stripes; yellowstriped armyworm is dark brown or black with yellow and orange stripes. The surface of the armyworm skin is smooth and is not bumpy or hairy when viewed under a hand lens.

Where eggs are laid. Armyworm eggs are laid in fluffy masses on crowns of seedlings and on leaves of older plants. In staked tomatoes, yellowstriped armyworm often lays its eggs on the stakes.

Other monitoring. Look for egg masses. Before planting, look for them on redroot pigweed, nettle-

Beet armyworm eggs are laid in masses covered with hairlike fluff.

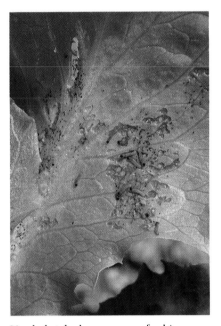

Newly hatched armyworms feed in groups.

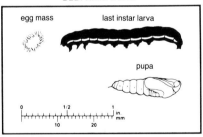

BEET ARMYWORM

egg mass   last instar larva

pupa

leaf goosefoot, and other weeds. On larger plants such as tomatoes, you can use a beating sheet, tray, or sweep net to collect larvae. In tomatoes and beans, check fruit or pods for shallow gouges and armyworms nearby. Check for armyworms parasitized by *Hyposoter* by pulling them apart—the parasite larva will pop out.

Control. Handpick. Natural control may be effective on caterpillars, especially virus diseases, parasites (*Hyposoter* and other species), and general predators.

Do not disturb parasitized or diseased individuals. Eggs are protected from parasites such as *Trichogramma* by fluff. Migrating armyworm populations can be stopped with an aluminum barrier or plowed ditch with the steep side toward your field or garden plot. *Bacillus thuringiensis* can give some control of armyworms. Smaller caterpillars will be easier to control, especially with less toxic insecticides, than larger ones. Ignore armyworms in sweet corn, where they do not usually cause economic damage.

The more mature beet armyworm caterpillar is green with distinct lengthwise stripes. The surface of the skin is smooth with only a few bristles. They commonly have a black spot on the segment above the second leg.

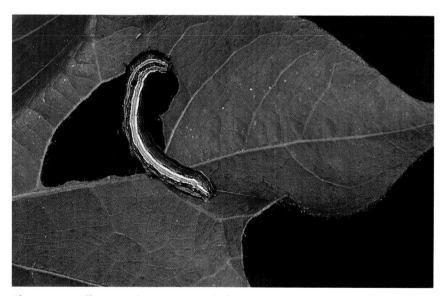

The western yellowstriped armyworm is dark brown or black with orange, yellow, or cream colored stripes.

This beet armyworm has been killed by a naturally occurring viral disease.

The parasitic wasp *Hyposoter* is laying an egg in a caterpillar.

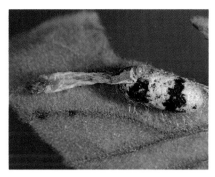

The black and white pupa of the parasite *Hyposoter exiguae* is shown here attached to the skin of the beet armyworm it consumed in its larval stage.

This yellowstriped armyworm has been pulled apart to reveal the larva of the parasitic wasp *Hyposoter* within.

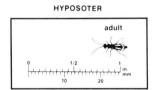

HYPOSOTER

Adult moths of many garden caterpillars look similar. Shown here are (a) variegated cutworm, (b) the beet armyworm, (c) the cabbage looper, (d) the tomato fruitworm, (e) the tobacco budworm and (f) the western yellowstriped armyworm.

Loopers can be distinguished by their distinctive looping movement.

Loopers parasitized by the wasp *Copidosoma truncatellum* curl into an 'S' shape after spinning a cocoon and fail to pupate. Numerous small wasps will emerge from each parasitized looper.

CABBAGE LOOPER

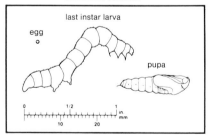

last instar larva

egg
o

pupa

0   1/2   1 in
                    mm
   10   20

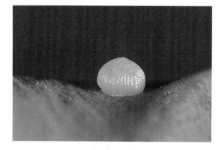

Looper eggs are laid singly and are dome shaped with distinct ridges.

Loopers
Trichoplusia ni

Crops attacked: Lettuce, cole crops, celery, tomatoes.

Damage. Loopers eat irregular holes in leaves, most often older leaves. They may also bore holes in lettuce and cabbage heads or damage tomato fruit.

Identification of caterpillar. Caterpillars are green with several white stripes down their backs. They arch their backs as they crawl; this "looping" movement gives them their name.

Where eggs are laid. Eggs are laid singly on undersides of older leaves; eggs are dome-shaped with ridges.

Other monitoring. Look for caterpillars and fecal pellets on undersides of leaves with holes. Use a beating tray for larger plants. Also look for parasitized eggs and larvae and disease-killed larvae; if substantial numbers are parasitized or diseased, populations may be reduced below damaging levels by these natural control factors alone. Pheromone traps are available for detection of adults.

Control. Handpick. Natural control by virus diseases, general predators, and parasites (*Hyposoter, Copidosoma, Trichogramma*) is often effective. *Bacillus thuringiensis* is effective. Mass releases of *Trichogramma* have worked experimentally in tomatoes.

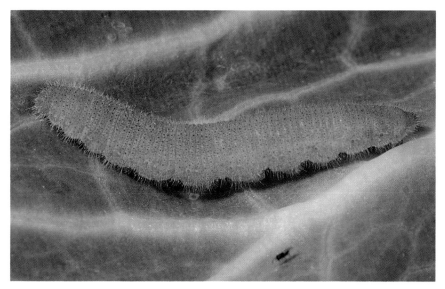

The imported cabbageworm has many short, fine hairs and faint yellow stripes down its sides and back.

Imported cabbageworm eggs are elongated and ridged; they are laid singly on leaves.

The cabbageworm butterfly is white to yellowish white. There are four black splotches on the upper surface of its wings.

Imported Cabbageworm
Artogeia rapae

Crops attacked: Primarily cole crops.

Damage. Caterpillars eat irregular holes in leaves and often contaminate edible portions with feces.

Identification of caterpillar. Cabbageworms are more sluggish than most other caterpillars found on leaves. Their skin has a velvety surface with many fine hairs. Older caterpillars are green with faint yellow stripes down their sides and back.

Where eggs are laid. Eggs are pale yellow to orange and are shaped like a football standing on its end; they are laid singly on leaves.

Other monitoring. Look for holes, frass, and caterpillars on undersides of leaves. Also look for parasitized and disease-killed caterpillars; high numbers of these can indicate that natural control will reduce numbers rapidly. Watch for the white to yellowish-white cabbage butterflies fluttering around; they have one to four black spots on their wings and are probably laying eggs. You can expect eggs to hatch and damage to begin a few days to a week after butterflies are sighted.

Control. Handpick. Natural control by virus and bacterial diseases and parasites (*Pteromalus puparum* on pupae, *Apanteles, Micropletis,* and tachinid flies on larvae, *Trichogramma* on eggs) can sometimes be effective. *Bacillus thuringiensis* is very effective. Mass releases of *Trichogramma* have been used with some success.

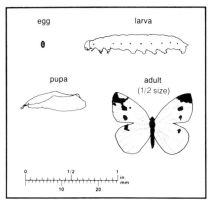

IMPORTED CABBAGEWORM

egg   larva

pupa   adult (1/2 size)

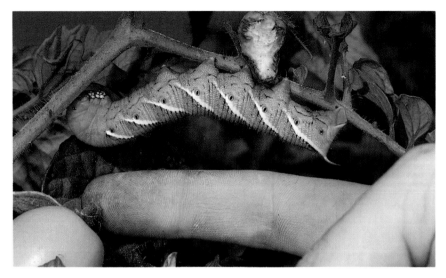

Hornworms are likely to be the largest caterpillars you see in the vegetable garden. Their striping pattern makes them hard to spot despite their size. Note the horn or thorn at the rear end.

Hornworms
Manduca species

Crops attacked: Tomatoes.

Damage. Hornworms consume entire leaves and small stems and may chew large pieces from green fruit.

Identification of caterpillar. Hornworms of all sizes have a distinctive horn or thorn at the rear end. Mature caterpillars are very large—up to 4 inches long.

Where eggs are laid. Smooth, round, pale green eggs are laid singly on leaves.

Other monitoring. Hornworm damage is obvious when the infestation is moderate to heavy because of the large amount of defoliation. Search for the large caterpillars. Large, black droppings on the ground beneath

The hornworm egg is smooth, round, and pale green.

tomato plants usually indicate the presence of hornworms.

Control. Handpick or snip hornworms with shears. Natural enemies (*Trichogramma, Hyposoter*) are common. *Bacillus thuringiensis* is effective on smaller larvae. Disking or rototilling after harvest destroys pupae in soil. Black light traps have been used to reduce populations in the southeastern United States.

Saltmarsh Caterpillar
Estigmene acrea

Crops attacked: Beans, lettuce, tomatoes, grapes.

Damage. Caterpillars skeletonize leaves.

Identification of caterpillar. Caterpillars are very hairy with conspicuous tufts of long hairs on each segment.

Where eggs are laid. Eggs are laid in clusters with no hairy covering and are spherical (versus stink bug eggs which are drum shaped). The pest commonly lays its eggs on weeds and migrates to vegetable crops.

Other monitoring. Watch for migrating groups of larvae in adjacent weedy areas before they enter the cultivated area.

Control. The saltmarsh caterpillar has many natural enemies. It seldom requires special control practices. Migrating larvae can be stopped with a physical barrier of heavy aluminum foil, a ditch of water, or a plowed ditch. *Bacillus thuringiensis* is effective against very small saltmarsh caterpillars.

Saltmarsh caterpillars have very long hairs with orange, white, and black tufts.

Corn Earworm
(Tomato Fruitworm)
Helicoverpa (Heliothis) zea

Crops attacked: Corn, lettuce, beans, tomatoes.

Damage. Corn earworm damage varies with the crop attacked. In tomatoes, it leaves deep watery cavities in fruit, which are contaminated with feces and cast skins. In lettuce, it decimates seedlings and bores into heads, causing damage and contamination. In beans, larvae feed on leaves, buds, flowers, and within pods. In corn, caterpillars are found eating down through kernels of ears. Prior to tasselling, the pest feeds on the developing tassels in the whorls of the plant. The seriousness of the damage depends on the time of year. Early tomatoes, beans, or sweet corn are not as seriously damaged as those planted for August or September harvest.

Identification of caterpillar.
The color of this species varies greatly and is not reliable for identification. However, older larvae usually have distinct stripes along sides and many short, whiskerlike spines over the body surface. Use a hand lens to check for warts along the back; each will have hairs or bristles growing out of it.

Where eggs are laid. Corn earworm eggs are spherical and are laid singly. Newly laid eggs are white, but they develop a reddish brown ring after about 24 hours. When the egg is about to hatch,

This corn earworm is boring into a tomato, where it is called the tomato fruitworm.

The corn earworm is a common pest of corn. The dark warts with bristles on them, common to all forms of the corn earnworm, are particularly visible in this color version.

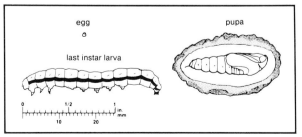
CORN EARWORM

egg

last instar larva

pupa

0   1/2   1 in.
mm
10   20

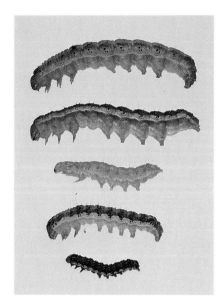

Do not rely on body color to identify corn earworms; they may come in many colors.

If you look at the earworm body surface with a hand lens, you will see many very short, whiskerlike spines over the body surface that distinguish them from armyworms.

the dark head and lighter body of the caterpillar can be seen inside; do not confuse about-to-hatch-eggs with those parasitized by *Trichogramma*, which are completely black within the egg shell. When available, corn silks are one of the earworm's preferred egg-laying sites; fresh silks are favored over older ones. On tomatoes, eggs are laid on terminal leaflets; in lettuce, a preferred site is the crowns of young seedlings.

Other monitoring. Slice open tomato fruit, corn ears, bean pods, or lettuce heads. Note size of larva on leaves and fruit; caterpillars larger than about a half inch are hard to control with *Bacillus thuringiensis*. Look for black eggs—a sign of parasitization. Pheromone traps are available for detection. Monitoring programs and treatment guidelines are available for commercial tomatoes and lettuce (see *IPM for Tomatoes* and *IPM for Cole Crops and Lettuce* in References).

Control. Home gardeners should consider avoiding sprays for this pest and discarding portions of fruit that have earworms in them. *Bacillus thuringiensis* kills 40 to 60% of the population, which should be good enough for crops such as tomatoes, beans, and lettuce in gardens and many small farms. *Bacillus thuringiensis* is not considered effective enough for use on sweet corn under most circumstances. However, some gardeners report success dusting silks with home-use dust formulations at 5 to 10% silk formation and continue treating every few days until silks turn brown; where corn is silking over a long period of time, this would require numerous applications.

A more reliable and environmentally sound control method for gardeners is to add 20 drops of mineral oil with a medicine dropper to the corn silk just inside each ear 3 to 7 days after silks first appear.

Because hatching larvae must be killed before they crawl into

A *Trichogramma* wasp lays its egg in the egg of a corn earworm. The corn earworm is only one of many caterpillar species *Trichogramma* parasitizes.

Eggs parasitized by *Trichogramma* normally turn darker than nonparasitized eggs as the wasp develops within.

the ear, any conventional insecticide, such as carbaryl, used against corn earworm in corn must be applied on the silks within 3 days after first silks appear; to achieve substantial control, applications must then be repeated at about 3 day intervals until silks turn brown. Again, breaking off and discarding wormy ends of ears may be the best control method for the gardener with only a few rows of corn.

In tomatoes, pick and destroy wormy fruit before caterpillars mature. Fruit harvested before late August usually escapes damage.

Important natural enemies include virus disease, *Hyposoter* and *Trichogramma* parasites, and general predators. Mass release of *Trichogramma* may also be helpful. Disking or rototilling plants and culls immediately after harvest eliminates some of the overwintering population and prevents migration to neighboring crops.

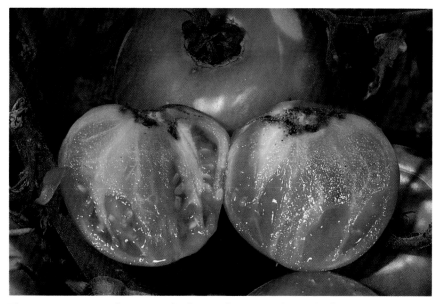

The tomato pinworm bores into the fruit at the stem end. Cut the fruit open to see the damage.

Tomato Pinworm
Keiferia lycopersicella

Crops attacked: Tomatoes (primarily coastal areas, southern California and San Joaquin Valley from Fresno south).

Damage. Pinworms bore into fruit at the stem ends, creating narrow, blackened tunnels, which expose fruit to decay. Often you must cut open the fruit to see the damage. Pinworms also mine leaves, but this activity does not usually produce economic damage.

Identification of caterpillar. Leaf mines and folded shelters are easier to spot than pinworm larvae. Caterpillars are tiny—when full grown they are less than ⅓ inch long. Their color varies from grey to yellowish with red or purple around each segment.

The tiny pinworm caterpillars have a mottled pattern; sometimes they are quite dark almost bluish black.

TOMATO PINWORM

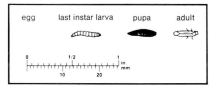

Where eggs are laid. Eggs are very tiny and too small to look for. They are laid on lower surfaces of leaves.

Other monitoring. Search foliage for folded leaf shelters and leaf mines that may contain larvae. Look under the green leaflike calyx of fruit for larvae, entry holes, or frass. Slice open fruit to find tunnels. In commercial fields, pheromone traps work well to detect reinvasions and subsequent generations after tomatoes have not been grown for 3 to 4 months (see *IPM for Tomatoes* in References).

Control. A host-free period of more than 3 months is essential for control of pinworm; the host-free concept requires that no tomatoes be grown in the area, including all neighboring fields and farms within several miles. Disk all plants and culls immediately after harvest; adults will not emerge from pupae buried more than 2 inches below soil surface. Natural enemies are often not effective. Pinworm is difficult to control with insecticides. Research indicates that the confusant technique using pheromone-filled dispensers (Scentry Attract 'n Kill, Tomato Pinworm Fibers and other formulations) may provide effective control if enough dispensers are used.

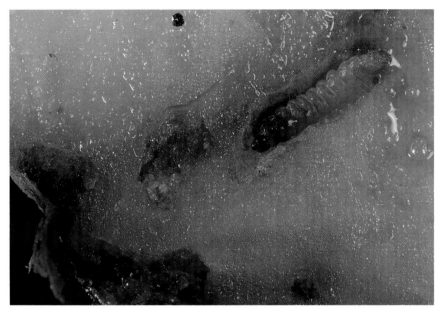

This potato tuberworm is burrowed deep into the flesh of a potato.

Potato Tuberworm
Phthorimaea operculella

Crops attacked: Tomatoes, potatoes, eggplant, peppers.

Damage. In tomatoes, tuberworms burrow into fruit and bore into terminal stems causing them to die. In potatoes, they tunnel in tubers, stalks, and leaves. Tuber eyes turn pink with excrement and silk. If potatoes are not stored properly, tuberworms can continue to cause damage in storage.

Identification of caterpillar. Tuberworms are dull white to pinkish, and do not grow longer than ½ inch. They have dark heads.

Where eggs are laid. The tuberworm's inconspicuous eggs are laid on tubers, foliage, plant debris, or soil. Moths will crawl through soil cracks to lay eggs on tubers beneath the soil.

Other monitoring. Check samples of harvested potatoes before storing. In commercially grown potatoes, tuberworms can also be monitored with water pan traps baited with pheromones. (See *IPM for Potatoes* in References for guidelines.)

Control. Damage to tomatoes, eggplant, or peppers can be avoided by not planting these crops near infested potato fields or following a potato crop in a garden. In potatoes, keep plants deeply (at least 2 inches) hilled with soil.

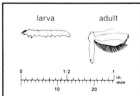

POTATO TUBERWORM

larva   adult

0   1/2   1 in
                              mm
         10   20

Do not allow soil to crack; soil cracks will allow entry of tubermoths to lay eggs. Sprinkler irrigation can help reduce cracks. Destroy infested tubers or store them at temperatures below 52°F to prevent further tuberworm development.

Garden Webworm
Achyra rantalis
Beet Webworm
Loxostege sticticalis

Crops attacked: The garden webworm attacks beets, beans, corn, strawberries, peas, and other crops. The beet webworm is a general feeder attacking beets, cabbage, carrots, beans, peas, potatoes, spinach, and cucurbits.

Damage. Webworms consume and web together leaves.

Identification of caterpillar. Webworms are somewhat hairy and may be various colors with a stripe down their back and three dark spots on each side of each segment. One to three bristlelike hairs grow out of each spot. Caterpillars web leaves together as they feed, producing abundant silk in and over these leaf shelters.

Where eggs are laid. Garden webworm eggs are laid in groups of a few to 50 on the undersides of leaves. Beet webworm eggs are laid in rows on undersides of leaves.

Other monitoring. Look for heavily webbed shelters with caterpillars inside.

Control. Clip off webbed leaves and destroy caterpillars. Control weeds, especially pigweed and lambsquarters, which may be sources of webworm invasions. Webworms are not very susceptible to natural enemies or insecticides because they are protected by their shelters; sprays must be applied when caterpillars are young before substantial webbing has been produced.

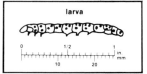

WEBWORM

larva

Webworms have numerous medium long bristles, a stripe down their back and dark spots. Colors vary. This one has two tachinid fly eggs on its head and thorax indicating that it is parasitized.

Cutworms

Crops attacked: Most garden crops.

Damage. Cutworms clip off seedling stems near or just below the soil level. A few species also climb up on foliage and chew holes or bore into heads of lettuce or cole crops.

Identification of caterpillar. Cutworms include various species, including variegated cutworm, *Peridroma saucia*, the black cutworm, *Agrotis ipsilon*, and the granulate cutworm, *Feltia subterranea*. They are dull-brownish caterpillars that curl into a C-shape when disturbed. Normally they are found on or just below

Cutworms are dull brownish caterpillars that curl up into a C-shape when disturbed. This one is the granulate cutworm.

CUTWORM

last instar larva

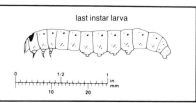

the soil surface or on lower parts of plants. They are smooth skinned and have various markings depending on species.

Where eggs are laid. Eggs are dome shaped and laid in groups or singly on stems and leaves near the ground.

Other monitoring. Cutworms are active mostly at night, so look for them with a flashlight. Look for cutoff seedlings, and dig around the base of injured plants to search for cutworms.

Control. To manage cutworms, thoroughly destroy crop residues, keep the garden weed-free and sod-free in winter, and destroy weeds in surrounding areas at least 10 days before planting. Cutworms often move to crops when weed hosts dry up in summer and are a special problem in new gardens planted in soil that was previously sod, pasture, or weeds.

Protect seedlings with cardboard collars, screens, or protective cloth. A ditch or partially buried metal fence around the garden may also deter immigrating individuals. Climbing cutworms can be kept out of perennial crops such as grapes with sticky collars. Insecticide baits can be used for cutworm control if applied before serious damage has occurred. Always plant a greater stand than desired to allow for some losses by seedling pests such as cutworms; dense stands can be thinned later. Some cutworm damage can be avoided by planting hardy transplants rather than sowing seed directly into the garden.

BORERS

The borers make up a diverse group of insects that tunnel into the bark and wood of trees and canes. Common boring insects include the larvae and adults of certain beetles, a few moth larvae and larvae of horntails, a group of insects related to bees, wasps, and ants. The species covered here include the three major boring insects that damage trunks and limbs of deciduous fruit trees: the peachtree borer, the Pacific flat-headed borer, and the shothole borer. Species such as the peach twig borer and the oriental fruit moth that attack fruit and twigs are covered in the fruit- and foliage-feeding caterpillar section. A number of boring insects attack blackberries and other caneberries and are briefly reviewed in the table for those crops in the back of the book.

On fruit trees, borers cause the most problems in newly planted trees and trees that have suffered damage from other pests or poor cultural practices. They are frequently a problem on young trees whose trunks have been sunburned; whitewash trunks of young trees with white latex paint from 1 inch below ground level to at least 2 feet above to prevent sunburn; tree protectors wrapped around the trunk can also be used. A program of careful monitoring for these pests and quick action when borers are discovered, coupled with normal good pruning, sanitation, adequate irrigation, and other recommended orchard management practices, should keep most of these pests from causing serious problems.

Peachtree Borer

The peachtree borer, *Synanthedon exitiosa*, attacks stonefruits and almonds. Most infestations of this pest in California have occurred in the Santa Clara Valley and Contra Costa County; however, it has also been reported as a pest in the northern San Joaquin Valley and may occur elsewhere. Damage is found primarily in the crown area or lower part of the trunk above or just below the soil line. These pests can girdle and kill a healthy young tree. Older trees are sometimes attacked but they usually tolerate the damage unless there are many larvae or a tree is attacked several years in a row.

The adult peachtree borer is a steel blue to black moth which emerges in the late spring and early summer. Female moths lay their eggs during the summer on bark at the base of tree trunks. Hatching larvae tunnel into the tree at or slightly below ground level. The larvae, light brown or pinkish with a darker head, feed in the crown area and burrow up into the tree, leaving piles of frass at their burrow entrances. In the late spring they pupate near the entrance of their burrows or in the soil. Adult moths emerge 20 to 30 days later. The peachtree borer has only one generation a year.

Management. If you live in an area where peachtree borers are a problem, examine your trees yearly. The best time to check is in the fall. Look for small accumulations of frass on the lower tree trunk, especially on trees that appear weakened. After the first rains, sap will exude from the larval tunnel. Remove 6 to 8 inches of soil from around the trunk and peel back the bark to observe the burrows. Pheromone traps are available for use in detecting peachtree borer in orchards.

Where only a few trees in a backyard situation are infested, you can remove borers through a technique called worming. Remove the soil from around the base of infected trees as described above, and use a pocketknife or some other pointed instrument to dig the larvae out. Be careful; it is possible to seriously injure the tree if too much wood is removed

Peachtree borer larvae can be controlled with applications of an insect parasitic nematode, *Steinernema carpocapsae*. Applications are most effective when larvae are feeding most actively and tunnel openings are largest in mid- to late summer. Apply nematodes into borer tunnels with a squeeze bottle applicator or 20-ounce oil can at a concentration of 1,000,000 or more invasive stage nematodes per ounce of distilled water. First clear the tunnel entrance of frass, then insert the applicator nozzle as far as possible into each gallery. Inject the suspension until the gallery is filled or liquid runs out another hole, then plug the tunnel entrances with rope putty or grafting wax. Agitate the applicator frequently to keep nematodes suspended in the liquid. Add a bit of latex pigment so you know which tunnels you have treated. Nematode treated larvae continue to feed and push frass from their tunnels for a few days. About 1 week after application, check that the opening of each gallery is plugged, replug any that have been opened, then spray the plugged openings

Peachtree borers are light colored larvae that can be found boring in the crown area of many fruit tree species.

When looking for peachtree borers, look for small piles of reddish brown frass at the base of the tree trunk as shown here.

with bright colored paint. Wait another week and check to see if these plugs are intact. If the gallery opening is no longer covered with paint, the larva has not been killed and you must retreat.

Insecticide sprays applied to trunks are also used to control adult moths in young commercial orchards with heavy infestations. Timing of application is critical and you need to choose a persistent material. Make one applica-

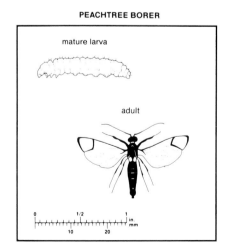
PEACHTREE BORER

mature larva

adult

tion in mid- to late-May to control adults as they emerge from pupae and another one in mid-July for later emerging moths and to prevent successful oviposition.

Pacific Flatheaded Borer

The Pacific flatheaded borer, *Chrysoborthris malis*, attacks most deciduous trees including stonefruits, apple and other pome fruit, walnuts, and almonds and occurs throughout California. Pacific flatheaded borers attack aboveground portions of the tree that have been previously injured by sunburn or other causes; this differentiates them from peachtree borers, which bore into undamaged trunks.

Tunnels excavated beneath the bark by the borer larvae cause sap to seep through the bark, creating spots that appear wet; there are no frass piles on the exterior as with the peachtree borer. Later, these areas may crack and expose the mines. Feeding by the Pacific flatheaded borer may cause a portion of the bark to die, or girdle and kill young trees.

The adult Pacific flatheaded borer—a hard, dark, mottled beetle with a bronzy cast—blends in well with the bark and is not commonly seen. The females lay their eggs on injured areas on the trunks of young trees or limbs of older trees. Young larvae feed

Larvae of the Pacific flatheaded borer have a distinctive flattened enlargement just behind the head.

under the bark in the rapidly growing outer wood and bore deeper into the trunk to pupate. Excavations are usually filled with finely powdered sawdust. The light-colored larvae have a distinctive shape with a flattened enlargement just behind the head.

Management. Flatheaded borers often invade sunburned areas on the trunk of newly planted first year trees. Wrapping or painting the tree trunk from 24 inches above to 1 inch below the soil line with white, indoor latex paint or whitewash will protect the trunk from sunburn and flatheaded borer invasion. Train your trees to avoid flat areas that will sunburn; when you prune, leave a few extra twigs to shade limbs below that may be prone to sunburn.

In older trees the best way to avoid infestations is to keep your trees sound and vigorous. Avoid

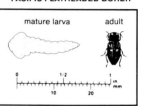

PACIFIC FLATHEADED BORER

mature larva   adult

0   1/2   1 in mm
    10   20

water stress and underfertilizing. Prune out all badly infested wood, and burn or remove it from the orchard before the growing season starts. Spraying for this insect is not recommended.

Shothole Borer

The shothole borer, *Scolytus rugulosus*, is a pest of many deciduous fruit and nut trees, including stone fruits, apples, pears, and almonds. It is a pest primarily in trees already weakened by root diseases, insufficient irrigation, sunburned limbs, Pacific flatheaded borer, or other maladies. Adults are tiny black beetles about 1/10 inch long. Females bore small holes, which look like shot holes, in the bark and lay eggs in a gallery 1 to 2 inches long running lengthwise down the cambium layer of the tree. Hatching larvae feed and excavate secondary galleries at right angles to the egg gallery, creating a gallery system that looks like a centipede. You must peel the bark off shothole-riddled areas of branches and twigs to see the galleries.

Healthy trees exude resin, which usually kills shothole borers before much damage can be done. If the tree has injured or weakened areas, this resin build-up does not develop and the invasion is successful. Ultimately the larval galleries may girdle the tree or a limb and cause its death. Two or three generations occur a year in California. Larvae spend the winter in their galleries beneath the bark.

Management. The key to preventing damage by shothole borers is to keep trees in sound and vigorous condition, with sufficient fertilizer and water, and free from disease and damage by other boring insects. The

Shothole borers are tiny black beetles.

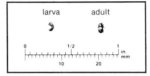

presence of shothole borers usually indicates that there are other serious problems. If shot-hole borers occur in one limb or scaffold, prune it out if the tree is vigorous, or remove the whole tree if it is weakened. Burn or remove all infested wood from the orchard before the growing season starts. Do not leave pruned limbs or stumps (healthy or infested) or firewood near orchards as these can be a source of infestations. Spraying is not recommended.

Sap leaking from many small holes on trunks or scaffolds is often a sign of shothole borer infestation.

LEAF- AND FRUIT-FEEDING BEETLES

Beetles are the most common and one of the most easily recognized insect groups. Fortunately, only a few are garden and farm pests. Many species, such as the familiar lady beetles, are predators. The species described here—flea beetles, wireworms, cucumber beetles, darkling beetles, vegetable weevils and green fruit beetles—are the most common pest beetles found in California vegetable gardens. Boring beetles, including the shothole borer and the Pacific flatheaded borer, which burrow into trunks and branches of fruit and nut trees, were discussed in the previous section.

All beetles have complete metamorphosis. Adult females lay eggs, which hatch into larvae of various shapes. After feeding and molting several times, larvae mature, pupate, and later emerge as adult beetles; the length of time it takes to complete the life cycle varies greatly from species to species and is also dependent on weather and other environmental conditions. Adult beetles are characterized by thick, hardened upperwings (elytra or forewings), which form a protective covering over most of the insect's body when it is at rest. When folded, the forewings usually form a straight line down the middle of the beetle's back. Beneath are the membranous hindwings, which are used for flying. Many adult beetles live for long periods of time and feed on the same crops as the larvae, as contrasted with other pest groups, such as caterpillars or maggots, with adult forms that do little, if any, damage.

Management practices vary, depending on species, insect stage, and crop.

Flea Beetles

Flea beetles are small, shiny beetles with back legs enlarged for jumping. The adult beetles feed mostly on leaves, leaving tiny pits or small holes at their feeding sites. Large numbers of flea beetles feeding together can cover leaves with bleached, pitted areas, ragged holes, or remove leaves altogether. Larvae usually feed on the roots of the same plants but normally do not cause injury; however, potatoes may occasionally be scarred by the shallow tunneling of flea beetle larvae.

Flea beetles are very common pests in newly planted vegetable gardens, especially on seedlings of tomatoes, potatoes, eggplants, peppers, cole crops, turnips, radish, and corn. Flea beetles normally fly (or jump) in from weedy areas; gardens surrounded by weeds that are drying up may be most seriously affected. Although flea beetles will feed on the leaves of many vegetable crops, serious damage is usually limited to seedlings with less than 6 leaves; on older plants flea beetles may be present, but remain mostly on the lower, older leaves. An exception is eggplant, which is particularly attractive to certain flea beetles and can be almost totally defoliated up through the fruiting stage. Flea beetles do not attack fruit.

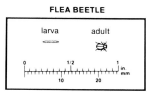

FLEA BEETLE

larva   adult

The tobacco flea beetle is brown with black markings across its back.

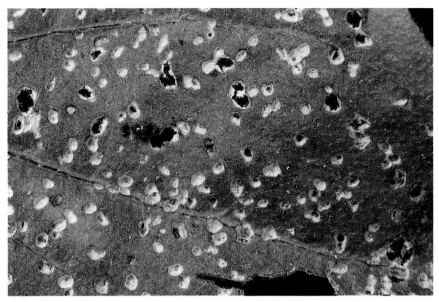

Flea beetles gouge many little holes on leaves as shown on this eggplant leaf.

Numerous flea beetle species occur in California. Among the most common are the tobacco flea beetle, *Epitrix hirtipennis,* which is brown with black markings across its back; the western potato flea beetle, *E. subcrinita,* which is shiny bronze or black; the western black flea beetle, *Phyllotreta pusilla,* which is shiny black to dark olive green; and the western striped flea beetle, *P. ramosa,* which is black with a white or yellow curved stripe along each side of the back. All are tiny, less than a tenth of an inch long, and jump like fleas when disturbed.

Management. Flea beetles overwinter as adults in debris and weeds, so it is important to remove these refuges by plowing or rototilling under weeds and crop debris in the fall after crops are harvested. Any management strategy for flea beetles must focus on protecting the youngest plants from attack, since older,

vigorous plants can usually tolerate substantial feeding. Seedlings can be protected with protective cloth or other coverings (see Figure 2-14 in Chapter 2) until they are in the 6 leaf stage (or older, if you are growing eggplant). Older transplants are less likely to suffer serious damage, so some losses can be eliminated by growing seedlings in greenhouses, under screens, or in other protected areas. Once heavy populations occur on seedlings in the cotyledon or first leaf stage and plants begin dying, there is little you can do, except wait it out or apply insecticides. Often only a few outside rows will require treatment. Among the organically acceptable materials, azadirachtin has been reported most effective.

Research has shown that crucifer flea beetle (*P. cruciferae*) populations are often lower on cole crop plants grown in a polyculture of other unrelated crops or healthy, vigorous weeds than in large concentrations of a single

type of plant grown in adjoining rows. Where flea beetles are a serious problem, such polycultures are a management alternative; however, other factors, such as potential for disease and other pests and competition for light, water, and nutrients, must also be considered. Some flea beetle species feed on a fairly wide range of vegetable crops and weeds, whereas others, like the crucifer flea beetle, are more restricted feeders.

Wireworms

Wireworms are the soil-dwelling larvae of click beetles, family Elateridae. Some species occur commonly in gardens and fields and injure seedlings by feeding on roots or boring into stems; most types of seedlings are susceptible to injury. Potato tubers and sweet potatoes may also be attacked by certain wireworms. Wireworms are slender, cylindrical, and usually yellowish and resemble mealworms. Damage is most common where the soil has a high organic content, especially fields that have recently been in or are adjacent to pasture, fallow land, or alfalfa. Check the soil in the root zone for wireworms to determine if they are the culprit. Prior to planting, soil fumigants and certain other pesticides can be used to kill wireworms, but such special efforts are seldom needed. Flooding a field for several weeks can also reduce populations.

Carrots are especially attractive to wireworms and reportedly can be used to trap wireworms out of a garden. Plant a nearly fully grown carrot (preferably with its top still on) in the soil every 2½ to 3 feet throughout the garden. Every 2 to 3 days, pull up the carrots, remove the wireworms and replace the carrots in the soil to trap more wireworms.

Wireworms are slender with 6 short legs close together near the head. This species is the sugarbeet wireworm.

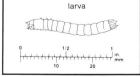

WIREWORM

larva

False Wireworms

False wireworms are the larvae of beetles in the genus *Eleodes*, family Tenebrionidae. They occasionally feed on sprouting seeds and seedlings of vegetables, usually in areas that have been fallow or in pasture and have not been intensely cultivated in recent years. The larvae may be dark brown to yellow and resemble true wireworms. False wireworm adults are dark brown to black beetles with smooth or ridged wing covers; adults can be distinguished from those of true wireworms by their longer, thicker antennae and front legs. When disturbed, they have the peculiar habit of placing their heads on the ground and elevating the hind part of their bodies as though they were standing on their heads. Adults do not normally cause serious economic damage. Control is the same as for true wireworms.

Darkling Beetles

Darkling beetles chew off seedlings or feed on foliage. Like wireworms, they are in the Tenebrionid family but are in the genus *Blapstinus* and are damaging principally as adults. Larvae resemble false wireworms but are much smaller. They usually invade from weedy areas or field crops, so damage often begins at the edge of the crop or garden. The beetles are most active at night but occasionally run on the ground in the daytime. They often hide under clods or debris during the hot parts of the day. To prevent beetle invasions from an adjacent field, fill a ditch full of water to keep them out. Baits placed around the edges of the field will also provide control.

Darkling beetles can be confused with predaceous ground beetles, but predaceous ground beetles are usually shiny, and the tip segments of their antennae are

Do not confuse the darkling beetle shown here with predaceous ground beetles. Ground beetles are shiny and do not have the tips of their antennae enlarged as do most darkling beetles.

DARKLING BEETLE

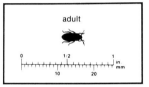

rarely as enlarged as those of darkling beetles. Some predaceous ground beetles have prominent patterns of lighter colors; they feed mostly on caterpillars and other insects.

Cucumber Beetles

Cucumber beetles are very common pests in vegetable gardens and may also attack ripening stonefruit. The most abundant species in California is the western spotted cucumber beetle, *Diabrotica undecimpunctata undecimpunctata*; however, the similar western striped cucumber beetle, *Acalymma trivittata*, may also cause damage.

Adults of both species cause the most serious damage. The beetles are shiny with black heads and about one-quarter inch long; the western spotted cucumber beetle is greenish yellow and has 12 black spots on its back; the western striped cucumber beetle is yellowish orange and has 3 black stripes. Adults feed on the leaves of melons, squash and other cucurbits, corn, potatoes, tomatoes, eggplant, beans, peas, beets, asparagus, cabbage, lettuce, and other vegetables as well as ripening peaches and apricots and other soft fruit. Shoots or blossoms may also be consumed. Cucumber beetles may also spread cucumber mosaic virus in cucurbits, although aphids are the primary vector.

The larvae of the two species are whitish and slender with three pairs of short legs; the head and tip of the abdomen are darker. Larvae feed exclusively on roots. Larvae of the western spotted cucumber beetle feed on the roots of corn, small grains, beans, sweetpea, wild grasses, and various weeds but do not usually cause noticeable damage in California gardens or crops. However, the larvae of a closely related subspecies, *Diabrotica undecimpunctata howardi*, known as the corn rootworm, cause serious losses to corn in the Midwest, but do not occur in California. The western striped cucumber beetle larva feeds exclusively on cucurbit roots and can damage these crops when infestations are heavy. The striped cucumber beetle larvae do not attack roots of other crops.

Life cycles of the two cucumber beetle species are similar. There are several generations a year. Beetles overwinter as adults in weedy areas and move into planted fields and gardens as soon as plants start to come up. They lay their yellow orange eggs at the base of plants or in soil cracks. Hatching larvae burrow into the ground seeking out roots, feed for 2 to 6 weeks, pupate, and emerge as adult beetles that attack the aboveground portions of the plant.

The western spotted cucumber beetle is a distinctive yellow green beetle with 11 or 12 black spots. Some people confuse them with lady beetles, but their antennae are much longer than lady beetle antennae.

The western striped cucumber beetle is similar in size and shape to the spotted cucumber beetle but has black and cream colored stripes on its wing coverings.

CUCUMBER BEETLE

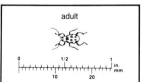

adult

Management. Management of cucumber beetles is difficult, and newly emerged seedlings can be destroyed in a few days. However, most older plants can support substantial numbers without serious damage. Insecticides can be used to kill them but applications must be repeated frequently to kill invading populations, especially if short residual botanical materials are used. The best strategy for most small farm and garden situations may be to place screens, protective cloth, or individual cups or cones over emerging plants and remove these protective devices after plants grow large enough to tolerate damage.

Little is known about the effectiveness of natural enemies against these pests in California. Tachinid flies and various general predators are known to attack them but rarely, if ever, provide economic control. Commercial preparations of entomophagous nematodes have been effective against the similar corn rootworm, but would have to be applied to the areas where larvae are developing—a strategy that would be feasible for the striped, but probably not the spotted, cucumber beetle in California.

Cucumber beetles are attracted to cucurbitacins, which are substances found naturally in all cucurbit species. Experimental studies have shown that if cucumber beetles can be attracted to substances mixed with cucurbitacin powder, the beetles will continue feeding until all the substrate is gone. Researchers in Illinois have combined insecticides with cucurbitacins and dusted the mixture on plants, reducing the quantity of insecticide required per unit area for control to about 1% of that required for conventional spray applications; however, no mixtures of cucurbitacin and insecticide are currently being developed commercially, and other strategies for use of cucurbitacin await development.

Vegetable Weevil

The vegetable weevil, *Listroderes costirostris obliquus,* is a sporadic pest in vegetables. It may attack potatoes, tomatoes, turnips, carrots lettuce, and other vegetables. Both adults and larvae feed on the buds, foliage, and roots, sometimes cutting young plants off at ground level or completely defoliating them.

The larvae are green, legless grubs about ⅜ inch when fully grown. Adults are small (⅜ inch) brown or gray snout beetles with an inconspicuous V-shaped spot at the tip of the wings. Both adults and larvae are most active at night and can be difficult to find in daylight. Adults do not fly, so infestation of new areas takes place slowly and damage within the field may be spotty. There is only one generation a year, but vegetable weevil adults may live 2 years or more.

Insecticides can be used effectively, but little is known about nonchemical controls. Young plants can be protected with protective cloth or cones. You may be able to prevent localized infestations from spreading by destroying infested rows or pick-

The vegetable weevil has a long snout and dull coloring.

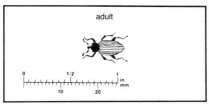

VEGETABLE WEEVIL

ing off all weevils at night. Sticky barriers can be used to keep beetles from migrating into new areas of the garden.

Green Fruit Beetle

The green fruit beetle, *Cotinis mutabilis,* is a large (1¼ inch long) metallic green beetle that attacks maturing soft fruit such as tomatoes, peaches, plums, figs, and apricots. Because of its scarab shape, it is often mistaken for the Japanese beetle, but it is much larger. It is primarily a backyard pest and not common in commercial orchards.

The larval stage is a C-shaped, creamy white grub, which feeds on decaying organic material in the soil. The larva does not damage plants. Manure and partially decomposed compost are favorite feeding sites and adult beetles seek out these areas to lay eggs. Adults also tend to feed in fruit trees near these sites.

The best control is to remove all manure from areas near fruit trees and turn compost piles frequently to expose small grubs. If grubs are found they may be killed by flooding the infested area for at least 2 days.

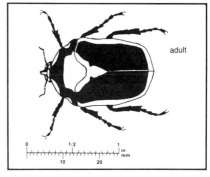

GREEN FRUIT BEETLE

A less effective strategy is to try to trap the adult beetles. Solutions of equal parts of grape juice and water or peach juice and water have been reported to be attractive. Place bait in the bottom of a 1 gallon container with a funnel of small mesh wire inserted in the top. Beetles will be attracted into the jar and once inside will be unable to escape. Combine this technique with early harvesting and removal of damaged fruit. Insecticides are not recommended against the adult beetles.

STINK BUGS, LYGUS BUGS, AND OTHER TRUE BUGS

Stink bugs, lygus bugs, and squash bugs belong to the insect order Hemiptera. The name Hemiptera or "half-winged" is a reference to their forewings, which are partly thickened and partly membranous. When folded, the tips of the wings overlap forming a fairly well defined X on the back of the body. This X-formation along with the distinctive inverted triangle, called the scutellum, behind the head make species within the Hemiptera group fairly easy to recognize. This group is often called the "true bug" group.

Although many of the true bugs are plant-feeding pests, there are also a number of important predators in the group. These predators include assassin bugs, ambush bugs, damsel bugs, minute pirate bugs, bigeyed bugs, and certain stink bugs. Pictures of several of these general predators appear on pages 40-44. Be sure you can distinguish them from plant feeders before taking control actions against true bugs.

True bugs go through incomplete metamorphosis (Figure 3-6). The young, called nymphs, resemble their parents but are wingless until they are about half grown when they begin to form wing pads. There is no pupal stage.

All the true bugs are sucking insects. The plant-feeding species insert their strawlike piercing-sucking mouth parts into tender plant tissues and suck out cell contents. Damaged fruit and buds often drop or become mishapened. Certain species inject a toxin while feeding that causes yellow spots on leaves and ripening fruit. Often spots and distortion do not appear until long after the true bugs have left, making it difficult to identify the cause of the damage.

The most common California true bug pests—stink bugs, lygus bugs, and squash bugs—are discussed individually below. Other true bug pests sometimes found in California gardens and orchards include false chinch bugs, *Nysius* spp., the western boxelder bug, *Leptocoris rubrolineatus*, and the leaftooted plant bug, *Leptoglossus clypealis.*

The boxelder bug is dark brown to black with red orange markings. Its eggs, laid in groups, are also red orange.

Stink Bugs

Stink bugs are shield shaped with a large scutellum or triangle on their backs. They are wider than most other true bugs and have 5-jointed antennae. The name stink bug comes from the offensive-smelling defensive substance many species give off when disturbed. A few species feed on insects. Two of the most common pests in California gardens and farms are the consperse stink bug, *Euschistus conspersus*, and the harlequin bug, *Murgantia histrionica*.

Consperse stink bugs attack a variety of fruits from stone fruit to pears to tomatoes, often leaving blemishes, depressions, or brown drops of excrement. On green tomatoes, damage appears as dark pinpricks surrounded by a light discolored area that remains green or turns yellow when fruit ripen. Areas beneath spots on tomatoes or depressed areas on pears and other fruit become white and pithy but remain firm as the fruit ripens.

The harlequin bug is a common pest of vegetables, especially those in the mustard family including cabbage, broccoli, kale, radishes, and turnips. Harlequin bugs leave yellow or white blotches on areas of leaves where they have been feeding; heavy infestations can cause plants to wilt, turn brown, and die.

Stink bug eggs are barrel or keg-shaped with distinct circular lids; they are laid in clusters of usually 10 or more on leaf surfaces. Nymphs are nearly round, often brightly colored, and disperse a few days after hatching.

The consperse stink bug is variable in color but usually gray brown to green with black speckled legs.

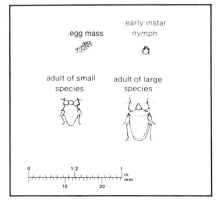

STINK BUGS

early instar
egg mass   nymph

adult of small   adult of large
species   species

Adults of the consperse stink bug are variable in color but usually gray brown to green with yellow to orange black speckled legs. Do not confuse the consperse stink bug with the rough stink bug, *Brochymena quadripustulata*, which also occurs in orchards but is an insect predator; the rough stink bug is steely gray and covered with white specks. Adults of the harlequin bug are bright

The harlequin bug is black with distinctive red markings.

The say stink bug is green with a white rim around its borders. This individual is a nymph and lacks fully formed wings.

Stink bug eggs are barrel shaped and laid in groups. These are say stink bug eggs. Other stink bug eggs vary in color but are a similar shape.

On fruit such as pears, stink bug feeding leaves deep depressions.

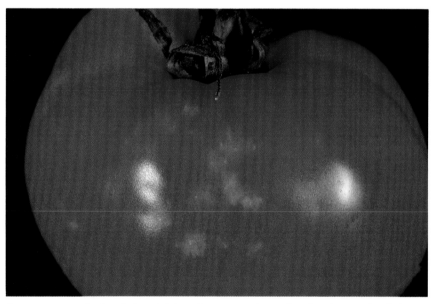
On tomatoes and other fruit, stink bug damage can leave discolored areas that fail to ripen with the rest of the fruit.

red and black. Coloration in other species varies. Stink bugs usually have three or more generations a year in California.

Management. In orchards, stink bugs breed and spend most of their time in ground covers or adjacent weedy areas. If stink bugs are a perennial problem, eliminate these areas in early spring before stink bug populations have built up; destroying weeds once populations are heavy will only drive more of the pests into the trees. Management of weedy areas often helps curb damage in vegetable crops, too. Wild blackberries are a favored host of the consperse stink bug and mustards often harbor the harlequin bugs.

The most important natural enemies of stink bugs are parasitic wasps, *Telenomus* spp., that attack the eggs and parasitic flies in the Tachinid family that attack nymphs and adults. Such general predators as bigeyed bugs and collops beetles may also attack eggs and nymphs.

In small vegetable gardens the egg masses and bugs can be picked off and destroyed by hand. Trap crops of mustards have been used to attract harlequin bugs in the early spring or late fall after the main crop has been harvested. Once the bugs have concentrated, trap crop and bugs can be destroyed with pesticides or other appropriate methods. However, be sure to destroy the fast moving bugs before disturbing them; otherwise you will just drive them to new hiding places. If this happens, your trap crop will have done more harm than good.

Lygus Bugs

Lygus bugs attack a large variety of crops but are a particular problem in beans, strawberries, and orchard crops. Adults are green, straw yellow, or brown with a conspicuous yellow or pale green triangle on their backs. Nymphs are light green and when tiny can be confused with aphids; however, they move much faster than aphids and have red-tipped antennae. Lygus bugs lay their narrow, cylindrical eggs singly just below the surface of leaf tissue; they are very difficult to find; eggs may be laid in many different crop and weed hosts. *Lygus hesperus* is the most common species.

On beans, lygus attacks buds and flowers, causing them to drop and thus lowering yields. After pods form, lygus feeding on seeds causes pitting and blemishing of pods. In strawberries, lygus bugs deform berries, causing gnarling and twisting—a symptom called catfacing—and enlarged, hollow, straw-colored seeds. In pears, tomatoes, and other fruit, lygus causes discoloration, bumps, or depressions on fruit, but without the white, pithy areas beneath characteristic of stink bug damage.

Many broadleaf weeds are lygus hosts, including redroot pigweed, lambsquarters, and related plants as well as knotweed, sunflower family weeds, and mustard family annuals such as wild radish, shepherdspurse, and London rocket. Lygus often moves into new crops or orchard trees when weeds are mowed or disked in orchards or vineyards

Adult lygus bugs have a conspicuous yellow or pale green triangle on their backs.

This photo shows three stages of lygus bug nymphs. Full-size wings do not develop until the adult stage.

Lygus bugs cause gnarling and twisting in strawberries, a symptom called catfacing.

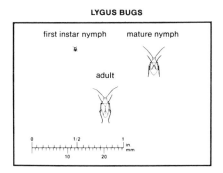

or when weedy fields of such annual crops as sugarbeets or tomatoes are harvested. Alfalfa is a favored host of lygus, and harvesting in that crop often stimulates major lygus migrations. Other legumes grown as crops or cover crops, including vetch, lupine, or fava beans, also support lygus populations. Maturing safflower is also a common source of lygus infestations. Rangeland weeds such as Russian thistle, lupine, and tarweed also harbor lygus, especially near the foothills.

Management. The best strategy for managing lygus is to remove sources of infestation whenever possible before damage is likely to occur on the crop. Disk or rototill weeds along field and orchard borders before the crop is planted or before susceptible fruit or buds appear on trees. If the ground cover in your orchard is the major source of lygus, consider removing it. An alternative approach in orchards is to keep your cover crop so lush that the lygus will not migrate up into the fruit to cause damage; however, maintaining this type of a cover crop can be difficult. Where nearby alfalfa fields are the principal source of lygus, strip harvesting or alternate harvesting of halves of the field may decrease migration into more susceptible crops.

Some growers have experimented with using borders of shasta daisies around commercial strawberry fields to keep lygus out of strawberries; these trap crops work if you can keep the daisies lush and constantly flowering. Once flowering ceases, lygus rapidly move into the strawber-

Squash bug adults are frequently found mating in the field.

ries. Also in commercial strawberries, growers have experimented with giant vacuum machines to suck lygus out of fields.

Insecticides can be used to control lygus around borders or in commercially grown beans but are probably not worthwhile in orchards or backyard gardens. A few lygus bugs can be tolerated since they primarily cause cosmetic damage. Lygus has developed resistance to many insecticides.

Lygus bugs have many natural enemies, which may keep their numbers under control under some circumstances. Predators include bigeyed bugs, damsel bugs, and collops beetles, which attack eggs and nymphs. Several parasites also attack lygus bug eggs and nymphs.

Squash Bugs

The squash bug, *Anasa tristis*, is a common pest of all cucurbit crops, especially squash and pumpkins. Adults are about ⅝ inch long, grayish or yellowish brown, flatbacked, and somewhat speckled, often with a dense covering of black hairs. The edges of the abdomen, which protrude from beneath the wings, are orange or orange and brown striped. Eggs are laid in the spring through midsummer on stems or on the undersides of leaves, often in the angle formed by two veins coming together. The orange yellow to bronze eggs are elliptical, and are deposited on their sides in groups. Young squash bugs are wingless and pale green to almost white, later turning darker brown. Very young nymphs feed close together. When squash bugs are crushed, they give off a disagreeable odor

A large population of squash bug nymphs feeds on a zucchini flower. Markings on the backs of nymphs become more distinct as they get older.

Squash bug eggs are bronze, eliptical, and laid in groups.

Squash bug feeding causes parts of vines to wilt and die.

and sometimes the term "stink bug" is incorrectly applied to them. Nymphs grow into adults by late summer or fall but these adults do not lay eggs until the following spring. There is only one generation a year.

Leaves fed on by squash bugs first develop small specks, which turn yellow and later brown; vines wilt from the point of attack to the end of the vine, and affected parts become black and crisp. Small plants may be killed. The squash bugs themselves may be hiding and not apparent; search under dead leaves and clods of earth to find them. All cucurbits are attacked, but pumpkins and squash are the most

seriously damaged. Zucchinis are among the squash varieties least susceptible to damage.

Management. Squash bugs are difficult to control. It is important to remove all debris from the field or garden once the crop is harvested by composting or thoroughly disking or rototilling it under. Squash bugs like to hide under sheltered areas such as boards and piles of trash, so remove all these overwintering sites.

Squash bugs can be trapped and killed in gardens or small acreages. In early spring place boards around and within the garden. Turn the boards over every morning and kill or vacuum up bugs that have collected beneath them; a rechargeable minivacuum works well. Hand-pick adult bugs and nymphs;

search for and destroy eggs in the spring and early summer. Squash bug numbers can be temporarily reduced by application of insecticides, but their populations usually return within days or weeks.

A parasite of the squash bug, a strain of the tachinid fly *Trichopoda pennipes*, attacks squash bugs in the eastern United States, causing up to 90% mortality. This parasite has recently been successfully introduced into California. Parasitized nymphs and adults may have tiny white eggs on their undersides. Another California strain of *T. pennipes* attacks the bordered plant bug.

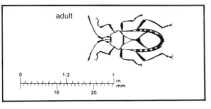

SQUASH BUG

adult

0   1/2   1 in
                        mm
      10   20

LEAFHOPPERS

Leafhoppers are small, slender insects that disperse rapidly when disturbed. They have short, bristlelike antennae and slightly thickened forewings. Both adults and nymphs run sideways and are good jumpers or hoppers. Adults also fly when disturbed. Leafhopper species vary somewhat in size, shape, and color. Most are basically wedge-shaped, less than ¼ inch long as adults, and generally varying shades of green, yellow, or brown and often mottled.

Leafhoppers go through incomplete metamorphosis in their development. Eggs are usually laid within plant tissues. Nymphs resemble adults except they lack wings, although later stage nymphs have small wing pads. There is no pupal stage and there are several generations a year.

Although leafhoppers can at times be found on almost every type of plant, their presence is of concern on only a few. Control actions are not normally needed on most crops. Leafhoppers have piercing-sucking mouthparts and suck plant juices from green parts of plants, often giving leaves a whitened, mottled appearance. On some plants, leafhopper feeding will cause a drying and yellowing or browning of the leaf margins or the whole leaf. Damage is most frequently seen on beans, cucurbits, potatoes, eggplants, grapes, and apples. Look for leafhoppers or their cast skins at their feeding sites on the undersides of affected leaves. Leafhoppers also often leave tiny varnishlike spots of excrement where they have been

The potato leafhopper, *Empoasca fabae*, is a common leafhopper on many crops.

These two nymphs are the white apple leafhopper, *Typhlocyba pomaria*, a pest of apples.

The grape leafhopper has mottled wings.

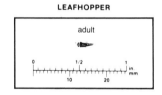
LEAFHOPPER

adult

feeding. These spots lower the quality of commercially grown table grapes, but can be washed off apples.

Leafhoppers transmit a few plant pathogens, most notably curly top virus, spread by the beet leafhopper, *Circulifer tenellus*, and aster yellows, caused by a mycoplasma and spread by the aster leafhopper, *Macrosteles fas-*

cifrons. Both of these diseases affect a wide range of vegetable crops. Because even a small number of leafhoppers can spread these pathogens, disease management focuses on resistant varieties and limiting reservoirs of viruses and mycoplasmas in crops, weeds, and foothill areas rather than controlling leafhoppers in the vegetable field or garden.

MANAGEMENT

Because of their mobility and abundance, leafhoppers are not easy to control. Fortunately, low to moderate populations of leafhoppers can be tolerated in most garden and small farm situations; a small amount of leaf browning will not generally cause serious crop losses. You may be able to prevent buildup of leafhopper numbers by careful management of ground covers and field borders. For instance, removing alternate hosts well before the preferred crop has emerged may reduce populations that could otherwise later migrate into the crop. Alternatively, planting and maintaining borders of more attractive hosts may keep leafhoppers out of your crop during sensitive growth stages.

Preliminary research with insecticidal soap has shown that these materials can at least partially control leafhoppers in grapes if applied when leafhoppers are small and during the development of the first brood in spring. Other insecticides available for leafhopper control are also more effective when leafhoppers are small.

Leafhoppers have many natural enemies, including general predators such as lacewings, damsel bugs, lady beetles, minute pirate bugs, and spiders. Various leafhopper species also have specific parasites or diseases that keep their populations low under most circumstances. Little is known about the efficacy of natural enemies in the control of most leafhoppers in California. An important exception is the grape leafhopper, *Erythroneura elegantula*, the eggs of which are parasitized by a tiny wasp, *Anagrus epos*, in Thompson seedless grapes. Check for *Anagrus* in your vineyard by looking for leafhopper eggs that are red or brown rather than the normal translucent color. Nonparasitized eggs are fairly difficult to spot but the red or brown parasitized ones stand out against the green leaf tissue; 90 to 95% of eggs may be parasitized in a Thompson seedless grape vineyard where *Anagrus* is well established. The parasite has not been as effective on other varieties.

Anagrus overwinters as a parasite in eggs of other leafhopper species in prunes, wild blackberries, and other hosts. It may be possible to improve parasitization of the grape leafhopper by *Anagrus* by providing refuges of these alternate hosts; however, attempts at establishing blackberry refuges have not been successful in increasing parasitization of the grape leafhopper on grapes. Researchers are now experimenting with border plantings of prunes near vineyards.

Grape leafhopper eggs are frequently parasitized by a parasitic wasp which turns them red brown as shown here rather than their normal translucent color.

Leafhoppers lower the quality of table grapes by leaving tiny black spots of excrement.

APHIDS

Aphids are small, soft-bodied insects with long, slender mouth parts with which they pierce stems, leaves, and other tender plant parts to suck out plant fluids. Almost every vegetable plant and fruit tree has one or more aphid species that occasionally feed on it. Fortunately, most plants can tolerate moderate numbers without significant damage. Curled, distorted leaves and sticky honeydew exudates on the leaves of plants often signal the presence of aphids.

There are dozens of species of aphids in California, some feed on several crops but many will attack only one species of plant or a few very closely related ones. Many aphid species are difficult to distinguish; however, identification to species is not necessary to control them in most situations. Some of the most commonly seen species are pictured here.

DESCRIPTION

Aphids may be green, yellow, brown, red, or black depending on species and food source. A few species appear waxy or woolly due to the secretion of a waxy white or gray substance over their body surface. All are small, pear-shaped insects with long legs and antennae. Most species have a pair of tubelike structures called cornicles projecting backwards out of the hind end of their body. The presence of cornicles distinguishes aphids from all other insects.

Generally adult aphids are wingless, but most species also

Aphids are pearshaped, soft-bodied insects with long legs and 2 tubelike projections (cornicles) coming out of the rear end of the body. Shown here are a wingless green peach aphid female and her offspring.

Most aphid species occasionally have winged forms such as this winged green peach aphid.

Aphids are often found in dense colonies. The rosy apple aphid shown here has a light waxy coating.

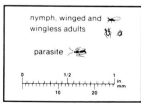

APHIDS

nymph, winged and wingless adults

parasite

0   1/2   1 in.
       10   20   mm

occur in winged forms, especially when populations are high or during certain times of the year. The ability to produce winged individuals provides the pest with a way to disperse to other plants when the food source gets scarce.

Although they may be found singly, aphids often feed in dense groups on leaves or stems. Unlike leafhoppers, plant bugs, and certain other insects that might be confused with them, aphids do not disperse rapidly when disturbed.

BIOLOGY

Aphids have many generations a year. Most pest aphids in California's mild climate reproduce asexually throughout most or all of the year with adult females giving birth to live offspring (often as many as 12 a day) without mating or laying eggs. Young aphids are called nymphs and molt, shedding their skins, approximately four times before becoming an adult. There is no pupal stage. Some species mate and produce eggs in fall or winter, which provides them a more hardy stage to survive harsh weather. In some cases, these eggs are laid on an alternative host, usually a tree, for winter survival.

When the weather is warm, many species of aphids can develop from newborn nymph to reproducing adult in less than two weeks. Since each adult aphid can produce up to a hundred offspring in a matter of a week, aphid populations can increase with great speed when not under control of natural enemies, weather factors, or actions taken by the gardener or farmer.

The woolly apple aphid is typical of aphid species that excrete a thick waxy coat which obscures the aphids.

Aphids excrete a sticky honeydew that frequently supports the growth of a dark sooty mold such as on this apple.

These apple leaves have been curled and distorted by the rosy apple aphid; this symptom is typical of many aphids and most crops.

DAMAGE

Damage by leaf-feeding aphids in most crops is similar. Low to moderate numbers are usually not damaging in gardens or on fruit trees. However, large populations cause curling, yellowing, and distortion of leaves and stunting of shoots; they can also produce large quantities of a sticky exudate known as honeydew, which often turns black with the growth of a sooty mold fungus. Seedlings of vegetables can be killed or severely distorted by aphids; susceptibility to damage decreases in most crops after seedlings have 5 or 6 true leaves, although full-grown squash or pepper plants can be killed. Some aphid species inject a toxin into plants, which further distorts crop growth. In commercially grown vegetable crops, the most important damage aphids do is contaminate the produce; aphids cannot be easily washed out and aphid-infested vegetables are generally not acceptable in the marketplace. This is a particularly serious problem in lettuce, spinach, broccoli, cabbage, and brussels sprouts.

In certain vegetable crops, aphids transmit viruses from plant to plant. Squash, cucumbers, pumpkins, melons, beans, potatoes, lettuce, beets, chard, and bok choy are crops having aphid-transmitted viruses commonly associated with them. The viruses cause mottling, yellowing, or curling of leaves and stunting of plant growth. (See disease chapter for photographs.) Although losses can be great, they are difficult to prevent through the control of aphids because infection occurs even when aphid numbers are very low. The best approach is to choose virus-resistant varieties and remove already-infected plants before planting a new crop. Aluminum foil mulches may help to deter aphids from landing. Be aware that many viruses are *not* spread by aphids.

A few aphid species attack parts of plants other than leaves and shoots. The lettuce root aphid is a soil dweller attacking lettuce roots during most of its life cycle; infestations cause lettuce plants to wilt and may kill plants when numbers are high; during the winter the lettuce root aphid takes refuge in aboveground parts of nearby poplar trees. The woolly apple aphid infests woody parts of apple roots and limbs, often near pruning wounds, and can cause overall tree decline if roots are infested for several years.

MONITORING AND MANAGEMENT

Check your garden or farm regularly for aphids—at least twice weekly when plants are growing rapidly. Once aphid numbers are high and they have begun to distort and curl leaves, it is often harder to control them because the curled leaf shelters them from pesticides or natural enemies.

Aphids tend to be most prevalent along upwind edges of the garden and close to other sources of aphids, so make a special effort to check these areas. Many aphid species prefer the undersides of leaves, so turn them over to check them. On trees, clip off leaves from several areas of the tree to check for aphids. Also check for evidence of natural enemies such as lady bugs, lacewings, syrphid fly larvae, and the mummified skins of parasitized aphids. Look for disease-killed aphids as well. Substantial numbers of any of these natural control factors can mean that the aphid population may be reduced rapidly and should be watched closely.

Ants are often associated with aphid populations, especially in tree crops. If you see large numbers of ants climbing up your tree trunks, check for aphids on the limbs and leaves above.

Biological control. Natural enemies can be very important in the control of aphids, especially on farms and gardens not sprayed with broad spectrum pesticides that kill natural enemy species as well as pests. Usually natural enemy populations do not appear in significant numbers until sometime after aphids begin to be numerous.

Among the most important natural enemies are various species of parasitic wasps that lay their eggs inside aphids. The wasp larva hatches, feeds internally on the aphid, killing it shortly before pupating. The skin of the dead aphid turns crusty and golden brown or black, a form called a mummy; the parasite larva pupates within, later cutting an exit hole to emerge as an adult wasp. The generation time of most parasites is quite short when the weather is warm, so once you begin to see mummies on your plants, the aphid population is likely to be reduced substantially within a week or two.

The brown colored aphids in this photo have been killed by a fungus; the green ones are healthy cabbage aphids.

The parasite *Diaeretiella rapae* lays its egg in a cabbage aphid.

Many predators also feed on aphids. The most well known are lady beetle adults and larvae, lacewing larvae, and syrphid (or hover) fly larvae. Naturally occurring predators work best, especially in a small backyard situation. Purchasing and releasing commercially available lady beetles can reduce numbers in some cases but most beetles will disperse before feeding and leave your yard. Green lacewings purchased and released as eggs or newly hatched larvae have shown mixed results.

Nectar-producing flowers increase food supplies for lacewings and parasitic wasps and may make your garden more attractive to these natural enemies. Supplementary food sprays

The crusty, golden aphid skins at bottom are the remains of parasitized aphids. The hole indicates the parasite has emerged. A healthy walnut aphid is at top.

Some parasite species turn aphids black. This photo shows a potato aphid parasitized by an Aphelinid surrounded by healthy pink forms of the potato aphid.

The convergent lady beetle is probably the best known aphid predator.

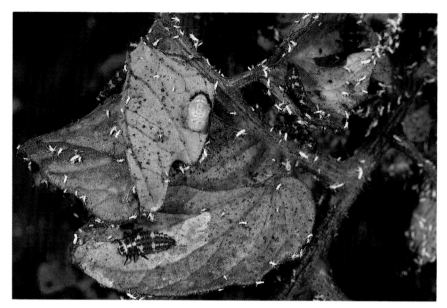

Larvae of the convergent lady beetle have cleaned up this aphid infestation on tomatoes. Note the orange lady beetle pupa and the many cast skins of aphids.

Lady beetle eggs are laid on end in groups.

consisting of yeast (or yeast hydrolysate) plus sugar have also been used experimentally to increase existing low populations of lacewings in the field. A yeast product (wheast) is commercially available (Merrick's, P.O. Box 307, Middletown, Wisconsin 53562-0307); however, it is fairly expensive; homemade mixtures of yeast and sugar could also be used. Procedures for use of these materials need to be further refined to use these products most effectively. Crop plants must be putting out volatile substances that are also attractive to lacewings to get them to land in the first place; research is ongoing to synthesize such substances. See References for more information (e.g., Hagen 1976).

Aphids are very susceptible to fungal diseases when it is humid.

Whole colonies of aphids can be killed by these pathogens when conditions are right. Look for dead aphids that have turned reddish or brown; they have a fuzzy, shrivelled texture unlike the shiny bloated mummies that form when aphids are parasitized. Fungus-killed aphids may sometimes have whitish mycelium growing over their surfaces.

Weather can have an impact on aphids. Populations of many species are reduced by summer heat in the Central Valley and desert areas, and aphid activity is also limited during the coldest part of the year. However, some aphids may be active year round, especially in the milder central coastal areas of California.

Ant control. In some situations ants tend aphids and feed on the honeydew aphids excrete. At the same time, they protect the aphids from natural enemies. If you see ants crawling up aphid-

infested trees or woody plants, put a band of sticky material around the trunk to prevent the ants from getting up. Prune out other ant routes such as branches touching buildings, the ground, or other trees.

Handpicking and pruning.
Where aphid populations are localized on a few curled leaves or new shoots, the best control may be to prune these areas out, drop the infested plant parts in a bucket of soapy water, and dispose of them. In large trees, some aphids thrive in the dense inner canopy; pruning these areas out can make the habitat less suitable.

Sanitation. Before planting vegetables, check surrounding areas for sources of aphids and remove them. Aphids often build up on weeds such as sowthistle and mustards, moving onto crop seedlings after they are planted.

Check transplants for aphids and remove them before planting. Each time you harvest part of your crop, remove any crop residues immediately.

Fertilizing. High levels of nitrogen fertilizer favor aphid reproduction. Never use more nitrogen than necessary. Use less soluble forms of nitrogen and apply it in small portions throughout the season rather than all at once. Or better yet use a urea-based, "time-release" formulation (most organic fertilizers can be classified as time-release products as compared to synthetically compounded fertilizers). Nitrogen fertilizers favor leaf growth and thus aphid buildup. Many gardeners and farmers use more nitrogen than necessary, yet neglect other important nutrients such as phosphorus.

Protective covers and transplants. Because many vegetable crops are primarily susceptible to serious aphid damage during the seedling stage, losses can be reduced by growing seedlings under protective covers in the field or in greenhouses or inside, and then transplanting them when they are older and more tolerant of aphid feeding. Protective covers will also prevent transmission of aphid-borne viruses.

Aluminum foil mulches. Aluminum foil mulches have been successfully used to reduce transmission of aphid-borne viruses in summer squashes, melons, and other vegetable crops. Aluminum foil mulches will also repel invading aphid populations,

reducing numbers on seedlings and small plants. However, as plants grow, aluminum foil mulches give mixed results for aphid control; they seem to repel natural enemies of aphids as well as aphids, and the few aphids that do drift into plants grow and reproduce with greater speed than those landing on plants on bare soil. This is mostly a problem where aphid contamination of produce is undesirable; yields of crops grown on aluminum foil mulches are usually increased,

despite higher aphid numbers, by the greater amount of solar energy reflecting on leaves.

To put an aluminum mulch in your garden, remove all weeds and cover beds with aluminum coated construction paper or use clear plastic painted silver. Bury the edges of the paper or plastic with soil to hold them down. After the mulch is in place, cut or burn 3- to 4-inch diameter holes and plant several seeds or single transplants in each one. You may furrow irrigate or sprinkle your

Syrphid fly eggs are elongate-oval and laid singly near aphid colonies. Note the characteristic striations on the egg surface.

The syrphid fly adult requires pollen to reproduce and superficially resembles a honey bee.

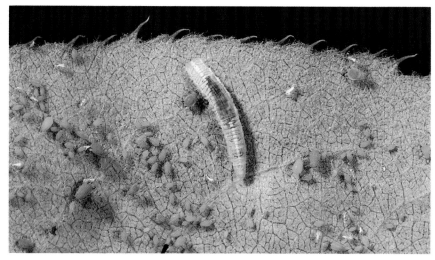

A syrphid fly larva feeds on a colony of apple aphids.

beds; the mulch is sturdy enough to tolerate sprinkling. In addition to repelling aphids, leafhoppers, and some other insects, the mulch will enhance crop growth and control weeds. When summertime temperatures get high, mulches should be removed to prevent overheating plants.

Water and soap sprays. One way to reduce aphid populations on sturdy plants is to knock them off with a strong spray of water. Most dislodged aphids will not be able to return to the plant and honeydew will be washed off as well. Using water sprays early in the day allows plants to dry off rapidly in the sun and be less susceptible to fungal diseases.

Water sprays can be made more effective by adding soap. A number of insecticidal soap products are available at garden supply centers.

The effectiveness of more concentrated soap sprays used as insecticides to kill aphids (rather than just knocking them off plants) varies with pest and crop species and environmental conditions. University of California research has produced conflicting results, although soap sprays generally lower populations to levels that can be more effectively controlled with natural enemies while causing less harm to the natural enemies than more conventional insecticides. Soap sprays have been used successfully for more than a decade along California highways to control melon, spirea, and other aphids on trees and bushes; research indicates that about 75% control is achieved with the sprays. In

brussels sprouts, soap sprays reduced cabbage aphids about 50% but discolored the inside leaves of the sprouts, so researchers do not recommend their use after sprouts begin to form. Two applications of insecticidal soap applied a week apart failed to control potato aphids on squash; part of this failure may be due to the hairiness of squash leaves; special surfactants are needed to effectively cover hairy leaves with soap sprays. Although more frequent applications may be more effective, they may reduce vegetable yields; for example, six weekly applications of insecticidal soap applied according to label directions reduced the harvest weight of cabbage by 23%. Soap sprays are reported to be less effective when mixed with water with a high calcium or magnesium content; such water is often called "hard water." Avoid use of soaps when conditions are hot or humid or plants are under water stress.

Oil treatments on dormant trees. Supreme- or superior-type oils will kill the overwintering eggs of aphids on fruit trees if applied as a delayed dormant application just as eggs are beginning to hatch in early spring. These treatments will not give complete control of aphids and are probably not justified for aphid control alone. Earlier applications will not control aphids. Common aphid species controlled include the woolly apple aphid, green apple aphid, rosy apple aphid, mealy plum aphid, and black cherry aphid.

Summer oil treatments. Summer treatments with supreme or superior-type oils also can be used to manage aphids on fruit trees. Good coverage of all areas of leaves and shoots infested with aphids is essential. See the introductory sections of this chapter regarding precautions for summer oil sprays.

Twice weekly sprays of mineral oil have reduced transmission of aphid-borne viruses on small vegetable plants such as squash and cucumber. Use a very fine mist of 4% water emulsion mineral oil (a petroleum based horticultural oil available as citrus soluble oil at your nursery) and cover all leaves completely. This oil may be toxic to some crop plants so test one or two plants before spraying your whole crop. Once plants get large, it is difficult to get thorough enough coverage to protect plants.

Other pesticides. Many other pesticides are available to control aphids in the garden and small farm including malathion and pyrethrum-soap combinations. Most are quite effective. Use them only when you are sure the aphids are about to cause irreparable harm to your tree or crop. Remember that moderate spring populations of many aphids attacking leaves of fruit trees often do not cause damage to the crop and low populations of aphids can be tolerated in most situations. Maintenance of low populations will also allow the buildup of natural enemies.

SCALE INSECTS

Scale insects can be serious pests on all types of fruit and nut trees and grape vines. Scales are so unusual looking that many people do not at first recognize them as insects. Adult female scales and many immature forms do not move and are hidden under a disklike or waxy covering. Scales have long, piercing mouth parts with which they suck juices out of the plant. They may occur on twigs, leaves, branches, or fruit. Severe infestations can cause overall decline and even death of trees. Most scales have many natural enemies that often effectively control them. Others are well controlled with oil sprays in the dormant season.

Two groups of scales, the armored scales and the soft scales, are important fruit tree pests. The armored scales lose their legs a day or two after hatching from eggs, settle down, and form a hard cover that is usually separate from the scale's body. Some soft scales (or unarmored scales) move around during their immature stages and retain their legs and antennae for life. Their covers may be smooth or cottony, but they are firmly attached to their bodies. Like aphids, soft scales excrete copious amounts of honeydew, attracting ants and causing growth of unsightly, sooty mold fungus. Common scales in each group are listed in Table 3-3. Excellent color keys for the identification of scale insects in California are available from the California Department of Food and Agriculture. See the References for titles.

TABLE 3-3.

Common Scales Found in California Orchards and Principal Hosts.

COMMON SCALES	PRINCIPAL HOSTS
ARMORED SCALES California red scale, *Aonidiella aurantii*	citrus, grape, olive
Italian pear scale, *Epidiaspis leperii*	walnuts, apples, pears, stone fruit
Olive scale, *Parlatoria oleae*	olive, almonds, stone fruit
Oystershell scale, *Lepidosaphes ulmi*	most deciduous fruit and nuts, especially apples and pears
Purple scale, *Lepidosaphes beckii*	citrus
San Jose scale, *Quadraspidiotus perniciosus*	most deciduous fruit and nuts
Walnut scale, *Quadraspidiotus juglansregiae*	walnuts
SOFT SCALES Black scale, *Saissetia oleae*	citrus, olives, almond, walnut, fig, apple, pear, stone fruit
Brown soft scale, *Coccus hesperidum*	citrus, avocado, stone fruit
Calico scale, *Eulecanium cerasorum*	walnuts, pear, stone fruit
Citricola scale, *Coccus pseudomagnoliarum*	citrus, walnuts, pomegranate
Cottony cushion scale, *Icerya purchasi*	citrus, other trees
European fruit lecanium (also called brown apricot scale), *Parthenolecanium corni*	walnuts, almonds, stone fruit, pears, grapes
Frosted scale, *Parthenolecanium pruinosum*	walnuts

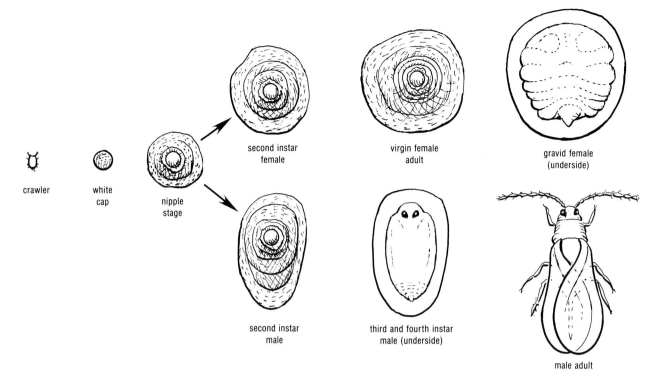

crawler

white
cap

nipple
stage

second instar
female

virgin female
adult

gravid female
(underside)

second instar
male

third and fourth instar
male (underside)

male adult

FIGURE 3-9. Life cycle of a typical armored scale, the California red scale. Eggs hatch into tiny crawlers which soon settle and secrete a cottony (white cap) cover and later a more solid cover (the nipple stage). After the first molt, males begin to develop an elongated scale cover whereas female covers remain round. Females molt three times with the final stage, the mated female with a rounded scab cover with a legless, wingless, immobile female beneath. Males molt four times. Males develop eyespots, which can be seen when scales are turned over in the third and fourth instar. The adult male has legs and two wings.

BIOLOGY

Most armored scales have several generations a year, while soft scales often have only a single generation. Eggs are often hidden under the mother's cover and are not commonly seen. Eggs hatch into tiny, usually yellow crawlers with legs. Crawlers walk around the plant surface or are blown by wind to other trees. Armored scales settle down permanently after a few days of the crawler stage, molt, and begin to form their characteristic covers. Soft scales move around for a while longer but also eventually settle at permanent feeding sites; half-grown individuals of some soft scale species move once again in

fall from leaves to wood, where they overwinter. Adult female scales are immobile and have a characteristic scale cover. Adult male scales are tiny, winged insects that superficially resemble parasitic wasps. They do not feed and usually live only a few hours. Among some scale species in California, males are rare or unknown and unnecessary for reproduction. The life cycle of an armored scale is shown in Figure 3-9.

DAMAGE

Trees heavily infested with armored scales often look water stressed. Leaves turn yellow and drop; twigs and limbs may die;

bark cracks and gums. Most armored scales attack leaves or fruit as well, leaving blemishes and halos on fruit. Damage to fruit is often just cosmetic but is not always acceptable in the marketplace. Armored scales can kill trees and must be controlled when their populations rise.

Soft scales also reduce tree vigor but do not commonly kill trees. The major concern with these scales is their production of abundant quantities of honeydew, which drips over leaves and fruit, encouraging the growth of black sooty mold. Honeydew also attracts ants, which protect soft scales from natural enemies and are a good indication of a soft scale infestation. Most soft scales infest leaves and twigs and do not attack fruit directly.

MANAGEMENT

Most scales are often well controlled by natural enemies. Among the most important predators are various lady beetles, most of which feed on scales both as adults and larvae. Many parasitic wasps attack scale insects; you can estimate the extent of their activities by checking scale covers for the round exit holes made by the emerging adult parasites. Often very high percentages of scale populations will be parasitized, reducing what at first appeared to be a serious infestation to a very low population. Major outbreaks of scale often follow treatment of trees with broad spectrum insecticides that kill natural enemies. Horticultural oils are less toxic to natural enemies than other insecticides.

Dormant season applications of specially refined horticultural oils called supreme- or superior-type oils are effective against several important scale pests of deciduous trees, especially San Jose scale, walnut scale, and the lecanium scales. Avoid oils called dormant oil or dormant oil emulsions; these are not as effective. When applied as a delayed dormant, just before bud swell, oil treatment can also kill a portion of overwintering mite, aphid, or caterpillar eggs on the woody portions of the tree. Treatments should always be made before budswell and when trees are not water stressed to avoid injury to trees. A good time to apply them is right after a period of rain or foggy weather. Do not apply them during fog, rain, or during or prior to hot or freezing weather (over 90°F or under 32°F). On deciduous trees, oils should not be applied within 30 days before or after applications of sulfur, captan, or certain other fungicides to avoid damage to trees. A combination of oil and organophosphate insecticide should be considered when scale levels are high, especially with armored scales.

Oils applied during the dormant season are not effective against oystershell or olive scales because susceptible stages are not present during the winter. Dormant season applications are not appropriate for scales on citrus or avocado because these trees do not enter a winter dormancy. However, horticultural oils are also used against scales in citrus; a superior-type narrow range (NR) 415 oil spray is suitable for the cooler fall and spring, whereas an NR 440 oil gives longer control in the warmer summer months. Summer treatments are also effective against walnut scale on walnuts. Always follow label precautions; oils can injure trees.

Oil applications in the spring and summer months must be carefully timed to kill crawlers. They are not very effective against mature scales or eggs. Double-sided sticky tape wrapped around twigs in spring will trap crawlers and indicate when sprays will be effective.

More specific guidelines are listed under each scale pest.

San Jose Scale
Quadraspidiotus perniciosus

Crops attacked: Most deciduous fruit and nut trees.

Description. The San Jose scale is an armored scale. Its life stages are similar to the California red scale pictured in Figure 3-9. It has no visible egg stage. The crawlers are bright yellow and about the size of the sharp end of a pin. After settling down, crawlers secrete a white, waxy cover (white cap stage). About a week later, a dark brown or black wax begins to surround and cover the white. This is the black cap stage. Adult female scales are round, about ¹⁄₁₀ inch in diameter, gray brown with a tiny, white knob or "nipple" at the center. If you remove the cover, the scale beneath is bright yellow. Adult males are tiny, yellow-winged insects. There are three to four generations a year.

Monitoring. Look throughout your orchard for scales, especially during the dormant season. Examine prunings, especially from tops of trees where sprays often do not reach. Twigs and branches heavily infested with scales retain their leaves during the winter and are easy to spot. Scales overwinter in the black cap and adult female stages primarily. Pull covers off scales to confirm that

The large, round scale is the adult female San Jose scale; the smaller pear-shaped one is the male. The tiny yellow crawlers and first stage white caps are also visible.

If you remove the cover, you can see the yellow body of the San Jose scale female.

they are living. Dead scale covers may remain on the tree for a time. Also look for emergence holes in scale covers that indicate the presence of parasites.

In commercial orchards, pheromone traps and sticky tapes can be used for monitoring and timing in-season application of pesticides.

Natural enemies. The San Jose scale has a number of natural enemies, which can provide substantial control when not killed by pesticides applied in the summer for other pests. The most important natural enemies include the twicestabbed lady beetle, *Chilocorus orbus*, and another

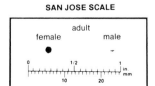

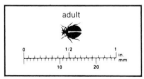

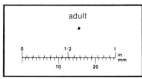

San Jose scale populations may cover limbs in poorly managed orchards.

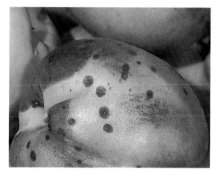

San Jose scale may get on the fruit, causing cosmetic damage.

small beetle, *Cybocephalus californicus*. A number of tiny parasitic wasps, including *Aphytis melinus*, also reduce populations.

Control. If San Jose scales are present on your trees, make a dormant treatment with supreme or superior-type oil in the late winter before new buds begin to swell. Be sure to thoroughly cover all limbs. Oils will not harm most natural enemies and should provide satisfactory control for the rest of the season. In-season sprays of insecticides may be required in commercial orchards. Use pheromone traps and degree-day calculations to time in-season applications; they are effective only during the crawler period in May. (See appropriate IPM Manuals listed in the References for further details.)

Walnut Scale
Quadraspidiotus juglansregiae

Crops attacked: Walnuts.

Description. The walnut scale is an armored scale and has a life cycle and appearance similar to the San Jose scale. Tiny eggs are often laid together in chains and hatch into yellow crawlers within a few days. After settling down, crawlers secrete a white, waxy cover (white cap stage). About a week later, a brown or gray wax begins to surround the white. This black cap stage is lighter in color than the San Jose scale. Eight to ten or more of the oval-shaped immature male scales often settle under the edge of the margin of their female parent making larger daisy-shaped scale cover formations. Adult females have a less pronounced nipple than the San Jose scale. If you remove the cover, you will see that the yellow body of the adult female is more indented than that of the San Jose scale. Adult males are tiny, yellow-winged insects. There are two generations a year.

Monitoring. Follow the same guidelines for monitoring as for San Jose scale. Concentrate monitoring efforts in the dormant season and examine prunings for scales. Look for dieback on limbs. Walnut scales are difficult to see because they blend in well with the bark. Populations tend to be spotty and clumped together. Once scales are found, pull off their covers and examine the shape of the adult female body to see if they are San Jose or walnut scales and confirm that scales are living.

Natural enemies. All the natural enemies listed for the San Jose scale also attack walnut scale. They are often quite effective if not disrupted by insecticide treatments.

Control. Applications of supreme or superior-type oils at the end of the dormant season will control walnut scale. However, oils can be damaging to walnut trees and are not recommended for use on trees that are stressed by drought or other factors. Oils during the dormant season are also not recommended for use in the central San Joaquin Valley south of Merced. In general, the later treatments are applied, the less likely damage will occur; the safest time is as bud swell begins (delayed dormant) and after adult males begin emerging. Check under the elongated male covers for pupal stages; the adults emerge about 3 days after pupation. Emergence usually occurs just before catkin elongation on early cultivars. At this time you can achieve good coverage of limbs and scaffolds.

In the central San Joaquin Valley, or if a high level of parasitization is observed (look for exit holes in scale covers), treatment with oils can be delayed until late spring after crawlers emerge. Oil sprays are not as damaging to walnuts at this time; however, it is more difficult to get good coverage.

California Red Scale
Aonidiella aurantii

Crops attacked: Citrus, grapes, olives.

Description. The California red scale is an armored scale. It has no visible egg stage. Crawlers are yellow and become immobile and begin to secrete a waxy white cover after a few days. A brown to golden cover forms in concentric rings within a few more days. Immature female scales are round whereas immature males are oblong. Adult males are tiny, yellow-winged insects almost identical to adult male San Jose scales. Adult female scales are round and light brown to golden with a nipple. Unlike other armored scales, the scale cover of the unmated California red scale cannot be removed to expose the scale body. (Once mated, the cover can be removed.) There are usually 2 to 3 generations a year depending on climate. Stages of the California red scale are shown in Figure 3-9.

Monitoring. The easiest way to spot California red scale is to look for scales on fruit. Fruit in commercial orchards should be inspected and sampled with a set sampling scheme 3 or 4 times a year. Look also for scales on leaves, twigs, and limbs. If scale populations are heavy enough to cause yellowing of leaves, leaf drop, and dieback of twigs and limbs, treatment is overdue.

Pheromone traps are available for monitoring California red scales in commercial orchards.

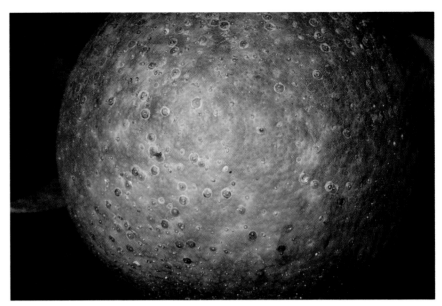
California red scale may be found on citrus fruit, leaves or twigs.

The adult male scale has one pair of wings.

The tiny *Aphytis* parasite lays an egg in a California red scale.

The most commonly used design is the paper clip card trap, consisting of a yellow or white sticky card, to which a rubber cap impregnated with the sex pheromone is attached with a paper clip. These traps catch the adult males as they emerge and take flight in search of females for mating. With a knowledge of the expected period of time between mating and hatching of nymphs,

CALIFORNIA RED SCALE

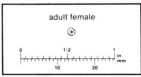
adult female

these traps can give you an idea of when treatment with oil or parasite releases may be most effective. For more information on how to use these and other sampling methods in commercial orchards, see *Integrated Pest Management for Citrus* in the References.

Natural enemies. Natural enemies are usually quite effective against the California red scale in unsprayed commercial groves in coastal areas and on backyard citrus trees in many parts of the state. Various species of parasitic wasps, including *Aphytis melinus*, *Aphytis lingnanensis*, *Encarsia perniciosi*, and *Compperiella bifasciata*, are important. Predators, including a small black lady beetle, *Lindorus lophanthae*, also feed on the scales. Where natural enemies have been temporarily disrupted by pesticides or harsh weather, mass releases of *Aphytis* can be made to augment dwindling populations. Releases should be made before adult male scales emerge and mate or just as they begin to be trapped in pheromone traps. *Aphytis* primarily attacks unmated female scales.

Parasites are very sensitive to many insecticides sprayed to control scales and other insect pests. Rely on less disruptive pesticides such as oils, soaps, and microbials where possible. Also minimize dust by oiling or watering paths and roads.

Control. If natural enemies are not providing adequate control, make releases of the *Aphytis* parasite just as male scales begin to fly and before they have mated, or use oil sprays. *Aphytis* does best in coastal areas and does not survive as well in the harsher climatic conditions of the interior valleys. Oil sprays should be applied between July and September (one spray should be enough). Oil sprays kill adult as well as immature scales, so precise timing is not too important. Organophosphate and carbamate insecticides kill only immature stages, so applications of those materials must be timed to reach the tiny, yellow crawlers before they have formed their protective scale covering. Good coverage is important, especially with oils. Foliage and fruit can be damaged when oil is applied under dry, hot conditions, so spray after an irrigation when soil moisture is high and temperatures will not go above 95°F.

Oystershell Scale
Lepidosaphes ulmi

Crops attacked: Most deciduous fruit and nuts, especially apples and pears.

Description. The oystershell scale is an armored scale. The cover of the adult is dark brown and shaped like a mussel or oystershell. This scale is not found on leaves or fruit but feeds on bark, sometimes causing leaves to turn yellow and dry up. Usually only one or a few limbs are affected at first. It is often a pest in fruit trees located near poplar, willow, or walnut trees, which seem to be preferred hosts. The scale has one generation a year.

Monitoring. No special monitoring procedures are needed. Survey your trees regularly and look for oystershell scales on bark below yellowing or dried up leaves.

Natural enemies. Little is known about natural enemies of this scale in California.

Control. Prune out infested limbs. If many limbs are affected, a spot treatment with an insecticide may be necessary. Sprays are effective only in late May or early June when crawlers are present. Dormant sprays are not effective against this scale because it overwinters in the egg stage protected by the old female covers.

Oystershell scales resemble a mussel shell in shape and inhabit only woody parts of the tree.

OYSTERSHELL SCALE

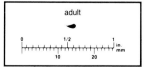

Brown Soft Scale
Coccus hesperidum

Crops attacked: Citrus, avocado.

Description. Young brown soft scales are oval, rounded, and mottled yellow brown. Mature scales are flattened and dark brown. All stages may be present at the same time since there are 3 to 5 overlapping generations a year. Brown soft scales are usually found on leaves and young twigs and rarely infest fruit. Moderate to heavy infestations will be accompanied by honeydew, sooty mold, and ants.

Management. Brown soft scales are usually under good biological control and outbreaks are generally a result of disruption of natural enemies. Insecticide sprays applied for other pests are a common cause; however, if only one spray was made, natural enemies usually recover within a few months. High populations of ants will also disrupt biological control, so keep ants out of trees. The most important natural enemy is the parasite *Metaphycus luteolus*; the lady beetle *Chilcorus cacti* is also important. Manage dust to encourage natural enemy activity. Spot treatments with oil may be made where immediate control is needed.

Young brown soft scales are yellowish, mottled, and convex.

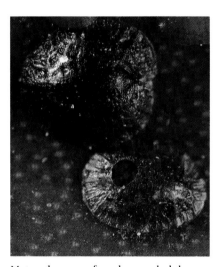
Mature brown soft scales are dark brown and somewhat flattened. A mature parasite has emerged from the hole in the scale at the bottom.

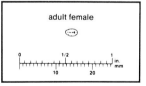
BROWN SOFT SCALE

adult female

Citricola Scale
Coccus pseudomagnoliarum

Crops attacked: Citrus, walnuts, pomegranate.

Description. Citricola scales look similar to brown soft scales except immature scales are mottled dark brown and mature ones are gray. However, the citricola scale has only one generation a year, so usually only one life stage is found at a time. Look for adults in the spring and early summer on twigs; and in the summer and fall, check the undersides of leaves for immature scales. This pest is a soft scale, so moderate to heavy infestations will be accompanied by honeydew, sooty mold, and ants.

Management. Biological control is effective against the citricola scale in southern California. The parasites *Metaphycus luteolus* and *Metaphycus helvolus* are most important. However, the parasite-host life cycle is not well synchronized in the San Joaquin Valley and the citricola scale may occasionally be damaging. Citricola scales are easiest to control with oils when they are small in the late summer or fall.

Black Scale
Saissetia oleae

Crops attacked: Citrus, olives.

Description. After the crawler stage, the immature black scale is mottled dark brown to gray with ridges in an H-shaped pattern on its back. Older scales are darker and hard, and the ridge may disappear. One or two overlapping generations occur in California, depending on climate. The black scale is a soft scale, so moderate to heavy infestations will be accompanied by honeydew, sooty mold, and ants.

Management. The black scale is normally under good biological control in citrus, primarily due to the parasite *Metaphycus helvolus*. However, biological control is often poor in olives in the San Joaquin Valley. Protect parasites from ants and avoid disruptive sprays. Naturally occurring populations may be augmented by releases of *Metaphycus* obtained from commercial insectaries; releases should be made in late summer or early fall. Oil treatments also control black scale. Apply oil in summer soon after crawlers have hatched and before young scales form the H-shaped ridge. Ridge formation usually occurs in September.

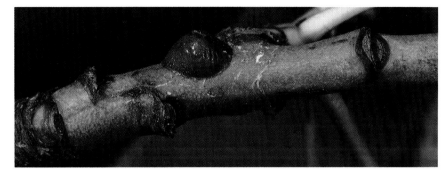

The domed, shiny brown shell of the European fruit lecanium has several ridges along its back. The frosted scale is similar but has a frostlike deposition of wax in the spring.

European Fruit Lecanium
Parthenolecanium corni
Calico Scale
Eulecanium cerasorum
Frosted Scale
Parthenolecanium pruinosum

Crops attacked: The European fruit lecanium (sometimes called the brown apricot scale) and the calico scale attack most deciduous fruit and nuts. Frosted scale is a pest mainly on walnuts and pistachios.

Description. These pests are soft scales, so infestations are accompanied by honeydew, sooty mold, and ants. All have one generation a year. They overwinter on twigs as small, flat, oval, brownish nymphs. In late winter (about February 1), females mature, exude a rounded shell, and are no longer effectively controlled with insecticides or oils. The domed shell of the European fruit lecanium is shiny brown, about ⅜ inch in diameter, and has several ridges along the back. The calico scale is mottled with a white and brown calico pattern as an adult. The cover of the

Newly hatched crawlers of the frosted scale are visible when the female cover is removed in early summer.

frosted scale has a frostlike, waxy coating in the spring from early March to mid-April. After this time the wax erodes and the insects appear shiny dark brown. Eggs hatch into crawlers beneath their mothers' covers in May through July. You can often flip the covers over and find many tiny crawlers beneath. Crawlers move out to the undersides of leaves in midsummer and young nymphs move back to twigs in early fall where they spend the winter.

Management. Parasites, especially those in the genus *Metaphycus*, play an important role in controlling these soft scales. Natural control, however, can be disrupted by adverse weather conditions, ants, or by insecticides. Parasitized nymphs are almost black and have convex covers; unparasitized nymphs are flat. Several parasites may emerge from a single adult scale, leaving a cover that looks perforated. Look for these signs of parasitism because heavily parasitized populations usually need no further control. Keep ants out of trees with sticky trunk barriers to protect natural enemies.

To monitor for scales, examine last year's growth in the dormant season looking for nymphs and signs of parasitism. Dormant oils are effective but must be applied by mid-January. Adults, which are difficult to control, begin to appear in February.

Cottony Cushion Scale
Icerya purchasi

Crops attacked: Citrus primarily.

Description. The cottony cushion scale is very different in appearance from the other scales described in this chapter. Like other soft scales, it excretes honeydew and is often accompanied by sooty mold growth and ants. The tiny, newly hatched nymphs are red with dark legs and antennae. They are found mainly on twigs and leaves. Older stages settle on branches and trunks and are covered with a thick, cottony secretion. Adult females secrete a fluted cottony egg sac, which remains attached to their bodies. Males are winged. There are 3 generations a year.

Management. Unless disrupted by insecticides, natural enemies provide excellent control of cottony cushion scale. The two important natural enemies are the vedalia beetle, *Rodolia cardinalis*, and the parasitic fly, *Cryptochetum iceryae*. If you find cottony cushion scale, look for vedalia beetles and their red eggs and larvae on top of scale egg sacs or the beetle's pupal cases. If your cottony cushion scale population is high and you can find no vedalia beetles, you can reestablish them by bringing them in from other orchards. Insecticides should not be needed.

An infestation of the cottony cushion scale covers this citrus twig and fruit. Note the honeydew dripping off the fruit.

The red and black vedalia beetle is a predator of the cottony cushion scale; its red egg and tiny larva are resting on the white fluted egg sac of a mature female scale.

COTTONY CUSHION SCALE

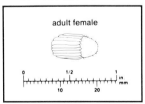

adult female

0   1/2   1 in
      mm
   10   20

VEDALIA

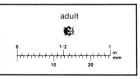

adult

0   1/2   1 in
      mm
   10   20

MEALYBUGS

Mealybugs are soft, oval, distinctly segmented insects that are usually covered with a white or gray mealy wax. They suck sap from stems, leaves, and shoots and, like their close relatives the soft scale insects, produce copious quantities of sticky, sweet honeydew. Mealybugs occur on a variety of fruit trees—mostly citrus, apples, pears, apricots, and grapes as well as on many ornamental shrubs and trees. They feed in dense colonies, producing wax and honeydew; when numbers are high, the colonies occur as white, sticky clusters among leaves and fruit and are easy to spot. An excellent identification key to California mealybugs is available from the California Department of Food and Agriculture (see References).

DAMAGE

Mealybugs, such as *Pseudococcus* and *Planococcus* species, that feed on stems and leaves lower fruit quality by covering it with wax or sticky honeydew upon which black sooty mold grows. These excretions are difficult or impossible to wash off during commercial handling of fruit. Another type of mealybug, the ground mealybugs in the *Rhizoecus* genus, injures roots. They may occasionally be found in strawberry, grape, or citrus plantings as well as on grasses and other annuals. Ground mealybugs spend their entire life underground and do not secrete the white waxy filaments common to the aboveground species.

Mealybugs feed in dense colonies apparent as white, sticky masses on leaves, twigs, or fruit. They are usually accompanied by honeydew with a dark, sooty mold growth over it.

The citrus mealybug shown here has a yellow orange body visible through powdery wax.

The mealybug destroyer lady beetle is an important mealybug predator. Here the adult and the large, cottony white larvae feed on a colony of smaller citrus mealybugs.

MANAGEMENT

Aboveground mealybugs are commonly adequately controlled by natural enemies, especially in citrus. Parasitic wasps are very important in orchards which have not been sprayed with insecticides. Native predators, including lady beetles, lacewings, and syrphid flies, help with control. An introduced beetle known as the mealybug destroyer, *Cryptolaemus montrouzieri*, is present

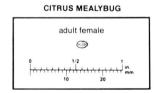

CITRUS MEALYBUG

adult female

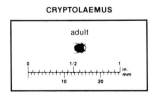

CRYPTOLAEMUS

adult

in areas without harsh winters and is also available commercially for release. The mealybug destroyer is particularly important in the control of the citrus mealybug but it attacks other mealybug species as well. Management of ants is extremely important in enhancing the activities of natural enemies because ants feed on honeydew and protect mealybugs from parasites and predators in order to increase this food supply. See the ant section for information on ant management.

On deciduous trees and vines, mealybugs overwinter in loose bark and removal of these overwintering areas can limit the number that survive the winter. When insecticide treatments are required in these crops, a delayed dormant timing is most satisfactory; mealybugs overwinter as nymphs or eggs and young nymphs do not have the waxy coat that protects older nymphs and mature mealybugs from toxic sprays in the summer. Mealybugs are sensitive to heat and populations often drop substantially in summer in warmer parts of California.

WHITEFLIES

Whiteflies are tiny, sap-sucking insects that are frequently found in vegetable and ornamental plantings. When infested plants are disturbed, clouds of the winged adult whiteflies fly into the air. Large numbers of whiteflies can cause leaves to turn yellow or appear dry. Like aphids, whiteflies excrete honeydew, so leaves may be sticky or covered with black sooty mold. The silverleaf whitefly, *Bemesia argentifolii*, injects a toxin that can cause distortion, discoloration, or serious losses in some vegetable crops. Some whiteflies transmit viruses to certain vegetable crops. With the notable exception of citrus, whiteflies are not normally a problem in fruit trees.

Whiteflies develop rapidly in warm weather and populations can build up quickly in situations where natural enemies are destroyed and weather is favorable. Most whiteflies, especially the most common pest species—the greenhouse whitefly, *Trialeurodes vaporariorum* and the silverleaf or sweet potato whiteflies—have a wide host range that includes many weeds and crops. They breed all year round, moving from one host to another as plants are harvested or dry up.

Whiteflies lay their tiny, oblong eggs usually on the undersides of leaves. The first stage larvae (called crawlers) are barely visible even with a hand lens; after hatching they crawl around for a while, then molt, losing their legs and antennae. Later immature stages are oval and flattened like small scale insects and do not move. They feed by sucking plant juices from the leaves, extracting much more than they can digest, and excret-

An adult greenhouse whitefly rests among late stage nymphs and pupae, which have long filaments. The T-shaped hole in the pupae indicates mature whiteflies have emerged.

ing the excess as sticky honeydew. Whiteflies go through complete metamorphosis and have a distinctive pupal stage. Adults of most species are whitish yellow with dull white wings. Some species have darker blotches

The round exit hole in this sweet potato whitefly pupal skin indicates that a parasite rather than a healthy whitefly emerged from it.

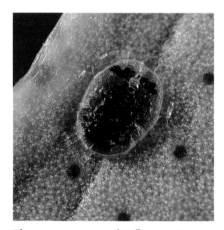

The most common whitefly parasites, *Encarsia* species, are black and can be seen within whitefly pupae before emerging. Other species of whitefly parasites, however, may be yellow and parasitized pupae can be difficult to distinguish from healthy ones.

WHITEFLY

pupa   adult

0   1/2   1 in.
                        mm
        10   20

on wings. A color photo key for identifying whiteflies in California is available from the California Department of Food and Agriculture.

MANAGEMENT

Whiteflies have many natural enemies, and outbreaks frequently occur when these natural enemies have been disturbed or destroyed by pesticides, dust buildup, or other factors. General predators include lacewings, bigeyed bugs, and minute pirate bugs. Whiteflies have a number of naturally occurring parasites. *Encarsia* spp. parasites are commercially available for release in greenhouse situations; however, they are not generally recommended for outdoor use since they are not well adapted for survival in temperate zones. An exception is the use of parasite releases for bayberry whitefly in citrus in southern California. You can evaluate the degree of natural parasitization in your crops by checking empty whitefly pupal cases. Those that were parasitized will have round or oval exit holes; those from which a healthy adult whitefly emerged will have a T-shaped exit hole. Whiteflies can often be quickly checked for parasitization before emergence by noting their color. Greenhouse whitefly and most of the whiteflies that occur on citrus have evenly white or yellow pupae; parasitized ones are commonly black or dark orange. However, many of the whiteflies that occur on ornamentals are black and should not be confused with parasitized ones.

Hand removal of leaves heavily infested with the nonmobile lar-

val and pupal stages may reduce populations to levels that natural enemies can contain. Control of dust and ants, which protect whiteflies from their natural enemies, can also be important, especially in citrus.

In vegetable gardens this strategy can be backed up with posting of yellow sticky board traps around the garden to trap adults. Make traps out of ¼ inch plywood or masonite board, painted bright yellow and mounted on pointed wooden stakes that can be driven into the soil close to the plants that are to be protected. Whiteflies do not fly very far, so many traps may be needed. You may need as many as one trap for every two large plants, with the sticky yellow part of the trap level with the whitefly infestation. Place traps so the sticky side faces plants but is out of direct sun.

Although commercially available sticky substrates such as Stickem or Tanglefoot are commonly used as coatings for the traps, you might want to try to make your own adhesive from one part petroleum jelly or mineral oil and one part household detergent. This material can be cleaned off boards easily with soap and water, whereas a commercial solvent must be used to remove the other adhesives. Periodic cleaning is essential to remove insects and debris from the boards and maintain the sticky surface.

Insecticidal soap is effective against whiteflies when properly applied. Be sure to cover undersides of all infested leaves; usually these are the lowest leaves and the most difficult to reach. Use soaps when temperatures are cool to prevent possible damage to plants. Use of other pesticides to control whiteflies should be avoided. Not only do most of these kill natural enemies, whiteflies quickly build up resistance to them, and most are not very effective.

A small, hand held, battery operated vacuum cleaner has also been recommended for vacuuming adults off leaves or sticky traps. Vacuuming must be done in the early morning or other times when it is cool and whiteflies are sluggish. Kill vacuumed insects by placing the vacuum bag in a plastic bag and freezing it overnight. Contents may be dumped the next day.

Aluminum foil mulches or silver-painted clear plastic mulches will repel whiteflies, especially away from small plants.

MITES

Mites are common pests on many fruit and vegetable crops. Although related to insects, mites are not insects but members of the arachnid class along with spiders and ticks. Unlike insects, mites lack antennae and body segmentation. Whereas insects characteristically have 6 legs, most adult mites have 8 legs; an exception is the family of eriophyid mites, which have only 4 legs and includes blister, gall, and rust mites. Mites feed by inserting their piercing mouthparts into plants and sucking up the liquid contents. Most mites cause some sort of yellowish stippling at the feeding site and general plant decline; others also cause distortion of leaves, fruit, or blossoms, or leaf drop.

Mites are tiny and often difficult to detect. You will need a good hand lens to see most of them. Usually plant damage will be noticed before you spot the mites themselves. To see them more clearly, shake a few off the leaf surface on to a white sheet of paper. Once disturbed, they will move around rapidly. Be sure mites are present before you treat. Sometimes the mites themselves will be gone by the time you notice the damage; plants will often recover after mites have left.

Two groups account for most of the plant-damaging species of mites. The tetranychid family includes the webspinning spider mites, the red mites, and the brown mites. The eriophyid family includes the rust, bud, and blister mites. One other mite family, the tarsonemids, contains a few serious plant pests, including the cyclamen mite, a major pest of strawberries, and the broad mite on lemons along the southern coast and in greenhouses.

Webspinning Spider Mites

The spider mites, also called webspinning mites, are the most common mite pests and among the most ubiquitous of all pests in the garden and farm. They can be found feeding on almost all fruit and nut trees and many vegetables in the garden and are a particular concern on asparagus, beans, squash and other cucurbits, sugar peas, and strawberries.

The webspinning spider mites include Pacific spider mite, twospotted spider mite, strawberry spider mite, and several other species. Most common ones are closely related species in the *Tetranychus* genus. They cannot be reliably distinguished in the field but there is no need to since their damage, biology, and management are virtually the same.

Description. To the naked eye, spider mites look like tiny moving dots; however, you can see them easily with a 10x hand lens. Adult females, the largest forms, are less than 1/20 inch long. Spider mites live in colonies, mostly on the lower surfaces of leaves; a single colony may contain hundreds. The names "spider mite" and "webspinning mite" come from the silk webbing most

Adult spider mites usually have a dark blotch on either side of their body. This twospotted mite is magnified 45 times.

Heavy infestations of spider mites are usually accompanied by webbing and stippling of leaves.

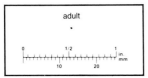

SPIDER MITES

species produce on infested leaves. The presence of webbing is an easy way to distinguish them from all other types of mites.

Adults have 8 legs and an oval body, with 2 red eyespots near the head. Females usually have a large dark blotch on each side of the body. Numerous long bristles cover the legs and body. Immatures resemble adults, except they lack the dark blotches and have only 6 legs during the first instar stage. Eggs are spherical and translucent, like tiny droplets, becoming cream colored before hatching.

Biology. In most parts of California, spider mites may feed and reproduce all year on crops, such as strawberries, and weeds that retain their green leaves through the winter. On deciduous fruit trees that drop their leaves, webspinning mites overwinter as red or orange females under rough bark scales and in ground litter and trash. They begin feeding and laying eggs when warm weather returns in the spring.

Spider mites reproduce rapidly in hot weather and commonly become numerous in June through September. If temperature and food supplies are favorable, a generation can be completed in less than a week. Spider mites are generally favored by hot, dusty conditions and are usually found first on trees or rows adjacent to dusty roadways or at margins of fields. As foliage quality declines on heavily infested plants, mites catch wind currents to disperse themselves to other plants. High mite populations generally undergo a rapid decline in late summer when predation overtakes them, host plant conditions deteriorate, and the weather turns somewhat cooler.

Damage. Mites cause damage by sucking cell contents from leaves. At first, the damage shows up as a stippling of light dots on the leaves; as feeding continues, the leaves turn yellow and drop off. Often leaves, twigs, and fruit are covered with large amounts of webbing. Damage is usually worse when compounded by water stress.

Loss of leaves will not cause yield losses in fruit trees during the year of infestation unless it occurs in the spring or very early summer, but it may impact next year's crop. On annual vegetable crops, such as squash, loss of leaves can have a significant impact on yield and lead to sunburning. On crops such as sugar peas and beans, where pods are attacked, spider mites can become an even greater concern.

Management. Spider mites have many natural enemies, which limit their numbers in many orchards and gardens, especially when undisturbed by pesticide sprays. Some of the most important are the predatory mites, including the western predatory mite, *Metaseiulus occidentalis*, and many *Amblyseius* and *Typhlodromus* species. Various insects are also important predators—the sixspotted thrips, the larvae and adults of the spider mite destroyer beetle, *Stethorus picipes*; the larvae of certain flies including the cecidomyid *Feltiella acarivora*; and various general predators such as minute pirate bugs, bigeyed bugs, and lacewings. These predators are pictured here or in the introductory section of this chapter.

Cultural practices can have a significant impact on spider mites. Dusty conditions often lead to mite outbreaks. Oil orchard roads or water pathways at regular intervals. Maintain ground covers in orchards and limit traffic, especially in hot spots where mites are frequently a problem. Trees and crop plants stressed for water will be less tolerant of spider mite damage. Be sure to provide adequate irrigation.

Spider mites frequently become a problem after application of insecticides for control of insect pests. Such outbreaks are commonly a result of the insecticide killing off the natural enemies of the mites, but also occur when certain insecticides stimulate mite reproduction. For example, mites exposed to carbaryl, methyl parathion, or dimethoate in the laboratory reproduce many times faster than untreated populations. Carbaryl, some organophosphates, and some pyrethroids apparently also favor mites by increasing the level of nitrogen in leaves. Insec-

ticides applied during hot weather usually appear to have the greatest effect on mites, causing dramatic outbreaks within a few days.

In gardens and on small fruit trees, regular, forceful spraying of plants with water will often reduce spider mite numbers adequately. Be sure to get good coverage, especially on the underside of leaves. If more control is required, use an insecticidal soap in your spray, but test the soap out on one or two plants to be sure it is not toxic to plants.

Sufficient control may also be obtained by using overhead sprinklers and sprinkling more frequently than would normally be required for irrigation alone. This strategy is used in commercial vineyards on the north coast and on grapes grown for raisins (for instance, sprinkling every 10 days for 12 hours in late June and July successfully controlled Pacific mites in one vineyard).

Selective miticides (e.g., propargite) are available for use in some commercially grown crops. These are less toxic to natural enemies than more broad spectrum materials, including dicofol, but can be skin irritants so are not registered for backyard use. Check registration status for materials and soap sprays on commercial crops and home gardens.

Release of predatory mites can be very useful, especially in orchard situations or larger plantings of strawberries and more permanent crops where naturally occurring predators are sparse. Predator mites currently commercially available for release are listed in Table 3-4. Often a combination of species is released for

A western predatory mite attacks a two-spotted mite.

The sixspotted thrips is an important predator on spider mites.

best results. These predators do not feed on foliage or become pests. If pest mites are not available, predators starve or migrate elsewhere. If pest mite populations in your orchard or garden are very high, use a soap spray or selective miticide to bring pest mites to a lower level and then release predatory mites to keep populations low. More than one application of predatory mites may be required if you want to

TABLE 3-4.

Commercially Available Predatory Mites.

PREDATORY MITE	DESCRIPTION
Metaseiulus (Galandromus) occidentalis	Most commonly released predatory mite, effective at high temperatures (above 90°F).
Phytoseiulus persimilis	Excellent predator under mild humid conditions up to 80°F. High temperature strains also available.
Amblyseius californicus	Often used in greenhouses as well as outside; useful at temperatures up to 85°F.
Amblyseius (Euseius) species	Feeds on all mites but particularly important against citrus red mite in citrus in southern California, does not normally enter areas heavily webbed by spider mites.

NOTE: See Hunter (1994) in the References for listings of suppliers.

The European red mite adult is bright red with a few long bristles coming out of its back.

reduce pest populations rapidly. Concentrate releases in hot spots where spider mite numbers are highest. Follow directions provided by suppliers. (See References for publications listing suppliers.)

Red Mites

The red mites are similar in shape to the webspinning spider mites but are bright red and produce little or no webbing. They are slightly larger than spider mites. The European red mite, *Panonychus ulmi*, is the most common of the red mites and feeds on a wide range of fruit and nut trees. Citrus red mite, *P. citri*, is primarily a pest of citrus, although it may also occur on almonds. Don't confuse these mites with red chigger mites that can be seen scurrying rapidly around plant surfaces and soil. Chiggers feed on various foods but are not plant pests.

Description. Both species have bright red globular eggs that are laid on bark or leaves. A slender stalk rises from the center of each egg. Immature mites are red, oval, and very tiny. Adults are also bright red with a few long hairs or bristles growing out of their backs. European red mites have white dots at the base of the bristles; citrus red mites do not.

Biology. European red mites overwinter in the egg stage on the bark of deciduous trees. On citrus, citrus red mites may be active all year. Both species become more active in the spring and in response to new growth flushes, producing a new generation every two weeks when temperatures are warm. Although European red mites may thrive in hot weather, citrus red mite populations are often reduced by high temperatures, especially in the Central Valley; their numbers increase again when temperatures cool down.

European red mite eggs have a slender spike rising from their center.

Damage. Red mites cause a characteristic stippling mostly on upper leaf surfaces. When damage is serious, leaves may bleach and burn at the tips. Mite populations must be extremely high to cause leaves to fall on deciduous trees but leaf drop may occur on citrus. Red mites do not produce webbing and are not as damaging as spider mites. They usually do not need to be treated in home orchards.

Citrus red mite attacks citrus fruit as well as leaves, causing a stippling and later silvering on the rind of mature oranges and lemons. Citrus trees suffer more damage when orchard conditions are hot and dry.

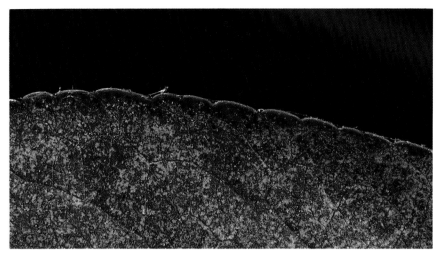

Red mites cause a stippling on the leaf surface similar to that caused by spider mites but produce no webbing.

Amblyseius species are predatory mites especially important in the control of the citrus red mite.

The spider mite destroyer is a tiny black beetle that feeds on many species of mites in both its adult and larval stages.

STETHORUS

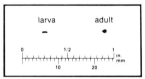

larva   adult

0   1/2   1 in
                            mm
     10   20

Brown lacewings can be important mite predators in orchards.

Management on deciduous fruit and nut trees. Numerous predators feed on red mites; the most important are brown lacewings, the spider mite destroyer, *Stethorus picipes*, which is a small lady beetle, and the western predatory mite, *Metaseiulus occidentalis*. The western predatory mite does not feed on European red mite eggs, so it is not as effective in regulating European red mite as it is on spider mites. However, it is effective enough to make mass releases worthwhile if you have a serious infestation. Follow procedures described in the webspinning spider mite section.

In-season sprays for red mites are not usually necessary or advisable in the home orchard. A supreme or superior-type oil spray during the delayed dormant period just as eggs are about to hatch should keep European red mites below damaging levels, if predators are not disrupted by sprays for other pests. Water, oils, or soap sprays can be used to knock mites off small trees.

Management on citrus. Citrus red mites have numerous natural enemies, which often keep the pest from reaching injurious levels on unsprayed trees. In addition to the predators listed for deciduous trees, predaceous mites in the *Amblyseius* (*Euseius*) genus are also very important. You can purchase these predators to augment your natural populations if necessary. Dusty wings, *Conwentzia barretti*, tiny insects related to lacewings, may also be important naturally occurring predators in some citrus groves. Another important biological control is a naturally occurring viral disease that often attacks mites under warm, humid, conditions. Affected mites walk stiffly, curl up, and finally die of diarrhea.

Treatment is rarely necessary on backyard trees. If stippling is severe on commercial trees, an oil spray applied in August or September in southern California and coastal areas should be sufficient. Spring sprays may be necessary in the San Joaquin Valley, but if hot weather is imminent, delay applications because populations are usually depressed by heat alone. Be sure to irrigate before applying oil, do not apply if temperatures are likely to go above 95°F, and follow label directions for citrus.

Brown Mite

The brown mite, *Bryobia rubrioculus*, is about twice the size of the European red mite and the largest of the common pest mites in California. Brown mites commonly feed on stonefruits, almonds, apples, and pears but rarely cause enough damage to warrant treatment in backyard gardens or small orchards.

Damage is similar to that caused by spider mites and appears as a stippling of whitish gray spots on young leaves, especially around the base of leaf veins. Brown mite produces no webbing and does not cause leaves to turn brown or drop. Adult brown mites are oval and slightly flattened, rusty brown to olive green in color with two very long legs in front. Eggs are bright red but lack the stalks that characterize red mite eggs; immatures are red and difficult to distinguish from red mites.

The life cycle of the brown mite is similar to that of the European red mite. However, it appears earlier in the spring and populations decline in the summer when temperatures rise and usually do not return again until the next spring. Unlike red mites and spider mites, brown mites do not spend all their time on leaves, but retreat to twigs and wood for much of the day and can be hard to find.

All the natural enemies that feed on other mites attack brown mites. Brown mites are often considered beneficial because they appear in the orchard early in the spring, providing food for natural enemies so they can build their populations up before the more damaging spider mites begin to do serious injury.

Blister, Rust, and Bud Mites

The eriophyid group of mites includes blister, rust, bud, and gall mites. All are very tiny, much smaller than the spider mites or red mites; a microscope or 15 to 20x hand lens is required to see them. They are pests mostly on tree and vine crops; a rare exception is the tomato russet mite. Table 3-5 lists common eriophyid mite pests in California.

Description. The eriophyid mites are very tiny and have only four legs, which appear to be coming out of their heads. Most are light pinkish or yellow to white. Some species are wedge shaped—including the rust mites, the silver mites, and the tomato russet mite; these species feed on leaf or fruit surfaces. The pearleaf blister mites, the citrus bud mite, and the grape erineum mite are cigar shaped and feed mostly in protected areas in buds or blisters. Damage is usually noticed long before these microscopic mites are detected.

Damage. Most of the eriophyid pest mites cause distortion or russeting of leaves or fruit. Their names give an indication of the damage. The rust mites cause leaf or fruit surfaces to turn brown or rusty. Usually the damage is not serious unless fruit is sold fresh commercially and scarring is unacceptable. On tomatoes, tomato russet mites can cause leaves to dry out and fall, leading to sun burning of exposed fruit. Peach silver mite produces a silvering of leaves on peaches, nectarines, and cherries. The prune rust mite attacks prunes and plums.

Blister mites cause red blisters to develop on pear leaves; these

TABLE 3-5.

Common Eriophyid Mite Pests in California.

Citrus bud mite	*Eriophyes sheldoni*
Citrus rust mite (Silver mite on lemon)	*Phyllocoptruta oleivora*
Grape erineum mite	*Colomerus vitis*
Grape rust mite	*Calepiterimerus vitis*
Peach silver mite	*Aculus cornutus*
Pearleaf blister mite (also attacks apples)	*Phytoptus pyri*
Pear rust mite	*Epitrimerus pyri*
Persea mite (avocados)	*Olygonychus perseae*
Redberry mite (caneberries)	*Acalitus essigi*
Tomato russet mite	*Aculops lycopersici*

These curled and dried out leaves show symptoms of a tomato russet mite infestation.

These peach silver mites show the wedge shape common to most eriophyid mites that feed on exposed leaves.

The russeting on these pears is typical eriophyid mite damage.

The elongated shape of the pearleaf blister mite shown here is similar to that of other eriophyids that feed in protected areas. These mites are magnified 45 times.

This lemon has been distorted by feeding by the citrus bud mite much earlier in the fruit's development.

blisters later turn black. Fruit attacked by blister mites have russetted spots and are bumpy and misshapen. The citrus bud mite feeds on buds on lemon, causing emerging flowers, fruit, and leaves to become grossly distorted. The grape erineum mite attacks leaves, producing a puckering on the upper surface and hairy patches on the underside of the leaf where the mites reside. In its bud mite phase, the erineum mite feeds on grape buds and destroys developing clusters.

Management. Most of the natural enemies that attack spider mites also feed on eriophyid mites. In most backyard situations, no further controls should be taken because damage by these pests is cosmetic and not very serious. Water or soap sprays and release of predators as described for spider mites are likely to provide some control of species that feed on exposed areas of leaves and fruit, although they would be less useful against species that remain in protected areas such as buds and blisters.

In commercial orchards, supreme or superior-type oil applied with lime sulfur in the dormant season are used against blister mites and rust mites. Citrus bud mite can be controlled with an oil spray in May and June and/or during the fall. Sulfur dusts are used against tomato russet mites (be sure temperatures are below 94° F), citrus rust mite (in winter), and grape erineum mites (in April or May). Follow appropriate precautions when applying these pesticides.

These puckered strawberry leaves are evidence of a cyclamen mite infestation. Look for the tiny mites on the newly unfolding leaves at the crown of the plant.

Cyclamen Mite

The cyclamen mite, *Steneotarsonemus pallidus*, a tiny mite in the tarsonemid family, can seriously damage strawberries. The pest is pinkish orange as an adult, translucent in immature stages, and smaller than a spider mite; you will need a 14x hand lens to see it.

The mite lays its eggs and feeds on young, unfolding leaves at the crown of the plant. When the leaves emerge, they appear stunted, crinkled and malformed. Leaf stems do not elongate. When the infestation is serious, the whole plant is dwarfed, leaves turn brownish green, and fruit are small, dry, and withered.

Adults overwinter in the crowns of strawberry plants and start reproducing when plants begin to grow in spring. A generation may be completed in 2 weeks, so populations can grow rapidly. Reproduction continues until October or November.

The best way to manage cyclamen mite is to remove and destroy infested plants as soon as they are spotted. Unlike spider mites, cyclamen mites can survive on few other plants and are introduced into the garden primarily on strawberries. Establish new plantings from mite-free stock and never plant new plants near infested ones. Avoid transporting mites on implements.

An alternative is to introduce predaceous mites into the strawberry patch. However, removing infestations is a more reliable way to handle the problem.

Leafminers

Leafminers are the larval stages of certain small insects that feed beneath the leaf surface. As they feed, they make distinctive winding trails or blotches that make their presence fairly easy to detect. On vegetables, the most common leafminers are the larvae of small flies in the genus *Liriomyza*, including the vegetable leafminer, the serpentine leafminer, and the pea leafminer. The major leafminer pest in apples and pears, the tentiform leafminer in the genus *Lithocolletis*, is the larva of a small moth.

Because most leafminers feed only on leaves, they do not commonly cause serious damage to crops grown for fruit or roots. Plants can usually tolerate substantial mining without loss of fruit yield or quality. On leafy vegetables, extensive mining can lower quality. Seedlings may be killed or their growth severely retarded by very high populations. Spinach and chard are among the most seriously affected crops.

Leafminers, especially those in vegetable crops, are often kept under good control by natural enemies; parasitic wasps are believed to be more important than predators. Leafminer outbreaks frequently follow insecticide sprays for other pests. Normally treatment is not needed where insecticide use has been avoided. Small seedlings can be protected with protective cloth or cold caps.

A new microbially derived insecticide, avermectin (Avid),

Leafminer adults, such as the vegetable leafminer shown here, are tiny yellow and black flies.

The most obvious evidence of leafminers is the twisting trails (or mines) the larvae leave as they feed beneath the leaf surface.

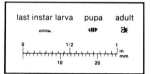

LEAFMINER

last instar larva	pupa	adult

A leafmine has been opened here to expose the vegetable leafminer larva feeding within.

which is effective against leafminers and has little toxicity to its parasites, has recently become registered for use in some vegetable and fruit crops.

Thrips

Thrips are tiny (less than ½₀ inch long), slender insects that are common on farms and in gardens. Adult thrips can be identified by the long fringe around the margins of both sets of wings. Some thrips species feed only on other insects and mites, but most are plant feeders that scar leaf or fruit surfaces with their rasping mouthparts. Although thrips damage to leaves is unsightly, it normally does not decrease yields unless plants are very young; thus insecticide treatment is not usually recommended in gardens. Some plant-feeding thrips are also predaceous; in cotton, the western flower thrips is considered more of a beneficial than a pest because of its role in reducing populations of spider mites. Common pest thrips include the western flower thrips, *Frankliniella occidentalis* on vegetables, grapes, and raspberries; the onion thrips, *Thrips tabaci*, and the bean thrips, *Caliothrips fasciatus*, on vegetables; and the citrus thrips, *Scirtothrips citri*, on citrus.

Thrips leave brown scars on injured leaves and leaves may be distorted. Damage is similar to that caused by windburn or blown sand. On vegetables, bean thrips damage can be distinguished by the presence of numerous black specks scattered over the damaged surface. These are thrips feces and remain long after the thrips themselves have left. On citrus, the citrus thrips produces a distinctive ring around the stem on the fruit skin. On grapes, thrips may leave dark scars surrounded

Thrips leave patchy white or brown scars on leaves. Note the black spots; these are thrips feces and a good indication of thrips activity. A tiny adult flower thrips is near the midrib of the leaf.

Adult thrips have two pairs of featherlike wings that are kept folded over the back. This dark one with black and white wings is the bean thrips; other species are lighter colors.

This western flower thrips nymph is feeding on mite eggs; sometimes its benefits as a predator outweigh the negative effects as a plant feeder, especially on older plants.

The scars on these nectarines are typical of thrips damage to fruit.

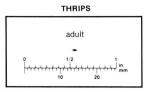

THRIPS

adult

0   1/2   1 in.
                  mm
    10   20

by lighter "halos" on the fruit where they have laid their eggs. On raspberries and nectarines, thrips feeding can deform developing fruit. On sugar peas, pods may be scarred or deformed.

Management. Thrips often migrate in from drying weedy areas or grasslands, so it is wise to avoid planting susceptible crops next to these areas. In small gardens thrips can be knocked off with a spray of water. Vigorous plants normally outgrow thrips damage; keep plants well irrigated. Insecticidal soaps are effective for temporary reduction of thrips populations. Sulfur dust has been used against thrips in some crops; however, watch for phytotoxicity. Some berry growers have been experimenting with the release of commercially available predaceous mites, *Amblyseius cucumeris* and *A. mackenziei*, which are used in Europe for control of thrips and mites in greenhouses. Aluminum foil and other disorienting mulches have been used with some success to reduce thrips infestations.

Maggots on Vegetable Crops

Three closely related species of maggots attack the underground parts of vegetable crops in California. They are the cabbage maggot, *Delia radicum*, which infests roots of cabbage, broccoli, cauliflower, brussels sprouts, radish, turnips, and other related vegetables; the seedcorn maggot, *Delia platura*, which damages seeds and germinating seedlings of many vegetables, including corn, beans, peas, cole crops, beets, potatoes, tomatoes; and the onion maggot, *Delia antiqua*, which can devastate bulbs of onions, leeks, and garlic. The three species are almost impossible to distinguish in the field. However, the crops attacked and the type of damage give a good indication of the probable species. Identification must be confirmed by an expert.

Adult maggots are dark gray flies that resemble the common housefly. Females lay small white eggs in plant stems right at the soil line or in cracks in the soil near plant stems. Eggs hatch in a few days and the maggots burrow through to roots or germinating seeds. The maggots are small, white, and legless—usually less than ⅓ inch when full grown. Their head end is pointed and the rear is blunt. After feeding for one to several weeks, maggots pupate in roots or surrounding soil. Pupae are brown and egg shaped. In most California growing areas, these maggots are active throughout the year and have several generations.

The head end of seed and root-infesting maggots is pointed and the rear end of their body is blunt. Look for maggots in seeds, roots, bulbs or surrounding soil.

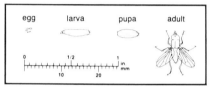

CABBAGE MAGGOT

| egg | larva | pupa | adult |

Although life cycles, appearance, and management of the three species of maggots is similar, each causes a somewhat different type of damage. All are usually found feeding in groups. The **seedcorn maggot** infests newly sprouted seed and seedlings, killing seedlings; they do not damage plants after the very early seedling stages. The **cabbage maggot** invades and destroys the root systems of crucifer crops. Heavily infested roots are often riddled with tunnels and rotted by decay organisms. Infested roots often break off if you try to pull them up. Affected plants are stunted, yellowed, and often wilted, especially during the hottest part of the day. The youngest plants are most susceptible; older plants can tolerate moderate infestations. The **onion maggot** bores down through the underground stem and into bulbs, mining out cavities or the entire bulb on small plants.

Management. Prevention is the best management strategy. These soil-dwelling maggots prefer to lay their eggs in moist, organically rich soil. If you are using manure, let it age and incorporate it well before planting. Avoid overfertilization with manure. Disk or rototill cover crops or weeds at least 2 weeks before planting. Female flies are attracted for egg laying by the moisture emitted from a newly planted seed row. Remove this moisture gradient by raking over the planted area or attaching a set of drag chains behind the planter during seeding.

Seedcorn maggot damage can be avoided by planting transplants or pregerminated seeds. However, even transplants have to be protected from cabbage maggot when crucifer crops are grown or the onion maggot when onions are grown because these maggots attack small plants as well.

Infestations can be prevented by covering seedbeds with a cheesecloth, screen cover, or a floating row cover immediately after seeds are sown. Be sure the cover extends at least 6 inches on each side of the seed row. Cheesecloth covers work best when they are designed to allow easy access to plants. A more permanent and sturdy cover can be constructed from standard window screen fixed to a wooden frame. Individual cone-shaped screen covers (Figure 3-10) can be useful in the home garden; they are easily constructed and can be stacked, stored, and used year after year.

Covers can be removed when plants grow big enough to tolerate damage.

Another barrier that has been used to protect cole crop transplants from the cabbage maggot is a 3-inch diameter disk of tarred paper, foam rubber, or other sturdy material placed flat around the base of each plant as it is transplanted; each disk has a hole in the center for the stem to go through (Figure 3-10). The disk will prevent flies from laying eggs near the plant stems and may provide places for aggrega-

tion of predatory ground beetles that eat the maggot's eggs and larvae. Paper collars around transplants have also been used but may not be as effective. Neither the disk nor the collar is as easy to use as a screen cone. Be sure transplants are maggot-free before planting them; cabbage maggots frequently move into the field on transplants. Some organic growers dip transplants into a solution containing commercially available predaceous nematodes to protect them during the critical first weeks after transplanting.

Once maggots are infesting a crop, there is no reliable way to control them. If plants are old enough, they may outgrow the damage, especially if they are maintained with a careful irrigation schedule. Disk or rototill under crop residues in infested fields immediately after harvest. Crop residues can provide an overwintering site.

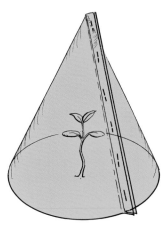

screen cone

foam or tar paper disk

FIGURE 3-10. Two devices for protecting young crucifer plants from the cabbage maggot. Sturdy window screen cones can be stored, stacked and used year after year. A foam or tar paper disk can be placed around the base of each transplant as it is planted. Either device is removed when plants grow big enough to tolerate damage.

Earwigs

Earwigs are among the most readily recognized and most commonly complained about insect pests in gardens and small farms. Although they can devastate seedling vegetables and often seriously damage more mature plants, they also have a beneficial role in the garden and recently have been shown to be important predators of aphids when released in apple orchards.

The common earwig species is the European earwig, *Forficula auricularia,* which was accidentally introduced into North America from Europe in the early 1900s. The adult earwig is about ¾-inch long and reddish brown, and has a pair of prominent forceps at the rear of its body; the male forceps are more strongly curved than the female. Contrary to popular myth and despite their ferocious appearance, earwigs do not attack humans.

Damage. Earwigs feed on a variety of dead and living organisms, including insects, mites, and growing shoots of plants. Damaged seedlings may be missing all or parts of leaves and stem. Leaves on older plants, including orchard trees, have numerous irregular holes or are chewed around the edges; this damage may resemble damage caused by caterpillars—look for webbing, frass, or pupae that would indicate the presence of caterpillars. Soft fruit such as apricots or strawberries may be attacked. On stone fruit, look for shallow gouges or holes that

The European earwig is easily recognized by its prominent forceps on the rear end of the body.

extend deeply into the fruit. On strawberries, distinguish earwig damage from that of snails and slugs by checking for the slime trails left by snails and slugs. On corn, earwigs feed on silks and blossoms, causing poor kernel development.

Earwigs feed most actively at night and seek out dark, cool, moist hiding places during the day. Common hiding places are under loose clods of soil, boards, dense growth of vines or weeds or even within fruit actually damaged by other pests such as snails, birds, or cutworms.

Management. In backyard gardens, earwigs can usually be adequately controlled by a persistent program of trapping. A low sided can, such as a catfood or tunafish can, makes an excellent trap; just add ½ inch of vegetable oil and a drop of bacon grease or tunafish oil and place numbers of them around the garden. Dump them and refill them with oil as they fill up with earwigs. Another common type of trap is a rolled-

Earwigs can be trapped in lowsided cans filled with vegetable oil and a drop of bacon grease or tunafish oil.

up newspaper, bamboo tube, or short piece of hose; place these traps on the soil near plants just before dark and shake accumulated earwigs out into a pail of soapy water in the morning. Continue these procedures every day until you are no longer catching earwigs. Complement the trapping program by removing refuge sites for earwigs, such as ivy, weeds, piles of rubbish, or leaves. Natural enemies, including toads, birds, and predators, may play an important role in some gardens.

In commercial orchards, keep weeds, brush, and suckers away from the base of trees throughout the year; these areas provide refuge for earwigs. If infestations are heavy in stone fruit orchards, you can apply pesticides to the lower 2 feet of trunks and soil around the base of each tree one month before fruit ripens and repeat the application in 2 weeks if the infestation is heavy. Keep pesticides off of upper parts of trees and fruit. Monitor populations with folded newspapers or burlap bags placed at the base of trees.

Sowbugs

Sowbugs are not insects but soil-dwelling crustaceans in the isopod order, more closely related to crayfish than insects. Sowbugs have a hard, shell-like covering that is made up of a series of segmented plates, superficially resembling an armadillo. Some sowbugs roll up into a ball when disturbed and are called pillbugs. Sowbugs have 7 pairs of legs and are dark gray or brown but may be almost purple or blue just after molting.

Sowbugs feed primarily on decaying plant material and are very important in the process of decomposing organic matter in the garden. However, they may occasionally feed on seedlings, new roots, lower (often partially decaying) leaves, and fruits or vegetables laying directly on the soil or near a damp soil surface. Sowbugs get blamed for more damage than they actually do, however, because they are frequently found in decaying fruit initially damaged by other pests such as snails or slugs.

Sowbugs have a shell-like covering that is made up of a series of segmented plates. The purple blue one has recently molted.

Like other crustaceans, sowbugs breathe through gills and require a moist environment. They are active at night and during the day hide in dark, moist, protected areas such as under mulches, boards, rocks, flower pots, ivy, or other large-leaved ground cover. They are especially common inhabitants of gardens with lots of organic mulch and are favored by sprinkler irrigation.

Management. The first step in managing sowbugs is to limit the moist and decaying matter environment in which they thrive. Try to water early in the day so plants and the soil surface dry out by the evening. Choose mulch materials that are coarse enough to let water pass through easily so the surface next to crop plants will not remain damp for long. Allow for air circulation as much as possible by providing trellises for vines, removing old decaying leaves and hiding places such as boards, flowerpots, and ivy. Elevate maturing melons and squashes on old strawberry baskets or pebbles. Strawberries are a particular problem when they are lying directly on soil or damp mulch. Any strategy you can devise to lift the strawberries off the ground will help. One suggestion is to use steplike tiers of narrow raised beds, one plant wide, which will allow berries to hang over the wooden sides of each stair. Black plastic mulches are also helpful in discouraging sowbugs because they get too hot to provide desirable shelter for the pests.

Crickets

Crickets occasionally invade gardens and fields, causing damage to vegetables or field crops. Normally they migrate in from weedy areas that are drying up and cause damage only for a short period. However, large numbers can destroy several rows of seedlings in a day or two. Crickets are nocturnal and require shelter during the day. They rarely are serious enough to require control in gardens. Various types of baits are available for large numbers of migrating crickets heading toward a newly planted field. Cones, row covers, and other protective devices can protect seedlings from damage.

This cricket is a nymph; it will have full sized wings in its adult stage.

Various species of grasshoppers may invade gardens and fields from neighboring foothills and rangelands. This one is the devastating grasshopper, *Melanoplus devastator*.

Grasshoppers

Grasshoppers are sporadic pests and an infrequent problem in many gardens. However, populations build up in foothills and rangelands, especially after a wet spring, and later migrate to nearby fields and gardens, often defoliating everything in sight. Damage is usually limited to a few weeks in early summer immediately after range weeds dry up. Sometimes grasshoppers can be controlled by spraying the edges around fields closest to the source of migration with carbaryl bait; repeat applications may be necessary. A microbial insecticide—*Nosema locustae*, which is a protozoan pathogen of grasshoppers—has been used as a bait with variable success to manage grasshoppers in rangeland in Utah, Colorado, and other western states; *Nosema's* limitation is that it does not provide quick knock-down and often kills only about half the resident grasshoppers, even when properly timed. Cones, floating row covers, and other protective covers provide some protection if numbers are not high. Grasshoppers will eat through row covers if they are hungry enough. A good strategy is to keep an attractive green field or border around your garden to trap the pests and continue to bait this area until hoppers are gone. Don't mow it, however, or you will send the hoppers straight into your garden.

Garden Symphylans

Garden symphylans, also called garden centipedes, feed on roots of many crops and weeds. They may damage seedlings before or after emergence and may slow the growth of larger plants. Symphylans are not insects; they have 12 pairs of legs and 14 body segments. They are associated primarily with moist, organically rich soil, and feed on decaying as well as living plant material. Areas that have had symphylan damage in the past commonly have problems with these little pests year after year.

Garden symphylans are difficult to manage. Flooding for 3 weeks can limit the problem in some situations if your soil will hold the water. Vigorously disturbing the soil when it is dry just prior to planting by rototilling may reduce populations through abrasion of the pest's soft body. Soil solarization has not been effective because symphylans are mobile and able to leave the solarized area. Soil drenches of various insecticides have been used to control symphylans in commercial crops.

Symphylan populations are usually quite localized within a garden or field. Look for them by taking soil samples as deep as you can and dropping each sample in a bucket of water to see if symphylans float to the top. Confirm their identity carefully, and be sure the organisms you find are not other soil dwelling invertebrates, such as true centipedes, which are important predators, or millipedes, which feed primarily on decomposing matter.

Garden symphylans are slender and white; they have 10 to 12 pairs of legs and a pair of antennae. They run rapidly when exposed to light.

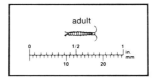

GARDEN SYMPHYLAN

adult

Ants

As a group, ants are important natural enemies of many insect pests. However, some species can be troublesome in the orchard and garden. The most common problem occurs with species that feed on the sweet exudates of honeydew-producing pests, such as aphids, soft scales, mealybugs, and whiteflies, and protect the honeydew producers from their natural enemies. Damaging populations of aphids, scales, and mealybugs are frequently associated with ant populations. Honeydew-feeding ant species can be identified by their swollen, almost translucent abdomens as they move down plants and trees. Other ants, such as the southern

The Argentine ant feeds on honeydew excreted by insects such as these brown soft scales. Note the swollen, almost translucent, abdomen, which identifies this species as a honeydew collector.

fire ant, *Solenopsis xyloni*, do not protect pests but may feed on tender twigs of small trees and can damage almonds drying on the ground after shaking; however, under most circumstances these nonhoneydew-feeding species probably do more good than harm because they feed on pest insects as well.

The most common honeydew-feeding species in California is the Argentine ant, *Iridomyrmex humilis*, which builds its nests under the soil surface. In fall, colonies move their nests to sunny locations; with the onset of hot weather, they build new nests under trees and other shady locations or under objects and structures like sidewalks, flower pots, or houses. The size of a colony varies from a dozen to many thousands and the number of queens from one to many hundreds. The ants usually aggregate in large colonies in the winter and break up into small colonies during the summer. Populations may increase enormously in midsummer and early fall. Southern fire ant colonies are smaller than those of the Argentine ant and usually have only one queen; their underground nests are marked by loose mounds of soil; often several colonies will be found at the base of a tree.

Management. Where ants are tending pests on trees, they may be managed by placing sticky barriers around the tree trunk to prevent them from reaching the honeydew-producers. Various commercial sticky barriers, including Tanglefoot and Stickem, are available. They must be applied frequently because they lose their stickiness when covered with debris. Temporary barriers can be made by wrapping a band of tape, plastic, or heavy paper around the tree trunk and coating it; however, spaces between the band and trunk must be plugged so ants cannot crawl under it. The temporary band can then be removed when ants are no longer a problem. When putting a barrier around the trunk, be sure there are no other routes ants can use to climb into the tree. Prune off branches that are touching the ground, buildings, or other trees.

Ants can also be controlled with pesticides in ant stakes or bait stations placed on the ground near nests or trails.

SNAILS AND SLUGS

Snails and slugs are among the most bothersome pests in many garden situations. The brown garden snail, *Helix aspersa*, is the common snail causing problems in California gardens. Many species of slugs may cause damage; the most common is probably the gray garden slug, *Agriolimax reticulatus*. Both groups are members of the mollusc class and are similar in structure and biology, except that slugs lack the snail's external spiral shell. Snails and slugs move by sliding along a mucus or slime trail secreted from their single foot. The slime trails dry to form silvery pathways that provide a record of their activities long after they have moved on.

Snails and slugs feed on a variety of living plants as well as on decaying plant matter. They cause the most serious damage to seedlings, tender, low-growing, leafy vegetables, such as lettuce, and ripening fruit, such as strawberries and tomatoes, that are close to the ground. Snails and slugs are most active at night and on cloudy or foggy days. On sunny days they seek hiding places out of the heat and sun; often the only clue to their presence is their silvery trails.

Management. The first step in a snail and slug management pro-

A brown garden snail feeds on a navel orange.

Slugs are closely related to snails but lack the shell. This is the gray garden slug, a common California pest.

gram is to eliminate, to the extent possible, all places where the pests can hide during the day. Boards, stones, debris, weedy areas around tree trunks, leafy branches growing close to the ground, and dense ground covers such as ivy are ideal sheltering spots. There will be shelters that cannot be eliminated, e.g., low ledges on fences, the undersides of wooden decks, and water meter boxes; make a regular practice of removing snails and slugs on these shelters or try to screen out migration from these areas into the garden.

Handpicking can be very effective if done thoroughly on a regular basis. At first it should be done daily; after the population has noticeably declined, a weekly handpicking may be sufficient.

To draw out snails, water the garden in the late afternoon. After dark, search them out (a flash light may be necessary), pick them up (rubber gloves are handy when slugs are involved), place them in a plastic bag, then crush and dispose of them in the trash. Alternatively, captured snails can be crushed and left in the garden.

Snails and slugs can be trapped under boards or flower pots positioned throughout the garden (Figure 3-11). You can make traps from 12" x 15" boards raised off the ground by 1-inch runners. The runners make it easy for the pests to crawl underneath. Scrape off the accumulated snails and slugs daily and destroy them. Crushing is the most common method of destruction.

Beer-baited traps have been used to trap slugs; however, they provide variable control.

Several types of barriers will keep snails and slugs out of vegetable gardens. The easiest to maintain are those made with copper flashing and screens. Copper barriers are effective even after rain and sprinkler irrigation and under certain circumstances, such as trunk banding of citrus trees, can last for several years. A well tested barrier for keeping snails out of vegetables is a vertical copper screen surrounding a snail free garden area. The screen should be erected 6 inches high and buried several inches below the soil to prevent slugs from crawling beneath the soil. Barriers of dry ashes or diatomaceous earth heaped in a band 1 inch high and 3 inches wide around the garden have also been shown to be effective. However, these barriers lose their effectiveness

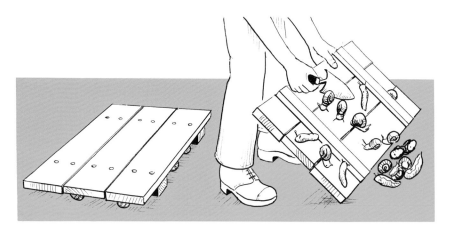

FIGURE 3-11. Make a snail and slug trap out of boards raised off the ground by 1 inch runners. Snails will collect underneath. Scrape off and destroy accumulated snails and slugs daily.

A copper foil barrier tacked onto raised beds can keep snails and slugs out.

after becoming wet and are therefore difficult to maintain.

Snails and slugs have many natural enemies, including ground beetles, pathogens, snakes, and birds, but they are rarely effective enough to provide satisfactory control in the garden. A predaceous snail, the decollate snail, *Rumina decollata,* has been released in southern California citrus orchards for control of the brown garden snail and is providing very effective biological control. Because of the potential impact of the decollate snail on certain endangered mollusc species, it cannot be released in many areas. Check with your local agricultural commissioner for regulations that apply to your area. It is also not recommended for garden situations since the decollate snails may feed on seedlings, small plants, and flowers.

Citrus trees may be protected from the brown garden snail by

The decollate snail shown here is a predator of the brown garden snail.

applying a band of copper metal sheeting around the trunk and pruning all branches that may touch the ground or other infested areas. A patented product, Snail Barr (Custom Copper, Ventura, California), works better than homemade barriers because it expands with the tree trunk as it grows. Other sorts of copper trunk barriers are available commercially, including some paper-like products.

Larger commercial vegetable fields are not usually plagued by snail and slug problems because they are far enough removed from sheltered areas and disked regularly. However, fields next to citrus groves, canals, and drainage ditches may have problems. Remove trash, weedy growth, and other daytime shelter to the extent possible. If damage occurs, disk under several rows of the crop between the shelter area and the crop, and keep the disked area as dry as possible to prevent further damage.

Snail and slug baits are available. They can be effective when used properly in conjunction with a cultural program incorporating the other methods discussed here. Snail baits can be hazardous and should not be used where children and pets cannot be kept away from the poisons. Never pile bait up in mounds or clumps; piling makes it attractive to pets and children. Scatter baits along areas snails and slugs have to cross to get from sheltered areas to the garden. Bait placed in ivy or other areas of heavy foliage is usually not very effective. Baits containing only metaldehyde are reliable when conditions are dry and hot but are rendered ineffective by rainy or foggy conditions. Before applying baits, water the area well if the ground is not already moist. This will draw out snails and slugs and moistens the bait to make it more attractive. Do not water heavily for at least 3 or 4 days after bait placement; watering will reduce effectiveness.

CHAPTER FOUR

Diseases

IT IS DIFFICULT TO MAKE IT THROUGH a season without evidence of disease showing up on some of the plants in any garden or farm. Many types of damage can be due to disease. Common symptoms include spotted, discolored, or distorted leaves; wilted, dying, or dead leaves and/or stems; blemished or rotting fruit; sap dripping from tree branches or trunks, and rotted or stunted roots. Because symptoms of some diseases are quite similar, it is often difficult to identify the actual cause of the damage; diagnosis is further complicated by the small (often microscopic) size of most disease-causing organisms. Disease symptoms can also be caused by noninfectious agents such as air pollution, herbicides, nutrient deficiencies, low temperature, salt or other toxic elements, or too much or too little water. Often professional help and laboratory tests will be needed to positively identify the cause of a disease.

Types of Pathogens

Microorganisms that cause disease are commonly called pathogens; fungi, bacteria, and viruses are the most common pathogen groups causing diseases in plants. Each type of organism has somewhat different disease cycles, dispersal methods, and disease symptoms.

Fungi rely on living hosts such as plants or dead organic material for a food source. Fungi usually grow through or on affected plant tissue as fine, threadlike structures (hyphae) that form a network or mass called mycelium. To reproduce or multiply, fungi produce spores, often on specialized structures. Spores are moved by wind, water, tools, machinery, insects, and anything else they come in contact with. When they land at a location where conditions are suitable for their growth, spores germinate to produce a new fungal mycelium. Some types of spores can survive adverse conditions for long periods of time, especially in the soil or in diseased plant tissue. Mycelium, spores, and spore-forming structures—when they can be seen with the naked eye or with a hand lens—help in identifying the pathogen in the field. For example, a mushroom is the spore-forming structure of some fungi. Fungi can cause many different types of symptoms, including soft rots of fruit, severe stunting of plants, curling of leaves, profuse gumming, smuts, rusts, leafspots, mildew, sooty molds, and wilts.

Bacteria are single-celled organisms. Like fungi, the bacteria that cause plant diseases require a host plant or some other organic material for a food source. Since plant pathogenic bacteria do not produce spores that can withstand adverse conditions, they must usually remain inside a plant host or plant debris to survive. Bacteria require warmth and moisture to multiply so are not as serious in areas where summers are dry; however, sprinkler irrigation can cause them to become problems even in arid areas. Bacteria are commonly spread by splashing water but can be dispersed by anything that moves and comes in contact with them or infested soil; this includes equipment, insects, and plant material. Common symptoms associated with bacterial diseases include shoot blights, leaf spots, soft rots, scabs, wilts, and galls on branches, twigs and tree trunks.

Viruses are particles that are much smaller than a bacterium.

They require a living host cell to multiply. Viruses are most commonly spread by plant feeding-insects—especially aphids, leaf-hoppers, and whiteflies. Nematodes, fungi, cultivation practices, pruning or grafting, pollination, and movement of infected seed also spread certain viruses. Common symptoms associated with viruses include stunting or reduction of growth and change in coloration, often expressed as a mottling or yellowing of leaves or fruit. Some viruses produce malformations or abnormal growth of various plant parts.

Monitoring and Diagnosing Diseases

Any plant disease involves a complex interaction between the host plant, the pathogen, and the environment. Disease symptoms, their rate of development and the damage produced are influenced by genetic characteristics of the plant, its stage of growth, when infection occurs, what other stresses the plant is under at the same time and environmental conditions—especially temperature and humidity. Under some conditions, some pathogens will have very little impact on a plant whereas under other environmental conditions, the same pathogen can be devastating. In general, it is much more difficult to determine the causes of disease symptoms than those caused by insects because of the greater variability in symptoms and the microscopic size of pathogens.

Check your crops regularly for

stress or disease symptoms. Record weather, soil conditions, and previous disease outbreaks to help in diagnosis; free water is especially important because many pathogens require water for germination of spores and infection.

When comparing symptoms in the field with illustrations and descriptions in this book, examine as many affected plants as possible. Look for plants with different stages of disease to determine how symptoms change as the disease progresses. The photographs in this book show only some stages. Do not rely on a single symptom, such as yellowing, to identify a disease. Look at all parts of affected plants. Be sure to pull plants up and check roots; many above-ground symptoms are actually caused by root diseases. Different diseases may produce the same or similar symptoms if the pathogens involved disrupt the same plant function. Observation of several different symptoms is usually needed to identify a disease. Carry a hand lens so you can check diseased plants for spores or other microscopic evidence of the pathogen.

It is not always possible to identify diseases with certainty in the field. Many pathogens require special laboratory techniques for identification; nutrient deficiencies and some other conditions require plant tissue analysis.

Disease Management

Most pathogens require certain conditions to spread and infect plants. Often you can make conditions in your garden or orchard less suitable for disease development by following recommended management practices. In many cases these procedures will limit the need for pesticide applications in the home garden, although sprays may still be needed for commercially grown crops.

Resistant varieties. Often the most effective way to avoid disease is to plant a resistant variety. Such varieties either cannot be infected by a pathogen or, if the pathogen can infect, it is not able to reproduce or cause damage. Varieties resistant to a number of diseases are available. In many cases, resistance can be used to avoid problems with powdery mildew, certain viruses, and vascular wilts. Check with your seed supplier or nursery to find out which ones are best suited for your needs and local conditions. In some cases you may choose to plant a more susceptible variety because of desirable horticultural characteristics. If so, be sure you know what diseases are likely to be troublesome. You may need to use pesticides to keep losses down, and should be prepared to accept more loss from disease.

Certified and disease-free planting material. Always select the highest quality seed and planting stock that is available. Many diseases can be transmitted in nursery stock, transplants, and on or in seed. Obtain your material from a reputable seed company, garden center, or nursery. Select certified virus-free nursery stock. Examine young trees, vines, and other planting stock for signs of root diseases, crown gall, and virus diseases. The best way to prevent disease is to keep it out of your garden or orchard from the start.

Planting site and spacing. Select a suitable location for your plants. Some plants prefer full sun while some do better in shady areas. Certain diseases—for example, powdery mildew, white mold, and gray mold—are more serious in shady areas. Avoid planting in soil that drains poorly; root diseases are likely to be a problem in such soil. Improve drainage conditions before planting. (See "Reducing Root Rots in Plants" listed in the References for more information on planning drainage for the backyard garden.) Avoid planting susceptible plants near fields with crops that may harbor harmful viruses. Do not overcrowd plants. Crowded conditions favor damping-off and gray mold of vegetables. Arrange plants according to their watering requirements; do not plant something that requires infrequent deep watering next to a plant that needs frequent light watering.

Irrigation and fertilization. Maintain an even water supply for your plants without overwatering. Too much water favors root diseases and several foliar diseases. Water early in the day or before daybreak so foliage has a chance to dry out quickly. In large gardens and commercial operations, use soil and plant tissue test results to determine fertilizer needs. Inadequate or excess fertilizer makes plants more susceptible to some diseases.

Roguing and pruning. Roguing refers to the removal of diseased plants. Remove diseased plants from your garden as soon as you see them; prompt removal will reduce the likelihood of the disease spreading to other plants. Bag diseased plants and discard them. Aerobic composting will destroy most disease-causing organisms, but a few, such as Fusarium and Verticillium wilts, may not be killed. Prune out and destroy diseased foliage from trees and shrubs as soon as it appears. Pruning is very important for fireblight control and is also useful for brown rot, shot hole, and some powdery mildews.

Sanitation. Keep garden implements clean to avoid spreading contaminated soil or disease organisms from infected plants. When working with diseased plants, sterilize your equipment with household bleach diluted 1:9 with water or 70% alcohol. Use this solution to sterilize pruning shears when removing diseased limbs from trees, but be sure to wash treated implements before storing them—bleach is corrosive. After working in an area where you know or suspect a soil pathogen is present, be sure to clean off your shoes as well as any implements before moving to another area. Wash your hands well after handling diseased plants. If you smoke, wash your hands before handling potato family plants (especially tomatoes and potatoes); tobacco mosaic virus may be spread through contact with tobacco.

Rotation. Rotating crops keeps many disease organisms from building up. Do not plant closely related plants—for example potato family plants or cucurbit family (cucumbers, melons, pumpkins, squash) plants—in the same place year after year. Corn is a good crop to rotate with many garden vegetables. If you have the space, change your garden location occasionally or allow a fallow (crop-free) period.

Weed and insect control. Weeds can be hosts for a number of disease-causing fungi and viruses. Keep them controlled in your garden and in adjacent areas. Be aware of weedy or wild plants that may harbor diseases that can spread to your cultivated plants. For instance, wild blackberries if you are growing blackberries or raspberries or wild plums or choke cherries if you are growing stone fruits. Follow recommended control practices for insect pests. They can spread viruses and some other foliar diseases too.

Aerobic composting. Aerobic composting is a technique for allowing plant refuse to decay with frequent mixing so that adequate air gets to the decaying material. If correctly done, many plant pathogens are destroyed by the process and the compost, itself, can suppress disease in the garden. Use a compost volume of about 1 cubic yard, mix equal amounts of green compost (such as grass clippings or other fresh plant material) and dry compost (such as leaves or other dead plant material), and thoroughly stir the compost every 2 or 3 days. Most but not all pathogens will be destroyed in a period of 3 weeks. See Chapter 2 for more information on composting.

Soil solarization. A number of soil pathogens can be destroyed or greatly reduced in the warmer areas of California by covering moist soil with clear plastic for 2 months during midsummer. For a more detailed discussion of soil solarization see Chapter 2.

Biological control. There are many organisms in nature that destroy pathogens or limit their growth and thus provide biological control of disease. For example, some soils, called *suppressive soils*, contain microorganisms that are antagonistic to certain pathogen species. These microorganisms, which normally have little or no impact on the crop plant, may produce antibiotics toxic to the pathogen, compete with the pathogen for food, or attack the pathogen directly. Although the existence of disease-suppressive soils has been well documented in the literature, little is known about how to manipulate these beneficial microorganisms artificially. However, crop rotation, regular addition of moderate amounts of organic matter such as compost, and limiting use of broad spectrum soil toxicants such as soil fumigants may help. Compost properly cured for 4 or more months commonly is disease suppressive when mixed with soil. Composts vary in the range of organisms they suppress. One soil fungus, *Trichoderma* species, is being used on farms to manage some of the fungi that cause damping-off in seedlings and many other diseases and may be available in the future.

Pesticides. Numerous pesticides are available to control plant pathogens. Most are fungicides for the control of fungi but a few bactericidal materials are available as well. Organically acceptable fungicides include sulfur, copper, Bordeaux mixture—a mixture of bluestone (copper sulfate) and lime (calcium hydroxide)—and fungicidal soaps. A particularly interesting new garden product is a formulation of fatty acid salts (surfactants such as in dishwashing soap) and micronized sulfur. This combination allows better coverage than conventional sulfur and may be easier for home gardeners to handle. When organically acceptable materials are known to be effective, they are mentioned in the text with some notes on proper timing. Very careful timing of application is required for successful use of these materials.

Because of the emphasis in this book on nonchemical and organic management methods, synthetic pesticides for the control of pathogens have not been detailed here. However, they are often more effective and easier to use than the old-fashioned organic materials; captan is one of the most common ones used in gardens and very small farms. In addition, synthetic materials are commonly less damaging to plants. Copper and sulfur can be especially toxic to some crops. Check with your nurseryman or farm advisor for currently available materials and their limitations.

With a few notable exceptions such as peach leaf curl on peaches and nectarines and powdery mildew on grapes, the diligent backyard gardener who is willing to sacrifice a few plants to disease should be able to get by with very little use of pesticides for the control of pathogens if he or she follows a careful cultural management program as described in this chapter. Commercial growers, especially those growing fruit trees, will have to rely more on chemical control to preserve a crop and protect the long-term health of the orchard.

Damping-off, Seed and Seedling Decay

In the field, garden, or planting box, seedlings often fail to come up, or die soon after they have emerged from the soil. Seeds may rot before they germinate, shoots may be decayed before they emerge, or stems of seedlings may be attacked near the soil line, causing young plants to collapse. These diseases often are collectively referred to as "damping off," and may be caused by a number of soil-inhabiting fungi.

Species of the soil fungus *Pythium* are most often responsible for damping-off, but several other fungi, including species of *Rhizoctonia*, *Fusarium*, and *Phytophthora*, can also cause decay of seeds, potato seed pieces, sprouts, or seedlings. Decay is

most likely to occur when old seeds or seed pieces are planted in cold, wet soil and is further increased by poor soil drainage, the use of green compost, and planting too deeply.

Symptoms. The first evidence of damping-off or seed piece decay is the failure of some plants to emerge. If seeds are attacked before they germinate, they become soft and mushy, turn dark brown, and decay. They may have a layer of soil clinging to them when they are dug up; the soil is interwoven with fine, threadlike fungus growth.

Germinating seedlings shrivel and may darken. If seedlings are attacked after they emerge, stem tissue near the soil line is decayed and weakened, usually causing plants to topple and die. When only roots are decayed, plants may continue standing but remain stunted and eventually die. As seedlings get older, they become less susceptible to damping-off fungi.

Biology. The fungi that cause damping-off and seed piece decay are present in virtually all soils. They survive on dead organic matter and also produce spores or other structures that survive for long periods of time. The young tissue of emerging seedlings is least resistant to infection by disease-causing fungi when growing slowly in cold, wet soil. Vigorously growing seedlings are fairly resistant to infection.

Management. Damping-off is worse when soil is wet or compacted. Prepare planting beds so that the soil has good drainage. Drainage can be improved by

The three tomato seedlings at right are suffering from damping-off. Note how the stem tissue is shriveled. These seedlings would topple over and die in the field. The one at left is healthy.

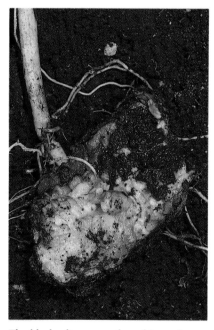

The black, slimy growth and irregular depressions on this potato seed piece indicate that it is suffering from seed piece decay.

using raised beds and soil amendments such as redwood shavings, peat moss, or fir bark. Use only well-decomposed compost. Green compost will encourage damping-off fungi. Use aerobic composting procedures to reduce the population of disease-causing fungi in the compost. Composted hardwood bark has been reported to reduce damping-off. Soil solarization during fallow periods will reduce pathogens in planting beds.

Plant when temperatures are favorable for rapid seedling growth. More shallow planting will speed up germination outdoors if conditions are marginal. If you want to start seedlings earlier, start them in the greenhouse or other protected areas and transplant them into the field when temperatures get warmer; do not transplant into cold, wet soil. Use only the highest quality seed available. If starting plants indoors, in cold frames, or in greenhouses, plant seeds in steam-treated soil or pasteurized potting mix. You can purchase treated potting soil or prepare your own. Soil mix must be held at 140° F (60° C) for at least 30 minutes. Soil can be heated in a conventional or microwave oven in a paper bag, by raising the temperature with boiling water, or by placing moist soil in a clear plastic bag in direct sunlight. Use a thermometer to make sure soil reaches the proper temperature for the correct time.

Use light sprinkler irrigations to encourage germination and emergence. Do not overwater. If growing potatoes, plant seed pieces in moist soil that is at least 50° F (10° C), and start them at

a time of year when irrigations will not be necessary before sprouts emerge from the ground.

After plants emerge, thin them so that there is good air circulation among the plants. Avoid putting on too much nitrogen fertilizer. Avoid planting the same crops in the same place year after year.

Powdery Mildew

Powdery mildew is a common disease on many types of plants (Table 4-1). Different species of fungi are involved depending on the plant affected. Powdery mildews do not require moist conditions to establish and grow and normally do well under warm conditions; thus they are more prevalent than many other diseases under California's dry summer conditions.

Symptoms. The disease can usually be recognized easily in most crops by the light-colored, powdery spore growth that forms on shoots, both sides of leaves, and sometimes flowers. An exception is the powdery mildew that affects tomatoes, eggplants, peppers, and sometimes artichokes, which produces yellow patches on leaves and often no powdery growth.

On vegetable crops, powdery mildew usually appears first as yellow spots on the upper leaf surface of older leaves; these spots develop the characteristic

Fruit, such as these apples, may develop weblike russetted scars after a powdery mildew infection.

Powdery mildew infections are characterized by a light-colored, powdery spore growth on shoots and both sides of leaves as shown on the apple shoot here.

TABLE 4-1.

Host Plants and Control Measures for Powdery Mildew Species.

FUNGUS SPECIES	HOSTS	CONTROLS
Erysiphe cichoracearum	composites: endive, lettuce, parsley, sunflower, chrysanthemum, dahlia, zinnia; cucurbits: cucumbers, melons, squash; potato; begonia; phlox	resistant varieties of lettuce, cucumber; water sprays; fungicides if necessary on squash and pumpkin
Erysiphe cruciferarum	cole crops	not usually required
Erysiphe pisi	peas	resistant varieties; sprinkler irrigation
Leveillula taurica	artichoke, eggplant, pepper, tomato	rarely required; fungicides if necessary
Podosphaera leucotricha	apple, quince, rarely pear and almond	tolerant varieties; pruning out infections; fungicides if necessary
Podosphaera oxycanthae	leaves and shoots of cherry, peach, plum	tolerant varieties; pruning out infections; fungicides if necessary
Sphaerotheca macularis	strawberry, a different strain infects caneberries	resistant varieties; removing infected tissue; fungicides if necessary
Sphaerotheca fuliginea	cucurbits, beans, okra, black-eyed peas	resistant varieties for some; fungicides if necessary
Sphaerotheca pannosa	leaves and fruit of apricot, peach, plum, rose	pruning out infections; fungicides if necessary
Uncinula necator	grapes, ivy	tolerant varieties; pruning out infections; water sprays; fungicides

powdery growth and symptoms spread to the undersides of leaves and stems. Affected leaves may turn completely yellow, die, fall off, and fruit beneath may become sunburned. Powdery mildew does not directly affect most vegetable fruits, although pea pods may get brownish spots, sugar content of fruit is often reduced, and the disease may affect the flavor of melons and squash.

The disease is more serious on grapes and fruit trees than most vegetables because on these crops it attacks new growth, including buds, shoots, flowers, and newly forming fruit as well as leaves. New growth is dwarfed, distorted, and covered with a white, powdery growth. Young fruit develop weblike russetted scars and sometimes a rough corky skin. Grapes may also crack or burst.

Biology. All powdery mildew fungi require living plant tissue to grow. On perennial hosts such as grapes, raspberries, and fruit trees, powdery mildew survives from one season to the next in infected buds or as fruiting bodies on the bark of stems and trunks. On strawberries the disease can survive on leaves that remain on the plants through winter. Year-round availability of crop or weed hosts is important for the survival of the powdery mildew fungi that infect cole crops and tomatoes, peppers, and eggplants. Special spores are produced that allow overwinter survival of the species that cause the disease in lettuce and peas and certain other crops.

Most powdery mildew fungi grow as thin layers of mycelium on the surface of the affected plant part. Spores, which are the primary means of dispersal, make up the bulk of the powdery growth of powdery mildew and are produced in chains that can be seen with a hand lens; in contrast, spores of downy mildew grow on branched stalks that look like tiny trees. Also, downy mildew spores are darker—a grey purplish color—and occur only on the lower leaf surface.

Powdery mildew spores are carried by wind to new hosts. Although humidity requirements for germination vary, all powdery mildew species can germinate and infect in the absence of water. In fact, spores are killed and germination and mycelial growth are inhibited by free moisture. Moderate temperatures and shady conditions are generally the most favorable for powdery mildew development. Spores and

The powdery mildew fungus has caused dark webbing and left a light dust of white spores on this grape. The disease may also stunt grapes, cause off flavor and low sugar content.

Powdery mildew spores grow in chains, which can be seen under a good hand lens.

mycelium are sensitive to extreme heat and direct sunlight.

Management. In most cases, planting resistant varieties or avoiding the most susceptible varieties and following good cultural practices will adequately control powdery mildew. Protection of susceptible varieties with fungicide sprays may be necessary where conditions are most favorable for mildew. Fungicide applications are most often needed on highly susceptible varieties of apples and grapes.

Resistant varieties. In many cases, varieties resistant to powdery mildew are available. Plant resistant varieties of cantaloupe, cole crops, cucumber, peas, and strawberries. Be aware of control needs when planting more susceptible varieties.

Cultural practices. Plant in unshaded areas as much as possible. Provide enough water and avoid excess fertilizer. Since spores cannot germinate when there is free moisture and may be

killed, plantings with overhead sprinkler systems or frequent water sprays may have reduced incidence of powdery mildew. As new shoots begin to develop on apples, grapes, blackberries, and raspberries, watch closely for the appearance of powdery mildew. Where infection is limited, prune out and bury or discard diseased tissue as soon as it appears. Keep grapes carefully pruned and trained to allow exposure of bunches to sunlight and good air flow through the canopy. Pruning and training is also helpful in controlling *Botrytis* bunch rot. If powdery mildew has been present during the season on perennial crops, prune out infected tissue during the dormant season. On apples, grapes, and peaches, look carefully for infected shoots and buds and remove them. Infected buds are flattened or shriveled in appearance compared to normal buds. The buds and infected shoots of apples have a thin layer of fuzzy white fungus on their surface that usually is easy to see. Remove and dispose of overwintering leaves on strawberry plants that are infected. If blackberries

or raspberries develop powdery mildew, remove the canes down to the roots after harvest.

Fungicide application. Planting the most resistant varieties available and using good cultural practices should eliminate the need for chemical control of powdery mildew in most crops. However, in some situations, especially in the production of grapes, fungicides may be needed. Powdery mildew is a problem in commercial apple orchards but does not usually require treatment in backyard trees. Sulfur and synthetic fungicides are available. A new formulation of sulfur, which is combined with surfactants similar to those in dishwashing detergent, has been recently released by Safer's. Proper timing of fungicide applications is critical to successful control; this is especially true with sulfur treatments, which only prevent rather than eradicate infections. Sulfur can cause injury to foliage and fruit if applied just before or on days when temperatures exceed 90° F. Timing for sulfur treatment in apples and grapes is described below. Fungicides can also be used on other tree fruits and vegetables but should rarely be needed for powdery mildew control. Two exceptions are squash and muskmelons. Be careful when using sulfur on muskmelons; some varieties can be damaged by sulfur while others are relatively sulfur tolerant.

Apples. Apply liquid lime sulfur or flowable sulfur at 2-week intervals, beginning when buds just start to open (green tip stage), until small, green fruit are present. Sulfur or lime sulfur sprays timed to control apple scab will also control powdery mildew.

Grapes. Begin applying wettable sulfur between budbreak and 2-inch growth. Thereafter, repeat applications with wettable or dusting sulfur at 7-to-10-day intervals until the sugar content of grapes is 12 to 15%, which is when they begin to approach ripeness and are no longer susceptible to infection. You can measure the sugar content with a refractometer, if you have access to one, or you can see if sample berries sink in a 15% sucrose solution. (Prepare the sucrose solution by dissolving 8½ teaspoons of table sugar in a half cup of warm water, then mixing in enough cold water to make the total volume one cup.)

Downy Mildew
Bremia lactucae and *Peronospora* spp.

In California, downy mildew is primarily a problem on a few vegetable crops when they are grown during cool, moist weather. Downy mildew gets its name from the downy masses of spores it produces on the undersides of affected leaves. These downy growths initially may be confused with those of powdery mildew. However, the two diseases differ in several important ways. Downy mildew produces spores mostly on the undersides of leaves and only after rain or very heavy fog. Spores disappear soon after leaves

Downy mildew spores grow on branched stalks that look like tiny trees.

Downy mildew often causes yellow to brown spots on the upper surfaces of leaves, such as these broccoli leaves.

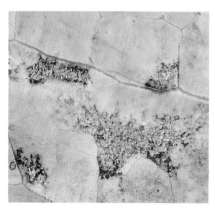
Downy mildew produces its fuzzy spores mostly on the under surfaces of leaves and only after rain or heavy dew; they disappear soon after sunny weather resumes.

dry out. Powdery mildew does not require water on the leaf surface for spore formation, and the powdery growth appears on both sides of leaves. As illustrated in the accompanying photographs, powdery mildew spores are produced in chains, whereas downy mildew spores grow on branched, tree-shaped fruiting structures. Powdery mildew thrives in warm weather whereas downy mildew is a cool weather disease. Dry, desiccating winds and clear, warm days inhibit growth and spread of downy mildew.

Although the characteristic downy growth appears primarily on the undersides of leaves, irregular yellow brown spots appear on both the upper and lower surfaces and remain through drier conditions after spores are no longer apparent.

The best way to prevent downy mildew is to avoid the environmental conditions that favor it. Freely circulating air, allowing plants to dry between irrigations, and keeping leaves as dry as possible are good ways to limit the disease. Cool, moist weather in early spring and late fall favor the disease, so adjusting planting times may help in some areas. Downy mildew survives from crop to crop on infected crop debris and susceptible related weeds. Always destroy these sources well before planting a new crop. Resistant broccoli varieties are available.

Brown Rot of Stonefruit
Monilinia spp.

Brown rot is the most common and serious blossom and fruit disease of almonds and stone-fruits. The disease also affects flowering cherries and plums, flowering quince, and some wild species of *Prunus*. Brown rot blossom blight causes withering of flowers and girdling of twigs. Later in the season the fungi cause rot of almond hulls and ripening stonefruit. A rot of ripe citrus fruit caused by *Phytophthora* is also called brown rot and is discussed in the next section; do not confuse it with the stone fruit disease.

Symptoms. The first symptom of the disease is the browning and withering of blossoms. Dead blossoms often cling to twigs for a long time. Sunken, brown areas called cankers may develop around twigs at the base of infected flowers, causing leaves at the tips of the twigs to shrivel up. Brown, sticky droplets of gum may exude from the base of dead flowers and the bark of infected twigs. During wet weather small, velvety, gray or tan tufts of spores are formed on diseased blossoms or twig cankers. These spores spread the disease to other blossoms. Large numbers of flower-bearing twigs are killed when the disease is severe.

On stone fruit or almond hulls, circular, brown or tan spots spread rapidly over the fruit surface and gray or tan, velvety tufts or lumps of spores are produced on the rotted areas. Rotted tissue remains relatively firm and dry. Fruit rot can develop on the tree or in storage.

Biology. The fungus that causes brown rot survives the winter in infected twigs, inside dead, blighted blossoms that remain on the tree, or in dried, rotted fruit (called "mummies") on the tree or on the ground. Spores produced on these sites in spring are carried through the air by wind and splashing water to infect flowers of the new year's crop.

Brown rot infection and disease development will take place over a wide temperature range and flowers can be infected from the time buds open until petals fall. Water must be present on the flower surface for infection to occur. Little or no blossom blight develops if there is no rain during the bloom period and irrigation does not wet flowers. Fruit is most susceptible to infection by the brown rot fungus when it is ripe. Most fruit rot develops during the month before harvest, although rot occasionally may develop on green fruit in early summer. Almond hulls may be inflected from the time they split until they dry.

Management. Prompt removal and destruction of diseased plant parts prevents the buildup of brown rot inoculum in isolated trees or in small orchards, and is often sufficient to keep brown rot below damaging levels. Prune your trees from the time they are planted to allow for good ventilation in the canopy. Use furrow irrigation or low angle sprinklers to avoid wetting blossoms, foliage, and fruit. Whenever possible, plant varieties that are least susceptible to brown rot.

These peach shoots and young fruit have been killed by brown rot fungus. Note the sticky gum exuding from the stem of the young fruit.

However, if brown rot builds up to high levels in your tree or orchard or is present in nearby orchards, trees, or ornamentals, application of a protectant fungicide may be necessary to prevent serious loss.

Variety selection. All varieties of almonds and stonefruits are susceptible to brown rot but some are less severely affected. You may want to choose the least susceptible varieties when planting new trees. Check with your nurseryman and other authorities such as farm advisors. If you are planting or already have susceptible varieties, you will want to pay close attention to brown rot control.

Sanitation. After leaves fall and before the first fall rains, remove all fruit and nut mummies and prune out branches with diseased

This whole almond branch has been infected with brown rot.

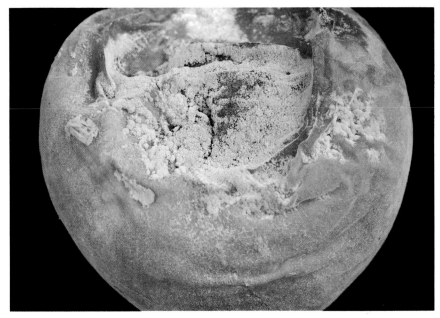
Ripe fruit can also develop brown rot fruit rot in storage. The rot spreads rapidly with the initial brown or tan spots developing gray to tan velvety tufts of spores within a day or so.

twigs and dead blossoms still clinging to branches; they signal the presence of overwintering brown rot infections. Destroy prunings and mummies by burning, burying, or bagging. If these infected plant materials remain exposed, they may produce spores that can infect blossoms or fruit. Pay close attention to flowering cherry, plum, and quince; they can harbor brown rot infections also.

During bloom, frequently check all trees, including ornamental hosts, for the appearance of brown rot symptoms. Prune out and destroy infected blossoms and twigs as soon as they appear. When green fruit are present, watch for symptoms of rot and remove affected fruit before spores begin to form on them.

As fruit begin to ripen, look carefully and frequently for fruit rot symptoms. Remove and destroy fruit as soon as symptoms appear. Watch for hull rot symptoms on almonds as soon as hulls begin to split. Remove and destroy spurs that develop hull rot symptoms before harvest.

Harvest your almonds as soon as 95 to 100% of the nuts at eye level have split hulls. If necessary, harvest your trees again to remove later maturing nuts.

After harvest remove all fruit or nuts remaining on your trees and destroy them. They are potential overwintering sites for other pests as well as brown rot. Dry prunes immediately after harvest. The drying process kills the brown rot fungus. Any delay in drying increases the likelihood of brown rot developing on the fruit.

During harvest and packing, handle fruit carefully to avoid injury and discard all fruit that show rot symptoms. Harvesting fruit before they are ripe decreases the incidence of brown rot; especially where brown rot blossom blight has been severe. Cool fruit as quickly as possible after harvest.

Fungicide application. In commercial orchards where brown rot incidence has been high or where neighboring trees pose a threat to your own, a protectant fungicide will probably be needed to avoid serious loss to brown rot. Apply Bordeaux mixture or other copper-containing fungicide or synthetic fungicides at budswell. If rainy weather occurs, additional applications at full bloom and petal fall may also be needed. Use additional lime in Bordeaux mixture after leaves begin to appear to decrease the likelihood of copper toxicity damage. If the blossom blight phase is severe, or if rains occur, application of a fungicide 1 to 3 weeks before harvest may be necessary to prevent severe fruit rot.

Phytophthora Brown Rot of Citrus

Ripe or ripening citrus fruit on the lower part of the tree may develop a light brown, water-soaked decay often called brown rot. Affected fruit develop a pungent odor. Fruit in the early stage of the disease may go unnoticed at harvest and infect other fruit during storage.

The disease is caused by the same *Phytophthora* fungus species that cause root or trunk rots, but brown rot often occurs when these symptoms are not present. Spores of the pathogen are water splashed from the soil onto tree skirts during rain storms, and infections develop under continued wet conditions, usually during fall and winter.

The only management option for brown rot of citrus is prevention. Apply a preventive copper fungicide, preferably Bordeaux mixture, in October or November before or just after the first rain.

Generally one spray provides protection through the wet season, but when rainfall is excessive, you may have to repeat the spray in January or February. Covering the tree skirts to about 4 feet above the ground is usually sufficient and does not harm natural enemies of insect pests. Spraying the ground underneath the trees also reduces brown rot infections.

Gray Mold and Bunch Rot
Botrytis spp.

Botrytis rot, often called gray mold on vegetables and bunch rot in grapes, occurs on a wide range of plants during damp, cool to mild weather. Commonly affected crops in California include strawberries, lettuce, squash and melons, artichokes, tomatoes, beans and peas, and grapes. Leaves, flowers, and green

or ripe fruit on or near the soil or in the dampest or densest areas of the canopy turn light brown and develop a gray or brown fuzzy growth of fungal spores. Symptoms often appear first on old flower parts and spread to fruit and foliage. Occasionally stems may be affected. Fruit or plants shrivel and rot and often develop flattened, hard, black masses, called sclerotia, under rotted parts. The disease may also develop in storage, causing rotting of harvested fruit or vegetables.

Biology. The fungus that causes gray mold, *Botrytis cinerea*, survives in decaying plant material in or on the soil and as sclerotia. It usually infects only flowers and senescing or injured tissue, especially plant parts in contact with damp soil. However, green, healthy tissue may also be affected. On lettuce, aging wrapper leaves or older leaves at the bottom of plants are usually first infected; whereas on beans, peas, strawberries, and tomatoes, flowers usually are the initial infection site, and the disease later spreads to pods or fruit. Grapes are infected at the flower stage, but the fungus remains dormant until fruit mature; bunch rot then develops if conditions become cool and damp.

Management. To minimize gray mold, use planting densities and training techniques that provide maximum air movement through the plant canopy. Avoid using excessive nitrogen fertilizer; it will encourage growth of foliage and limit ventilation. Two-wire trellising of grapes with midseason cane pruning ("hedging") reduces the incidence of bunch

The light brown, water-soaked discoloration on the ripening lemon on the right indicates it is infected with Phytophthora brown rot.

This strawberry fruit is infected with gray mold. The fungus is characterized by globular bunches of spores held at the ends of the fungal fruiting stalks. Use a hand lens to see them.

The gray, velvety coating on the calyx of these fruit and the browning or graying of fruit are typical of gray mold infections on tomatoes.

rot, but also delays ripening of berries. A better strategy is to remove leaves from around each grape cluster at late bloom. Keep strawberry plants thinned so that fruit are not completely covered with foliage, especially in cool, moist climates.

Plant on well-drained, raised beds and use furrow irrigation whenever possible and mulch, or stake plants so that fruit do not contact moist soil. Prop heavy ripening fruit up on old strawberry baskets. Try to keep surfaces of vegetable beds as dry as possible. When sprinkling plants, do so in the morning so that foliage will dry quickly during the day.

Harvest fruit in advance of rainy or humid, cool conditions whenever possible. Avoid storing fruit or vegetables that are susceptible to gray mold if they have been exposed to wet, cool conditions before harvest.

Remove diseased vegetable plants, strawberries, and grape bunches as soon as they appear; dispose of them or bury them deeply in the compost pile. Remove plant residue at the end of the season and discard it or compost it properly to reduce or eliminate mold inoculum. Any plant residue left in the field or garden must be buried several inches to prevent infection of the next planting.

Gray mold infections in grapes start with single berries, often in the middle of the bunch, and spread to adjacent berries.

White Mold
Sclerotinia spp.

The first sign of infection by white mold or lettuce drop on lettuce is wilting of outer leaves.

White mold is a distinctive disease of many vegetables that most often affects stems and foliage at the base of plants. Bean and pea pods, lettuce heads, cabbage heads, cucurbit fruit, and the stalks and flowers of cole crops may also be affected. Affected tissue develops a soft, watery rot and white, cottony mycelium forms on the surface. Plants may wilt if stems are girdled by the decay. As affected stem or leaf tissue dries up, it turns yellow to white, and hard, black sclerotia form on the surface or inside the dead stems. White mold usually affects plants nearing maturity that have developed a substantial canopy.

The white mold fungus, *Sclerotinia* spp., survives from one crop to the next as sclerotia in or on dead plant tissue. Because prolonged wetness is required for infection, disease usually occurs at the base of plants or beneath heavy plant canopies. As with *Botrytis,* infection and development is favored by cool, moist conditions.

Management. White mold is favored by a wet soil surface. Use of raised beds and careful furrow irrigation that does not overflow onto bed surfaces can help limit damage. Level your land to provide for even distribution of water and assure good drainage with beds as high as possible, especially where white mold has been a problem in the past. Space

On tomatoes, white mold kills and whitens stems, leaving hard, black sclerotia on the surface.

plants well enough to allow good air circulation, although as plants mature, their shade will allow more retention of water on the soil surface.

Remove and destroy entire infected plants as soon as you see them. Remove all crop residues from infested fields after harvest.

Sclerotinia species have a broad host range and can infect many broadleaved vegetables and weeds. Corn and other grains are not affected, but a three year rotation out of susceptible crops, combined with good control of broadleaved weeds, would be required to

A white, cottony mycelium and hard black clusters of dormant sclerotia develop on the lower stem of lettuce plants affected with the Sclerotinia fungus.

reduce survival of the pathogen in the sclerotium stage.

Soil solarization will control *Sclerotinia.*

Common Smut of Corn
Ustilago maydis

Common smut is a fungal disease of corn that frequently occurs in gardens and small farms. Symptoms are very distinctive. All actively growing parts of the plant may be affected including leaves and stems, but infections on ears are most obvious. Plant parts develop distinctive galls, which are at first a glistening, greenish white to silvery

white. Galls on ears and stems expand and fill with masses of powdery, dark olive-brown to black spores. Leaf galls remain small and become hard and dry. Ear and stem galls rupture, and wind, rain, or irrigation water spread spores through the farm and beyond. Hot, dry weather favors development of the disease.

All corn varieties are susceptible to the disease to some degree. Disease incidence is higher in soils that have had heavy applications of nitrogen or manure. Minor abrasions due to blown soil, cultivation, spraying, or detasseling increase opportunity for infection. In home gardens and small farms, remove and burn (or otherwise destroy) galls from infected plants before they rupture.

Common smut of corn causes kernels to expand and fill with masses of dark spores.

Bacterial Soft Rots of Vegetables

Ripe or ripening fruit, fleshy stems and tubers or bulbs are often attacked by bacteria that cause a watery, mushy, often foul-smelling rot. Almost any crop may develop this disease. Many of the bacteria that cause this type of damage are secondary invaders that attack parts of the fruit already damaged by other diseases or pests. Others, primarily *Erwinia* species, *Xanthomonas* species, and *Pseudomonas* species, will attack healthy plants in the field.

Symptoms normally develop similarly on any host. The disease starts as a small water soaked spot that enlarges rapidly and becomes soft and mushy. The surface of the fruit or affected part may remain intact while the tissues beneath disintegrate. Cracks often develop causing the contents to ooze out. A whole fruit or tuber may become a soft, watery, colorless, decayed mass within 3 to 5 days. A strong odor is often given off especially with cole crops, potatoes, and onions. No powdery spores or white fungus growth is produced as is the case with rots caused by gray mold or white mold. If you put an affected fruit under water, then squeeze and place some of the exudate under a microscope, you can often see the bacteria oozing out.

Certain cultural and sanitation practices can help reduce bacterial soft rot incidence. Always harvest in a timely manner and remove infected debris from the field. Plant in well-drained areas, avoid crowding, and avoid sprinkling maturing fruit of susceptible crops. Pesticides are not recommended. Sanitation is important in reducing incidence of bacterial soft rot in storage as well.

Shot Hole
Stigmina carpophila (Coryneum beijerinckii)

Shot hole, also called Coryneum blight, is an important disease of almonds, apricots, nectarines, and peaches in California. The disease is also referred to as peach blight in peaches. It rarely occurs on plums or prunes. The disease causes spots or lesions on buds, leaves, twigs, and fruit, and is most severe following warm, wet winters and when wet weather is prolonged in the spring.

Symptoms. Shot hole first appears in the spring as reddish or purplish brown spots about $\frac{1}{10}$ inch in diameter on new buds, leaves and shoots. The spots expand and their centers turn brown. Tiny, dark specks, visible only with a hand lens, form in the brown centers, especially on buds; these dark specks, which are the spores of the fungus, distinguish shot hole from other diseases. Spots on young leaves have a narrow, light green or yellow margin and their centers often fall out as leaves expand, leaving "shot holes." Leaves may fall from the tree, especially if affected when young. Blemishes on almonds or apricot fruit become rough and corky as they get older but rarely have the dark

The purplish brown spots on these young apricots were caused by the shot hole fungus.

TABLE 4-2.

Parts Most Commonly Affected by Shot Hole in Different Stonefruit Species.

Peaches	buds and twigs
Nectarines	buds and twigs
Apricots	leaves and fruit
Almonds	twigs and leaves

specks. Fruit spotting almost always occurs on the upper surface. The parts of the tree affected by shot hole vary among the different stone fruits (Table 4-2).

Biology. The fungus that causes shot hole survives the dormant season inside infected buds and in twig lesions. The spores produced on lesions can remain alive for several months. They are spread by splashing rain or irrigation water. Spores that land on twigs, buds, blossoms, or young leaves require 24 hours of continuous wetness to cause infec-

If you look at shot hole spots with a hand lens, you can sometimes see dark specks in the center, which are the spores of the fungus. This is a distinguishing feature of shot hole.

tion. Only the current season's growth is susceptible to infection. Under California conditions, twig and bud infections of apricot, nectarine, and peach can occur during rainy weather any time between fall and spring. The fungus can germinate and infect at temperatures as low as 36°F (2°C).

Management. Most stone fruit varieties are susceptible to shot hole. Check with your nurseryman or farm advisor about susceptibility of specific varieties.

Sprinkler irrigation increases the severity of shot hole. If you use sprinklers, make sure their angle is low enough that water does not get on the foliage. In almonds, fruit infections are a problem only when overhead sprinklers are used.

Remove and destroy infected twigs, buds, blossoms, and fruit as soon as symptoms appear. Look carefully for twigs and buds with symptoms after leaf fall; infected buds have a varnished appearance. Discard fruit and prunings following the recommendations for brown rot.

Where the incidence of shot hole is low, sanitation may be sufficient to prevent significant loss the next season. However, where disease incidence has been high, it is difficult to find and remove all infected buds and twigs in the fall. Application of a protectant fungicide may be necessary to avoid serious loss in commercial crops the next season. Apply Bordeaux or a fixed copper fungicide after leaf fall and before the first fall rains. This treatment also controls leaf curl under average rainfall conditions. Additional treatments may be necessary if prolonged wet weather occurs in the spring. Apply Bordeaux or fixed copper fungicide to peaches and nectarines at the pink bud stage. Make applications to almonds and apricots at pink bud and at full bloom; add additional lime to the fungicide mixture when leaves begin to appear to reduce the risk of copper damage to the leaves. Synthetic fungicides are also effective and eliminate copper injury hazards.

Leaf Curl
Taphrina deformans

Leaf curl, a fungal disease that affects the leaves and shoots of peaches and nectarines, is one of the most common disease problems for backyard gardeners. The distorted, reddened foliage that it causes is easily seen in the spring. When severe, the disease can reduce fruit production substantially. Control requires application of a fungicide in the fall and/or early spring prior to the emergence of new growth.

Symptoms. Leaf curl first appears in the spring as reddish areas on developing leaves. These areas become thickened and puckered, causing leaves to curl and become severely distorted. The thickened areas turn yellowish gray and velvety as spores are produced by the pathogen. Affected leaves later turn yellow or brown and fall off the tree; they are replaced by a second set of leaves that develop more normally unless wet weather continues. Affected shoots become thickened, stunted, and distorted and often die. Occasionally, irregular, wrinkled areas develop on fruit surfaces.

Biology. The leaf curl fungus, *Taphrina deformans*, survives the winter underneath bud scales and in other protected areas as resting spores called conidia. In the spring, conidia are moved by splashing water to newly developing leaves where infection occurs. The fungus grows between leaf

In the spring, trees infected with peach leaf curl develop distorted, reddened new leaves, which later fall off.

cells and stimulates them to divide and grow larger than normal, causing swelling and distortion of the leaf. Red plant pigments accumulate in the distorted cells. Cells of the fungus break through the surface of distorted leaves and produce another type of spore, called ascospores, that give the leaf a powdery or feltlike appearance. Development of leaf curl ceases when young tissue is no longer developing or when temperatures are above 79° to 87° F (26° to 30.5° C).

Management. Generally peach and nectarine trees should be treated every year in the fall after leaves have fallen to prevent peach leaf curl. Copper-based fungicides including Bordeaux mixture or synthetic fungicides can be used. However, to be effective, copper-containing compounds must have at least 50% copper; those containing less do not adequately control leaf curl despite advertizing claims. If timed properly (see previous section on shot hole), a single fall spray will normally prevent losses due to shot hole as well as peach leaf curl. In areas of high spring rainfall, it may be advisable to apply a second treatment in the spring when buds begin to swell, but before any green color appears. Although symptoms of leaf curl will be seen primarily in the spring as new leaves develop, there is little you can do to control this disease at this time. Some people remove diseased leaves but there is little evidence that this improves control. Normally, diseased leaves will fall off after a few weeks and be replaced by new, healthy looking leaves,

unless it is rainy. However, if trees have shown symptoms in spring, be sure to treat the following fall to prevent more serious losses the next year.

Bacterial Canker and Blast
Pseudomonas syringae

The bacterium *Pseudomonas syringae* attacks a wide variety of fruit trees. It causes bark cankers that develop during the dormant season and early spring on almonds and stone fruits. On almonds, pears, and occasionally on apples, cherries, and peaches, *Pseudomonas* causes a blight or "blast" of buds, blossoms, leaves, green shoots, and green fruit. Losses from bacterial canker and blast can be reduced by selecting suitable rootstocks, avoiding planting on shallow soil, providing adequate nutrition, and protecting trees from freezing temperatures during bloom.

Bacterial canker. Bark cankers occur primarily on stone fruit trees. They are irregularly-shaped, brown, water-soaked or gum-soaked areas that develop in the bark and outer sapwood of spurs, branches, and sometimes the tree trunk. Small cankers sometimes develop on twigs at the base of infected buds. When trees begin active growth in the spring, amber-colored gum may exude from the margins of cankers. Cankers are darker than the surrounding, healthy bark, and the

The wet, gummy accumulation on this almond branch marks a developing bacterial canker. The bacteria entered through the adjacent dead spur.

underlying diseased tissue is reddish brown, moist, and may be sour smelling. Cutting into the bark beyond the margin of cankers may reveal small, brown flecks in the inner bark tissue, especially in apricots and plums. Affected limbs may fail to leaf out in the spring or may produce new growth, which dies soon after temperatures increase in the summer. If trees are killed by bacterial canker, new shoots are frequently produced from the rootstock.

Bacterial blast. During cold, wet spring weather, blossoms of almonds, apricots, cherries, apples, and pears may be blighted or "blasted" by *Pseudomonas syringae*. Blast symptoms develop where temperatures are lowest, such as in low-lying areas and on bottom limbs. Affected blossoms turn brown, shrivel, and usually

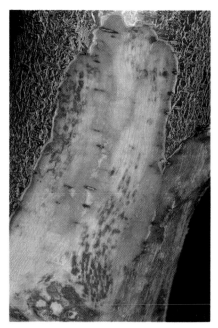

Beneath the bark, active bacterial canker infections cause red flecks in the wood just beyond the margins of the main canker.

These apple blossoms and shoots have been killed by bacterial blast.

cling to the tree. Leaves, young fruit, and green shoots also may be infected if cold rains occur after bloom. On leaves, dark brown or black spots develop and later drop out, leaving holes. Infected shoots turn dark brown or black. Small, dark brown or black, sunken spots may develop on young fruit.

Bacterial blast symptoms on apple and pear are similar to fireblight. However, blast seldom affects more than one or two inches of the shoot tip, where fireblight frequently extends a foot or more down the shoot. The bark of twigs affected by bacterial blast is light tan and has a papery appearance, in contrast to the dark brown or black, sunken, hard appearance of twigs with fireblight. Grayish-brown bacterial ooze is usually produced on fireblight lesions during humid or wet weather, but is never produced on tissue affected by bacterial blast.

Biology. *Pseudomonas syringae*, the bacterium that causes bacterial canker and blast, is always present on the surfaces of plants. When conditions are favorable, bacteria invade trees, usually through buds, leaf scars, wounds, or natural leaf and fruit pores. Infections occur during fall, winter, and early spring. Frost injury during bloom and the presence of moisture is required for infections that cause blossom blast. Cold, rainy weather increases the incidence of blast on leaves, shoots, and young fruit. Growth of cankers ceases in late spring and summer but may become active again in fall.

Susceptibility of trees to infection by *Pseudomonas syringae* is influenced by a number of factors. Trees grown on sandy, shal-low, or nitrogen-deficient soils are more susceptible. In California, feeding by high populations of ring nematodes is thought to be the most important factor making almond and stone fruit trees susceptible to bacterial canker. Late fall irrigation of young peach trees on sandy soil may increase the incidence of bacterial canker. Pruning and pruning wounds do not appear to influence susceptibility of trees to infection. Trees that receive adequate nitrogen fertilization are better able to recover from the effects of bacterial canker.

Management. Damage from bacterial canker and blossom blast can be minimized by certain cultural practices and careful selection of rootstocks and varieties. Avoid planting trees on sandy or shallow soils. If you are planting an orchard on a site where the soil is shallow because of a hardpan, break through the hardpan when preparing for planting. Have your soil analyzed for nematodes before planting an orchard. If potentially harmful species are present, they must be controlled before planting, and you may choose to plant rootstocks or species less likely to be damaged.

If overhead sprinklers are available, use them on apples and pears to prevent frost injury when low temperatures occur during bloom. Trees can be protected from frost injury and bacterial blast by using lattice or plastic shelters. For small plantings or backyard trees, consider covering them or using protective shelters if freezing conditions occur during bloom or early fruit growth.

If trees have been damaged by bacterial canker, remove entire affected branches in the summer, being sure to eliminate the entire canker and a few inches below.

Stone fruit and pear varieties differ in susceptibility to bacterial canker and blast. Choice of rootstock can influence the susceptibility of plums, prunes, and cherries to bacterial canker. Check with your nurseryman and other local authorities for the best recommendations before planting.

The scars on these apples were caused by apple scab. Note the distortion of the smallest fruit.

Apple Scab
Venturia inaequalis
Pear Scab
Venturia pirina

Scab is generally considered the most serious disease of apples in California, causing loss or severe surface blemishing of fruit. Apple scab is most severe in the coastal areas of California, where spring and summer weather is cool and moist; however, it can be a problem wherever apples are grown. Scab can also cause serious losses in pears.

Dark, velvety spots may grow on the underside of apple leaves affected by scab. Note the distortion of the leaves.

Symptoms and Damage. Scab infections are usually first noticed on leaves. Affected leaves become twisted or puckered and have black, circular, scabby spots on their upper surfaces. On the undersurface of leaves, the spots are velvety and may coalesce to cover the whole leaf surface. Severely affected leaves may turn yellow and drop. Scabby spots are also found on fruit; they are sunken and tan and may have spores around their margins.

Infected fruit become distorted and may crack, allowing entry of secondary organisms. Severely affected young fruit may drop.

Biology. The pathogen overwinters primarily in infected leaves on the orchard floor, although the pear scab fungus may also overwinter on twigs. In the spring, spores are carried by wind-driven rain or splashing water from the orchard floor to flowers, leaves, or fruit where they germinate and infect. New spores are produced on the infected leaf or fruit surface 10 to 20 days later, allowing further spread of the disease until conditions become hot and dry. Infection occurs most rapidly between 55° and 75° F and leaves must remain wet continuously for at least several hours for infection to occur. W. D. Mills at Cornell University devised a chart in

1954 that correlates the length of time it takes spores to infect apples using leaf moisture and air temperatures. (This chart is shown in Table 4-3.)

TABLE 4-3.

Mills' Chart: Temperature and Moisture Requirements for Apple Scab Infection.

AVERAGE REQUIRED TEMPERATURE ($°$F)	HOURS OF WETTING FOR INFECTION
78	13
77	11
61 to 75	9.5
60	9
57 to 59	10
56	11
55	11
54	11.5
52 to 53	12
51	13
50	14
49	14.5
48	15
47	17
46	19
45	20
44	22
43	25
42	30
33 to 41	—*

*Not known.

Source: Derived from research by W.D. Mills and A.A. LaPlante 1954. See References.

Management. All commonly grown varieties are susceptible to scab. Late-blooming varieties may have less scab if their development begins after spring rains. Some "heirloom" apple varieties are reportedly resistant to scab, and might be acceptable for backyard trees in cool coastal areas. In single backyard trees, removal of leaves from beneath trees in winter may be sufficient to limit the disease to tolerable levels. In commercial orchards, use of several techniques may be necessary. Fall foliar fertilizer (urea) applications hasten leaf fall and decomposition and will help reduce the number of spores in spring. If you are using overhead irrigation, operate sprinklers between sundown and noon of the following day to allow adequate leaf drying before significant infection can occur. Spores are released mainly during the day and require many hours for infection to occur, depending on temperatures. The Mills chart (Table 4-3) gives the number of hours leaves must be wet for infection to occur. Several synthetic fungicides and sulfur are available for control where spring rains are heavy. If sulfur is used it must be applied before the spores germinate and treatments must be repeated weekly as long as conditions are favorable for infection.

Viruses

Viruses are tiny (submicroscopic) particles that require a host cell for reproduction. They are transmitted mechanically in plant sap, by insects that feed on plant sap, by certain root-feeding nematodes, in seed or vegetative parts of plants used for propagation, and during budding and grafting. Most cultivated plants are susceptible to at least one or more virus diseases. Mosaic viruses, which usually cause yellow or light green mottling of foliage, are the most common. Other symptoms caused by mosaic viruses as well as other types of viruses include distorted or stunted growth, reduced yields, premature fruit ripening, poor quality fruit, and premature or abnormal coloration of foliage. In some cases virus-infected plants do not show visible symptoms, although their growth and yield may be reduced. Virus diseases are controlled by planting resistant varieties, using certified virus-free planting material and the best quality seed available, carefully removing diseased plants, isolating different aged plantings of susceptible crops, and controlling spread by mechanical means. Control of insect vectors is usually not practical.

Symptoms. The most striking virus symptoms occur on foliage. Mosaic viruses cause a mottling of light green, yellow, or white areas on leaves that may resemble symptoms of some nutrient deficiencies. This mottling can be very pronounced on apple leaves and in cucurbits, beans, and peas;

This yellow squash shows a green mottling caused by watermelon mosaic virus.

Leaf mottling is typical of many mosaic viruses, such as the watermelon mosaic virus infecting this squash plant.

it may be subtle and difficult to see in potatoes and tomatoes. Almond calico is a yellow mottling or banding of leaves caused by *Prunus* ringspot virus on almond trees.

Leaf distortion may accompany mosaic-type symptoms or may be present without mosaic patterns. Tomatoes affected by tobacco mosaic virus may have leaves so severely distorted they are string-

like in appearance. This damage is very similar to injury caused by the herbicide 2,4-D. In some cases, such as in potatoes and strawberries, leaves may have a rough, crinkled appearance because leaf tissue between veins becomes raised while veins remain sunken. Potato leafroll virus causes leaflets to curl upward at their edges. In some cases, distorted leaves may grow in bunches or rosettes. For example, leaves of almond trees affected by yellow bud mosaic, a strain of tomato ringspot virus, may form distorted bunches at the ends of branches.

Other types of virus symptoms include spots, streaks, rings, darkening of nutmeat conducting tissues, and vein banding—a discoloration along leaf veins. The leaves of strawberries, blackberries, grapes, and fruit trees may turn red prematurely.

Fruit may also be affected by virus infection. Cucumbers, melons, and other cucurbit fruit on plants infected with mosaic viruses are mottled with yellow, dark green, or pale green patches and may develop raised, warty areas. Melons may develop cracks or have poor flavor because their sugar content is low. Pods of beans and peas may become mottled and misshapen, or develop raised areas, and their seeds may become shriveled. Tomato fruit may develop dark green, yellow, or brown streaks or spots. Curly top virus causes tomato fruit to be distorted and ripen while still very small. Tubers of Russet Burbank (Netted Gem) potato plants infected with leafroll virus develop a dark brown speckling, often concentrated at the stem end; this symptom is called "net necrosis."

Certain viruses disrupt food-conducting vessels (called the phloem) at the graft union of some tree species. This results in aboveground symptoms that are similar to those caused by root rots—pale, sparse foliage, reduced shoot growth, and a steady decline of the tree. Symptoms may be visible as a dark brown or black horizontal line if the bark is cut

Lettuce plants affected by lettuce yellows or beet western yellows virus show general yellowing, which is most intense around the edges of the outermost leaves.

away from the sapwood at the graft union. Viruses that cause these symptoms include tomato ringspot virus, which causes prune brownline in plums and prunes and a similar reaction in Red Delicious apple on M106 rootstock, citrus tristeza virus, and cherry leafroll virus, which causes walnut blackline.

Biology. Virus particles move within the host plant in the food-conducting tissue. Transmission to another host plant requires contact with plant sap, which may occur when plants are rubbed together, handled, cultivated or pruned, or during budding and grafting. However, the most common means of spread is through the feeding of insects, especially aphids, leafhoppers, and whiteflies, that have fed on the sap of infected plants. Usually only one or a few species of insects transmit a specific virus.

Virus diseases also may be transmitted during vegetative propagation. Common examples

Tobacco mosaic virus has caused a mottled pattern of yellowing on this tomato leaf.

of this occur when infected strawberry plants or potato seed pieces are used or during grafting of fruit trees. Some viruses can be transmitted in seed or pollen. Bean common mosaic, lettuce mosaic, pea seedborne mosaic, and squash mosaic are seed transmitted. Walnut blackline and *Prunus* ringspot virus can be transmitted in the pollen of infected trees.

Management. The best way to reduce the effect of virus diseases is to plant resistant varieties and to use certified, high-quality planting stock and seed. Follow practices that minimize mechanical spread, protect susceptible crops from insect vectors, and avoid planting susceptible crops near fields that may harbor harmful viruses. It is especially important to avoid overlapping plantings of a single virus-prone crop in the same area. Virus inoculum which builds upon the first crop will rapidly spread to the second crop, devastating it before it has a chance to become established. Remove vegetables when they develop symptoms. In some cases trees with virus symptoms should be removed.

Resistant varieties. Whenever possible, use varieties resistant to viruses that are likely to be a problem in your area. Check with your nurseryman or seed supplier for the latest available varieties. For instance, resistant varieties are available to protect against some mosaic viruses in beans and peas, tobacco mosaic on tomatoes, curly top virus of beans, cucumber mosaic virus for cucumbers, and potato leafroll virus in potatoes.

Some viruses can cause rosetting or concentrating of leaves at branch terminals. Shown here are almond branches infected with yellow bud mosaic.

Some viruses cause disruption of the food conducting tissues at the graft union. This walnut tree has been infected with the cherry leafroll virus, which causes walnut blackline. The line of dead black tissue at the graft union is a distinguishing feature.

Virus-free or certified propagation material. Use certified seed or the highest quality seed available to avoid seedborne viruses such as bean common mosaic, lettuce mosaic, pea seedborne mosaic, and squash mosaic. Plant only certified bulbs, corms, and seed potatoes. Do not use potatoes from the previous season's crop for seed; they may be infected with virus even when the plants did not show visible symptoms. When planting trees, blackberries, raspberries, and strawberries, obtain nursery stock that is certified to be virus-free. Also use certified material for scion wood when budding or grafting.

Top growth is weak on trees affected by viruses affecting food conducting vessels at the graft union, such as the one causing walnut blackline. Suckers of branch growth commonly shoot up from below the graft union.

Control of insect vectors. Reducing the spread of viruses to susceptible crops by controlling aphids or other insect vectors is not easy and often not possible. For instance, insecticides are not usually effective because most viruses can be transmitted to a plant before the insect vector is killed. However, in gardens or small areas, spun-bonded polyester or other types of insect-proof covers layed over rows before seedlings emerge can protect crops from viruses during their most sensitive seedling stages. More novel protection methods include oil sprays and shiny aluminum mulches. Oil sprays (corn oil or mineral oil) may prevent the spread of viruses that are carried on the mouthparts of aphid vectors. These treatments need to be repeated weekly to remain effective. Reflective film mulches reduce virus spread by repelling aphids. Protection from insect vectors is particularly important for cucurbits (cucumbers, melons, pumpkin, squash) because they can be severely affected by mosaic viruses. In desert areas, protection of lettuce from whiteflies also may be necessary. Other crops that may require protection from insect vectors in some locations include beans, peppers, spinach, and tomatoes. Planting dates can sometimes be adjusted to avoid peak populations of insect vectors. (See Chapter 3 for more information on controlling insects that vector virus diseases.)

Cultural practices. Take precautions to reduce the mechanical spread of viruses. Avoid handling plants when they are wet, because viruses that are spread mechanically are spread more easily when plants are wet. Avoid excessive handling of cucurbits, peppers, potatoes, tomatoes, and strawberries. If you smoke or handle tobacco materials, wash your hands thoroughly before handling tomatoes, and do not smoke around these plants to avoid transmission of tobacco mosaic virus. Clean equipment used for pruning, budding, and grafting operations. If tomato ringspot virus is present in orchards, avoid practices such as flood irrigation and movement of contaminated equipment that would spread soil nematodes, which can transmit this virus.

Control weeds and rotate crops. Many weeds are hosts for viruses that infect garden plants and trees. Rotating crops reduces the buildup of viruses and their vectors. When growing strawberries, it is a good idea to replace the plants annually or at least every few years to reduce the buildup of viruses as well as certain other diseases. Destroy or compost crop residues immediately after harvest.

Diseased vegetable plants should be removed immediately. Be sure to remove plants when they are dry. Backyard trees may need removing only when they become unproductive. In commercial orchards, case-by-case judgments must be made. When replacing trees, be sure to use a rootstock resistant to the virus present, if one is available.

Avoid planting susceptible plants near plants or fields that may be infected with virus. Do not plant beans near clover or gladiolus. Avoid planting beans, melons, spinach, or tomatoes near sugar beet fields or tomatoes near alfalfa. Plant Chinese cabbage or bok choy well away from old cole crops fields, and do not plant strawberries near old plantings of strawberries. Destroy wild blackberry vines in the vicinity of blackberry or raspberry plantings, or else avoid planting these near stands of wild blackberry. Do not plant tomatoes when residue from previous tomato plantings is still present.

Vascular Wilts

When all the leaves on one branch or stem wilt and turn yellow or brown, it often means a plant is affected by vascular wilt. Vascular wilts of garden plants and orchard trees are caused primarily by species of the soil fungi *Fusarium* and *Verticillium*. These fungi invade the water-conducting tissue of the host plant and restrict water flow to the foliage, thereby causing wilting. *Fusarium* is mainly a problem in vegetable crops, whereas *Verticillium* may cause losses in fruit and nut trees as well as certain vegetables.

Symptoms. Wilt symptoms are generally similar in all plants affected. Leaves turn yellow, wilt, then brown, and die, usually starting on lower parts of the plant first, although Fusarium wilts may sometimes start with a yellow flagging of one or more shoot tips. Affected leaves may curl. Young shoots also wilt and die. Symptoms often start on one side of the plant first or on one stem of a tree or viney plant, especially on grapevines or trees affected by Verticillium wilt. If you make a cross section of infected stems, roots, twigs, or branches, you will usually see a partial or complete ring of yellow to reddish brown or brown, discolored vascular tissue.

In crops such as tomatoes and cabbage that are affected by both Fusarium and Verticillium wilt, it is extremely difficult to distinguish the two diseases in the field. A laboratory analysis by a plant pathologist is usually required to confirm diagnosis.

Biology. Fusarium wilts tend to be more common and destructive in warmer temperate climates; in most cases symptoms do not develop until temperatures are above 68° F (20 C). Verticillium wilt tends to be more serious in cooler areas; the fungus dies out in the branches of some woody hosts during hot summer weather.

Both *Fusarium* and *Verticillium* form resistant structures that can survive for years in the soil in the absence of a living host. These soil fungi are moved in soil water, on farm equipment, transplants, or tubers. In the presence of a host plant, the resistant structures germinate and penetrate the plant's roots either directly or through wounds. Once inside the root, the fungus grows until it reaches the water-conducting cells, inside of which it spreads upward through the plant, restricting water flow.

Management. Wilt diseases can be avoided in many cases by planting resistant varieties. When growing susceptible plants, problems sometimes can be minimized by rotating crops, following good sanitation practices, and using soil solarization.

Resistant varieties. Varieties resistant to wilt are available for most vegetables. Check with local authorities for the varieties that work best in your area. Some tree

Tomato plants affected by Fusarium wilt may first show symptoms on one side or just one branch, but eventually the whole plant will die.

The xylem or water conducting tissue in plants affected by Fusarium wilt turns brown. In tomatoes, this darkening may be slightly more intense than those infected with *Verticillium*.

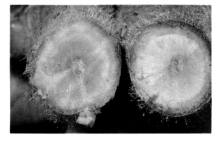

The tomato stem at left is infected with *Verticillium*; the one at right is healthy, showing no vascular discoloration.

rootstocks are less susceptible to *Verticillium* than others. Tables 4-4 and 4-5 list some crops for which varieties and rootstocks resistant or tolerant to wilt diseases have been developed; not all these varieties will be suitable for all areas or for commercial production. Root knot nematodes can break down the resistance of tomato plants to *Fusarium*. Use tomato varieties that are resistant to root knot nematodes as well as *Fusarium* and follow other prac-

In almonds, a Verticillium wilt infection usually becomes apparent when leaves on one or more branches suddenly wilt, turn light tan, and die. Dead leaves generally remain on the tree throughout the growing season.

TABLE 4-4

Crops Commonly Affected by Verticillium Wilt and Availability of Resistant or Tolerant Varieties or Rootstocks.

CROP	TOLERANCE OR RESISTANCE AVAILABLE?
Almond	yes
Avocado	no
Blackberry	yes
Cabbage	yes
Cherry	no
Cucurbits	yes (for some crops)
Eggplant	no
Grapes	no
Olives	yes
Persimmon	no
Pistachios	yes
Potatoes	yes
Raspberries	yes
Spinach	no
stone fruit	yes
Strawberry	yes, but lower yielding cultivars
Tomatoes	yes

tices useful for controlling nematode damage if root knot nematodes are present in your soil.

Sanitation. After working with a wilt-susceptible crop, clean off equipment and shoes before moving to another area to avoid spreading wilt organisms. Be sure surface water does not flow from an area where wilt has been present to an area free from wilt. Remove all residue, including roots of wilt-susceptible plants, at the end of the season and dispose of it. Aerobic composting can destroy the pathogens if compost temperatures reach 150° F.

Crop rotation. Crop rotation has limited value against the wilt diseases. *Fusarium* can persist in the soil for many years, so it is not practical to return to a host crop every few years. However, the Fusarium wilt pathogens have very

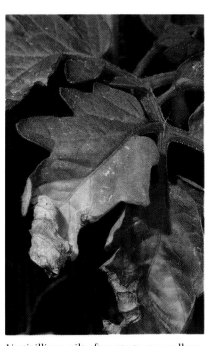

Verticillium wilt often starts as a yellowing between the major veins of the leaves. Eventually whole leaves and stems wither and die.

TABLE 4-5.

Crops Commonly Affected by Fusarium Wilt and Availability of Resistant or Tolerant Varieties.

CROP	TOLERANCE OR RESISTANCE AVAILABLE?
Asparagus	yes
Cabbage	yes
Celery	yes
Cucumber	yes
Melons	yes
Peas	yes
Radish	yes
Spinach	no
Sweet potato	yes
Tomato	yes

limited host ranges—usually only one, or at the most two, crops—so infested areas can be used to successfully grow many other crops. The *Verticillium dahliae* fungus, on the other hand, has a wide host range, which includes many crops, ornamentals, and weeds. Crops that can be infected include those in the tomato, cucurbit, crucifer, and spinach families as well as grapes, strawberries, caneberries, and stone fruit, avocado, olive, cherry, persimmon, and many other trees. Keeping land out of these susceptible crops for 3 to 5 years can reduce *Verticillium* to levels where a crop can be profitably grown for 1 to 2 years. Corn, other grains (especially flooded rice), carrots, celery, asparagus, sweet potato, lettuce, bean, pea, and alfalfa are some crops that could be used.

Soil solarization. Soil solarization has effectively reduced the amount of *Verticillium* and *Fusarium* in infested soil in some situations where summers are hot. Use solarization during a summer fallow period before planting a wilt-susceptible crop. (See Chapter 2 for details on using soil solarization.)

Phytophthora Root and Crown Rot

Almost all fruit and nut trees as well as most ornamental trees and shrubs can develop Phytophthora root and crown rot if soil around the base of the plant remains wet for prolonged periods. Red stele, a serious root rot of strawberries, is also caused by species of *Phytophthora*. Cauliflower and other cole crops may be affected by root or systemic rot, particularly in coastal areas on finer-textured (clay) soils. Tomatoes, peppers, and eggplant can also be affected. Asparagus is affected by Phytophthora crown and spear rot in wet years. In many of these crops, different species of *Phytophthora* are involved. Losses to *Phytophthora* are minimized by providing good soil drainage and selecting the most tolerant rootstocks or varieties available.

Symptoms. The leaves of plants affected by Phytophthora root or crown rot appear drought stressed. Trees or plants often wilt and die rapidly with the first warm weather of the season. Leaves may turn dull green, yellow, or in some cases red or pur-

Aboveground symptoms of plants affected by Phytophthora root or crown rot are typical of root and crown diseases. Leaves appear drought stressed and may wilt and die quickly with the first warm weather of the season.

plish. Often, only plants in the most poorly drained area of the field or garden are affected. In avocados, abnormally large numbers of small fruit are produced on infected trees.

Symptoms may develop first on one branch or stem then spread to the rest of the tree or plant. Trees may decline over a period of years before finally dying or they may be killed in a single season. Slow decline occurs when the roots are attacked; rapid decline occurs when the crown or basal stem is attacked and girdled in a single season. For instance, strawberries affected by red stele root rot may decline in productivity over a period of years.

Symptoms on roots and crowns depend on the species of *Phytophthora* involved, the plant being attacked, the resistance of

Phytophthora infections may cause leaves to take on a purple color and wilt as shown on this brussels sprouts plant.

Although Phytophthora root rot is a common disease in tomatoes, it often goes undetected because the aboveground symptoms—slow growing plants that wilt and die in hot weather—are not distinctive.

The lower crown in this photo shows discoloration typical of Phytophthora crown rot of strawberries. The upper crown is healthy.

More distinctive symptoms of *Phytophthora* occur on the roots, which dry out and rot off. The inner tissue of the root turns brown as shown on this tomato root.

the variety, and soil moisture and temperature. In general, trees affected by *Phytophthora* develop darkened areas in the bark around the crown and upper roots. Gum or dark sap may ooze from the margins of the diseased trunk area. If bark tissue is carefully cut away, reddish brown streaks or zones can be seen in the inner bark and outer layer of wood. No fungus mycelium is visible in the tissue affected by *Phytophthora*, distinguishing this disease from Armillaria root rot.

When tomatoes and eggplants are affected by Phytophthora root rot, roots of all sizes develop water-soaked spots that dry out and turn a chocolate brown as the disease becomes advanced. Early infections kill seedlings ("damping-off"); later infections reduce plant vigor and may cause collapse and death of the plant. If you cross section roots you will see that the stele or central core of conducting tissue, in the tap root, is yellowish or brown above the rot lesions. Stele discoloration may extend into the lower stem.

Biology. *Phytophthora* is a soil-inhabiting fungus that is favored by wet conditions. Species of *Phytophthora* produce resting spores that survive for years in moist soil in the absence of a suitable host. However, if the soil is completely dried out, these spores will survive for only a few months.

When a host is nearby and free water is present in the soil, resting spores germinate to produce motile spores that can directly penetrate roots, branches, or crowns as long as free water is present. Wounds are not required for infection.

Resting spores, decaying host tissue in the soil, and active cankers can all be sources for new infections. The fungus can be spread in splashing rain or irrigation water, in surface irrigation and runoff water, and by movement of contaminated soil, equipment, or plant parts. Flooded and saturated soil favor the spread of *Phytophthora* to healthy plants.

Some *Phytophthora* species are favored by warm weather, some by cool weather. Root rot of avocado, citrus, and tomato are favored by warm conditions, developing most extensively in late spring and early summer. Decay of crown, trunk, and branches of other tree species and red stele root rot of strawberry are favored by cool, wet conditions. These decays develop most rapidly in late fall and early spring.

Management. The most important factor in reducing the threat of Phytophthora root and crown rot is good water management. Avoid prolonged saturation of the soil or standing water around the base of trees or other susceptible plants. Irrigate only as much and as often as necessary; in an orchard, keep track of the soil moisture around each tree and water only when necessary. If you irrigate trees with sprinklers, use low-angle sprinkler heads and splitters to avoid wetting the trunk and lower branches. If using a drip system, place the emitters at least a foot away from the trunk. Avoid planting susceptible species on poorly drained or shallow soils.

For all vegetable and orchard plants, provide good soil drainage.

In trees, such as the almond one shown here, *Phytophthora* may cause gumming on bark and reddening of wood beneath the bark.

Good soil drainage is best provided before planting. Drainage should be good to the rooting depth of the plants, generally 3 to 6 feet for trees, 2 to 4 feet for shrubs, and 1 to 2 feet for bedding plants. You do not want the roots and crown of a plant to remain wet for the 6 to 8 hours that are required during favorable weather for *Phytophthora* to infect. Plant trees and shrubs on mounds made by working up soil to provide adequate drainage. Raised beds provide good drainage in garden situations. Group plants according to their irrigation needs: separate those needing frequent, light irrigations, such as potatoes and strawberries, from those needing infrequent, deep irrigations, such as tomatoes and melons.

Do not plant trees or other plants deeper than they were planted in the nursery; never cover the graft union with soil. If you are not sure where the graft union is, ask someone at the nursery to show you and mark it. Do not have irrigated turf around the base of trees, remove all weeds, and do not water the crown area directly.

At the first signs of aboveground symptoms, examine the tree at the soil line for crown rot. Carefully cut away bark that looks affected. If crown rot is present, trees can sometimes be saved by removing soil from the base of the tree down to the tops of the main roots and allowing the crown tissue to dry out.

Sanitation. It may be possible to slow the spread of *Phytophthora* within an orchard by avoiding movement of infested soil, water, and plant parts from an area where Phytophthora root and crown rot has developed. Surface and subsurface drainage water and anything that can move moist soil can carry the fungus to a new area. If drainage water flows from infested to uninfested areas within the orchard during wet weather, consider putting in watertight drains to channel the water away from healthy trees.

Selection of planting stock. Plant only certified nursery stock from a reputable source, and choose the most resistant rootstocks or varieties available for your area. Less susceptible rootstocks or varieties are available for almonds and stone fruit, apples, cauliflower, and strawberries.

Rotation. If tomatoes have been affected by Phytophthora root rot, avoid planting tomatoes or other susceptible plants such as egg-

plant in the same soil for at least one or two seasons. Plant a resistant crop such as corn instead, or leave the soil unplanted and unirrigated, but keep it well worked to allow the soil to dry down as deeply as possible. Different species of *Phytophthora* attack beans, and cole crops; so these plants can be used as well.

Armillaria Root Rot
Armillaria mellea

Armillaria root rot affects a wide variety of trees, shrubs, and herbaceous plants including almonds, apples, citrus, grapes, stone fruits, strawberries, and walnuts. The fungus that causes the disease, *Armillaria mellea*, is often called oak root fungus because it is frequently associated with oak trees or found where oak trees have grown.

Aboveground parts of trees and plants suffering from Armillaria root rot show the same general symptoms of decline as those with Phytophthora root rot; damage usually appears in localized areas within a field or orchard and spreads to other plants in a circular pattern. However, Armillaria root rot differs from Phytophthora root rot because the fungus is usually visible within infected tissues. Thin, white or yellowish mats of mycelium grow beneath the outer layer of infected bark; dark brown to black, root-like structures called rhizomorphs spread over the surface of infected roots and beneath the bark of severely decayed roots or crown; and clusters of light brown or honey-colored mush-

The Armillaria root rot fungus forms whitish mycelial plaques between the bark and wood in the crown region of trees.

rooms sometimes appear around the base of infected trees and shrubs during wet weather in fall or winter. Fungus mycelium and decayed host tissue have a distinct mushroom odor.

Armillaria mellea is a soil fungus that survives in the roots of host plants or in fragments of woody material, where it can live for many years as long as the soil remains moist. Management of Armillaria root rot requires removal of all potentially infectious host material as well as the same watering and drainage practices used for Phytophthora root rot.

When a susceptible plant is affected by Armillaria root rot, little can be done except removing the affected plant, allowing the soil to dry out, and replanting with a resistant species. If a tree is affected, remove it and, in orchards, all adjacent trees also. Remove and destroy all infected

The Armillaria root rot fungus invades healthy roots and crowns of neighboring trees with its dark, shoestringlike structures called rhizomorphs. The dark Armillaria rhizomorph is shown here above healthy, light brown apple roots.

In winter, *Armillaria* often forms clusters of mushrooms at the base of infected trees after a rain.

tissue, including all roots that are larger than one-half inch in diameter. Dig down at least 2 feet, and preferably deeper, to remove these roots. Burn all woody tissue at the site or bag it carefully for removal to avoid spreading the fungus to new locations. Air drying to the middle of infected tissue will also kill the fungus. Allow the soil to dry out for one or two seasons before replanting with a resistant species or rootstock. During the fallow period, a deep-rooted cover crop such as sudangrass that is not irrigated helps dry out the soil.

A tolerant tree species such as apple, citrus, prune, or plum may be saved by removing soil away from the trunk to a depth that exposes the upper lateral roots. Make sure the trunk is not wetted by irrigation water. Avoid planting susceptible tree species in locations where Armillaria root rot is known to have occurred or where *Armillaria* is likely to be present, for example old orchard sites, recently cleared forest land, and near rivers and streams.

Crown Gall
Agrobacterium tumefaciens

Crown gall occurs on a wide range of fruit tree and ornamental species. It is encountered most frequently on pome fruit, stone fruit, and almonds, walnuts, grapes, blackberries, and raspberries. Crown gall is most damaging to blackberries, raspberries, and young trees and vines. If you obtain good quality material from a reputable nursery and use careful planting and pruning techniques, you should not be troubled by crown gall.

Crown galls are rough, warty tumors caused by the soil-inhabiting bacterium *Agrobacterium tumefaciens*. This organism can survive in the soil for at least 2 years in the absence of host tissue or for several years in decomposing crown gall tissue. Bacteria are released into the soil when galls are wet or when older gall tissue disintegrates. Seedlings may be infected by *Agrobacterium* during germination; established trees and vines are infected only through fresh wounds such as those caused by growth cracks, pruning, damage by cultivation equipment, or freeze injury. Galls develop on trees most commonly on large roots at the crown. On blackberries, raspberries, and grapes, galls may form on stems also. Galls first appear as smooth swellings, and develop rapidly into large tumors with a rough, warty or cracked appearance.

You can greatly reduce problems with crown gall by obtaining planting material from a reputable nursery. Examine trees and vines for signs of galling before purchase. When caring for susceptible plants, avoid injury or pruning wounds that will come in contact with soil. If you need to remove suckers close to the soil line, rinse the soil away from the area before pruning. If you are planting in an area where crown gall has occurred, you may want to treat the roots of susceptible plants with a commercial biological control agent (Galltrol) before planting. Follow the label directions carefully. This treatment has not been effective on grapes.

Larger trees and vines can usually tolerate the development of crown galls. Younger plants can be more quickly girdled and killed by the developing gall tissue. Examine the crown area of trees and vines for presence of galls at the end of their first and third year in the field. If galls are present, apply bactericide formulated for treatment of crown gall tissue (Gallex) during the dormant season or in spring or early summer. Before treating galls, rinse soil away from them with a garden hose or other source of pressurized water and allow them to dry. If galls are very large, cut away some of the gall tissue before treating. After treatment allow galls to dry for at least one day before replacing soil around the roots. Gallex stops all flow of water and nutrients through gall tissue. Therefore, do not treat more than one third of a tree's circumference at a time. If more than this is galled, repeat treatments after 6 to 12 months.

Trees infected with crown gall commonly have warty tumors on large roots near the crown. A tree as large as this can usually tolerate the growths; however, young trees planted near it may be seriously affected unless treated before planting.

CHAPTER FIVE

Nematodes

Nematodes are tiny eel-like roundworms. The species that attack plants are so small that a microscope is usually required to see them. Most of the major nematode pests of fruit and vegetable crops in California are soil-dwelling species that feed on plant roots. Root feeders impair the plant's ability to take up water and nutrients and cause general decline of plants, often without readily recognizable symptoms and usually without killing them directly. Damage is often more pronounced when plants are under other stresses such as lack of water or nutrients, weed competition or damage from insects, pathogens or other pests.

Root Knot Nematodes— The Most Common Problem

Although there are many different species of root-feeding nematodes in California, by far the most important in gardens and small farms are the root knot nematodes (*Meloidogyne* species). Their importance and frequency of occurrence is largely due to the wide range of crops that they attack. The most serious problems with this group of nematodes occur in broadleaf plants grown in the warmest and sandiest soils.

Root knot nematodes usually cause distinctive swellings, called galls, on roots of affected plants. Root knot nematode galls may merge to grow as large as one inch in diameter on some crops such as okra and fig but are usually much smaller—not much larger than a pea or lima bean and never as large as the galls caused by diseases such as crown gall (see the Disease chapter). Root knot galls are truly swellings on the root. They do not rub off easily as do the beneficial nitrogen-fixing nodules that often occur on roots of legumes. On a few types of plants such as certain legumes and grasses, root knot nematodes may invade, reproduce, and damage roots but cause no galling.

There are several different species of root knot nematodes in California. Four or five species may be present in a single location, especially in warmer areas such as the San Joaquin Valley south of Stockton. Each garden or farm site may have a slightly different complex of root knot nematode species depending on what crops have been grown in the soil, sources of contamination and geographical region. Different root knot nematode species have different host preferences; growing certain crops will cause a particular species of root knot nematode to become dominant.

Root-feeding nematodes impair plants' ability to take up water and nutrients with a resulting decrease in aboveground growth. The small grapevines in this photograph are suffering from a root knot nematode infestation.

Root knot nematodes cause galling on roots. These tomato roots are severely infested.

This lettuce plant shows a more typical level of galling by root knot nematodes.

Because of their weakened root systems, plants infected with root nematodes often wilt during the hottest part of the day as shown in the lettuce plant at left.

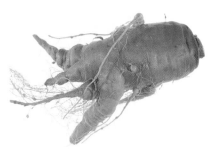

Root knot nematodes can cause galling as well as forking of roots in carrots.

Less Common Root-Feeding Nematodes

In addition to root knot nematodes, several other root-feeding nematodes may sometimes cause problems in gardens and small farms. These include the cyst nematodes (*Heterodera* species), which attack only certain vegetables including cole crops, beets, spinach, turnips, and related weeds; the root lesion nematodes (*Pratylenchus* species), which feed on a wide range of trees and some vegetable crops; and the citrus nematode (*Tylenchulus semipenetrans*), which may damage citrus or olive trees or grapevines. All of these groups feed inside of plant roots for major portions of their lives. Several species of ectoparasitic nematodes, which live freely in the soil and browse on root tips, can also occasionally be damaging; these include the stubby root (*Paratrichodorus* and *Trichodorus* species), dagger (*Xiphinema* species), and ring (species in the genus *Criconemella* and related genera) nematodes. With the exception of the cyst nematode, which produces distinctive egg-containing cysts on roots, invasions of these less common root feeders are difficult to identify without a laboratory diagnosis of roots and soil by a nematologist.

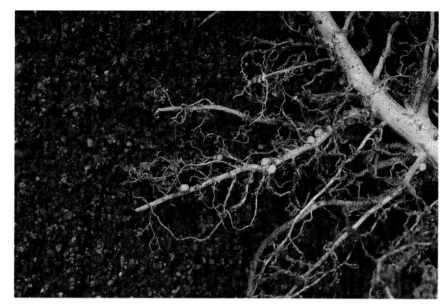

Females of the cyst nematode are visible as white bodies on the surface of these broccoli roots.

The white lemon-shaped body at left is a sugarbeet cyst nematode female and the shiny brown, egg-shaped cyst to the left contains the nematode's eggs.

Nematodes Attacking Aboveground Plant Parts

In California, only two non-root-feeding nematode groups are likely to cause damage in vegetables and fruits. The most common of these is the stem and bulb nematode, *Ditylenchus dipsaci*, which infests bulbs and cloves of onions, garlic, leeks, and related crops and causes discoloration or browning of leaf sheaths surrounding the bulb and rotting in the base of the bulb. Use of certified seed, rotation, and sanitation are recommended management practices for this pest. A group of foliar nematodes (*Aphelenchoides* species) may infest strawberry leaf and flower buds or growing points, african violets, ferns, and some other ornamentals; but damage is not common due to strict inspection by the nursery industry. Both foliar and bulb and stem-feeding nematodes are favored by cool, moist growing conditions, but can occur wherever infested planting stock is grown.

Table 5-1 lists the major nematode pests on commonly grown California vegetable and orchard crops. The remainder of this chapter will focus on the root-feeding nematodes.

The rotting at the base of these garlic bulbs and browning of leaves is typical of stem and bulb nematode damage.

Complexities of the Small Farm and Garden

Geographical location, cropping history, and soil texture are important determinants of a nematode problem. Garden situations are all the more complex because of the diversity of plants that generally occur. Trees, bushes, and vines develop deep root systems. If these roots are present in the garden or invading from the neighbor's garden, they can provide a continual source of nematode inoculum for the vegetables and other annually planted crops. Chopping down a tree, vine, or bush does not eliminate the problem because the roots can survive and nurture nematodes for many years after removal of the vine or tree trunk (at least 8 years in the case of grapevines). Nematodes such as root knot are capable of moving upwards 3 feet by the end of the growing season, so they can easily move from surviving roots to invade newly planted crops for many years. It does not make sense to use elaborate methods of complete soil treatment if infested roots or other sources of the problem nematode are to remain in or adjacent to the garden. If you plan to plant nematode-susceptible vines or trees near your vegetable garden, try to be sure you purchase ones on nematode-resistant rootstocks.

An additional complexity for gardeners is choosing appropriate management strategies. Most non-chemical management methods, including rotation, resistant varieties, soil solarization, and fallow-

TABLE 5-1.

California Vegetable, Fruit, and Nut Crops Likely to Suffer Significant Damage from Common Nematode Groups.

	ROOT KNOT (*Meloidogyne* spp.)	ROOT LESION (*Pratylenchus* spp.)	CYST (*Heterodera* spp.)	OTHER NEMATODES
FRUIT AND NUT TREES				
Almond	●[1]	●	○	ring nematode
Apple	●	●	○	
Apricot	●[2]	●[2]	○	ring nematode
Avocado	?	●	○	
Cherry	○	●	○	
Citrus	○	●	○	citrus nematode
Olive	T	●	○	citrus nematode
Peach and Nectarine	●[1]	●	○	ring nematode
Pear	?	H	○	
Plum and Prune	○	●	○	pin and ring nematodes
Walnuts	T	●	○	ring nematode
GRAPES AND SMALL FRUITS				
Grapes	●[3]	●	○	citrus, dagger, stubby root and ring nematodes
Blackberries/raspberries	?	●	○	dagger nematode
Strawberries	●	●	○	foliar nematode
VEGETABLES				
Asparagus	T	○	○	
Beans—blackeye	●[4]	●	○	
Beans—lima	●[5]	○	○	
Beans—snap	●[6]	●	○	
Beets	●	○	●	
Carrots	●	○	○	
Celery	●	○	○	
Cole Crops	●	○	●	
Corn	T	●	○	
Cucumbers	●	○	○	
Eggplant	●	○	○	
Lettuce	●	○	○	

continued

TABLE 5-1.

California Vegetable, Fruit, and Nut Crops Likely to Suffer Significant Damage from Common Nematode Groups, continued.

	ROOT KNOT (*Meloidogyne* spp.)	ROOT LESION (*Pratylenchus* spp.)	CYST (*Heterodera* spp.)	OTHER NEMATODES
Melons	•	○	○	
Onions and garlic	T	○	○	stem and bulb nematode
Peas	•	•	•[7]	
Peppers	•	○	○	
Potatoes (Irish)	•	•	○	
Potatoes (sweet)	•[8]	○	○	
Radish	•	○	•	
Spinach	•	○	•	
Squash	•	○	○	
Tomato	•[9]	○	○	
Turnips	•	○	•	

Key

• = Most varieties susceptible to at least one species in group, nematode likely to cause crop loss

•[1] = Resistant Nemaguard or Nemared peach rootstocks may be used against root knot

•[2] = Royal Blenheim rootstocks are resistant to root knot and root lesion

•[3] = Resistant grape varieties such as Harmony and Freedom are available

•[4] = California blackeye 5 cowpeas are resistant to *M. incognita*

•[5] = Ventura N white limas are resistant to *M. incognita*

•[6] = Nemasnap varieties are resistant to *M. incognita*

•[7] = Garden peas are susceptible to *Heterodera trifolii*, which is a different species from the ones attacking cole crops, beets and spinach

•[8] = Sweet potato varieties are available that prevent nematode reproduction, but are still damaged in heavily infested soils

•[9] = Tomato varieties designated N are resistant to most root knot nematode species (but not *M. hapla*) and prevent nematode reproduction.

T = Tolerant host, nematode can feed and reproduce, but does not cause significant damage

○ = Nematode unable to develop and reproduce in plant

? = Susceptibility not known

ing, must be instigated before trees or vegetables are planted.

Even so, if nematodes are discovered in the middle of the season, you can often limit the extent of damage. For instance, with some crops, it is possible to nurse the crop along to harvest with frequent irrigation and fertilizing. You will then have to take control action prior to next year's planting because nematode populations will be quite high.

Symptoms and Damage

Root-feeding nematodes damage roots, reducing the plant's ability to take up necessary water and nutrients. Aboveground symptoms include the same general signs of wilting and stress associated with poor irrigation or fertilization or other pests that damage roots. Usually damage is confined to certain areas of the garden or orchard at first, spreading through the area in subsequent years. In the case of root knot nematodes, damage is more serious in sandy soils and warm regions.

Vegetable plants infested with root knot nematodes grow more slowly than neighboring healthy plants, starting early to midseason; plants often wilt during the hottest part of the day and leaves may yellow slightly. Often fewer and smaller fruits and leaves are produced, and very susceptible, heavily infested plants damaged early in the season may die. On fruit and nut trees and grapevines, nematode damage may reduce vigor, retard growth, yellow leaves, decrease size of leaves and fruit, and cause premature autumn leaf loss. It may take several years for these symptoms to show up on trees and vines.

The root galls or swellings caused by the root knot nematodes make infestations of these nematodes easier to recognize than those caused by other root-feeding species. Substances produced by the nematodes while they feed stimulate the growth of plant cells and subsequent formation of root galls. Root knot nematodes feed and develop within the galls. If you cut a gall open, you may see the mature female nematode as a tiny (no more than $1/16$ inch long) white pearl within the tissue. More than one female may be in a single gall. Gall formation damages the water- and nutrient-conducting tissues in the roots, causing plant decline. Galls may crack or split open, especially on the roots of vegetable plants, allowing the entry of pathogens that cause root rots or wilts.

Root symptoms caused by other root-feeding nematodes normally will not be sufficient to identify the cause of the problem. Nematodes that move freely in the soil and browse on roots—such as the stubby root, stunt, ring, and dagger nematodes—may simply deform, shorten, or reduce root growth with no other recognizable clues to the source of damage. Citrus nematodes cause root reduction, but their presence can sometimes be detected at high population levels by their jelly-like egg masses, which can cause soil to stick to roots when they are pulled up; the soil cannot be readily removed with water. Root lesion nematodes produce dark, depressed areas or lesions on the root surface of many, but not all, damaged trees. Root lesion nematodes also reduce root growth, especially on young feeder roots.

When nematodes damage the growing point of roots, they may cause the plant to put out many side roots, giving the root system a "whiskery" appearance. This symptom is particularly common with cyst nematodes but may also be produced by root knot or stubby root nematodes when feeding on certain plant species. Cyst nematode infestations can be distinguished by the presence of the nematodes themselves. Check lateral roots with a hand lens to look for the lemon-shaped, white females and brown cysts, which contain the nematode's eggs.

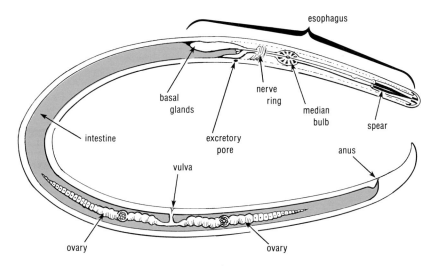

FIGURE 5-1. Anatomy of a typical nematode. Plant liquids are sucked out with mouth parts called the spear or stylet.

Life Cycle

Nematodes that feed on plants go through six stages: an egg stage, four immature stages—each of which is followed by a molt or shedding of the skin—and an adult stage. Many nematode species can develop from egg to egg-laying adult in as little as 21 to 28 days during warm summer-time temperatures. Immature stages, adult males, and sometimes adult females are long, slender worms (Figure 5-1) but are too small to be seen—even with a hand lens. Mature adult females of some species such as the cyst, citrus, and root knot nematodes have a swollen lemon or pearlike shape and remain embedded in infested roots. Nematodes have many generations a year in California's mild climates and potentially great rates of population increase depending on environmental conditions, particularly temperature.

The ectoparasitic nematodes (stubby root, ring, stunt, dagger, and related nematodes) move about freely in the soil around the roots throughout their life cycle, browsing around the root tip area; females lay their eggs singly in the soil around the roots. Eggs may provide a pro-

tected overwintering stage when suitable host plants or soil temperatures are not available. Root lesion nematodes, which can inhabit both soil and roots, lay their eggs singly within roots or in the soil near roots. Root lesion nematodes may overwinter in old roots that remain in the soil after harvest.

The swollen, immobile female cyst, citrus, and root knot nematodes lay their eggs into a jelly-like mass that extends out through the root surface. In the case of the cyst nematode, only a small percentage of the eggs are laid into the gelatinous mass, the rest remain in her body, which becomes a tough, brown protective cyst for the eggs. After the female dies, the cysts detach from roots, allowing wide dispersal of the nematode in soil; eggs may remain viable in cysts for several years.

It is believed that the root knot nematode survives primarily in the egg stage when crops are out of the ground; however, survival

is not great, and populations decrease by 80 to 90% in unplanted areas in the winter. Even so, if no control action is taken, enough root knot nematodes will survive on weeds and roots in the soil to damage subsequent crops for at least 2 years. The life cycle of the root knot nematode is shown in Figure 5-2.

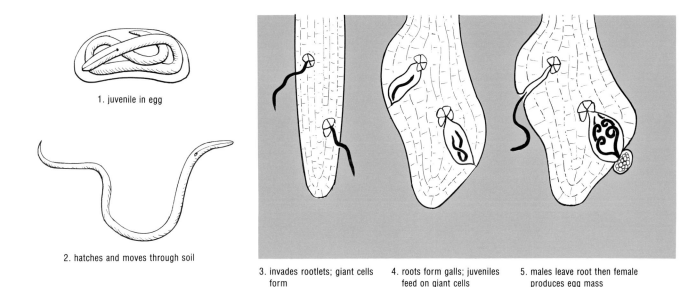

1. juvenile in egg

2. hatches and moves through soil

3. invades rootlets; giant cells form

4. roots form galls; juveniles feed on giant cells

5. males leave root then female produces egg mass

FIGURE 5-2. Root knot nematodes spend most of their life cycle in galls on roots. Second stage juveniles invade new sites, usually near root tips, causing some root cells to grow into giant cells where the nematodes feed. As feeding continues, the plant produces a gall around the infected area.

Management

The first step in managing nematodes is to study your crop or planting site to determine if you have or are likely to have nematode problems. While pest nematodes are undoubtably present in every California soil, many gardens and small farms will not have problems with them because climate, soil type, crops planted this year and/or in previous years, nematode species, and other biological factors are unsuitable. For instance, while root knot nematodes may occur in any soil type, they are a serious problem mainly in sandy or sandy loam type soils. On the other hand, cyst nematodes may cause damage in all kinds of soils but only on a few crops.

If you believe there have been nematode problems in your planting site in the past, it may be worthwhile to get the nematode species identified. General plant symptoms may be a clue, but because other pathogens and environmental conditions can cause the same overall symptoms of decline, they cannot be counted on to positively identify a nematode problem. Galled roots in plants can signal root knot nematodes, and cyst nematode females and cysts may be observed on roots of susceptible crops. However, other nematode groups and specific species within the cyst or root knot nematode groups are difficult for anyone but a professional nematologist to diagnose. Check with your farm advisor for names of reliable laboratories for identifying nematodes. When commercial fields and orchards are infested, several soil samples should be made and sent to a diagnostic laboratory to determine population levels as well as nematode species. See the University of California IPM manual for your crop (listed in References) for more information on sampling commercial fields. For some crops, including tomatoes, citrus, grapes, and cole crops, injury thresholds have been established.

If you determine that nematodes are present at potentially damaging levels, try to limit damage using a combination of methods over time. Figure 5-3 shows some sequences of practices that could be followed in a vegetable field or garden that is heavily infested with root knot nematodes. The area may be divided into several sections to allow a variety of crops to be grown while some parts of the field might remain bare (fallow), or be solarized, or fumigated at any given time. Study the examples in

	FIRST WINTER	FIRST SUMMER	SECOND WINTER	SECOND SUMMER	THIRD WINTER	THIRD SUMMER
Section A	Fallow	Fallow	Fallow	Summer Susceptible Crop	Winter Spring Crop	Early Harvest → Summer Resistant Crop (e.g. N tomato varieties, Nemasnap beans)
Section B	Fumigate (e.g. Vapam)	Summer Resistant Crop (e.g. N tomato varieties, Nemasnap beans)	Winter Spring Crop	Summer Solarize	Fallow	Summer Susceptible Crop
Section C	Fumigate (e.g. Vapam)	Summer Susceptible Crop — Early Harvest	Winter Spring Crop — Early Harvest	Summer Resistant Crop (e.g. N tomato varieties, Nemasnap beans)	Fumigate (e.g. Vapam)	Summer Susceptible Crop
Section D	Winter Spring Crop	Summer Solarize	Fallow	Summer Susceptible Crop	Winter Spring Crop	Summer Solarize
Section E	Winter Spring Crop — Early Harvest	Soil Amended Summer Susceptible Crop	Fallow	Fallow	Fallow	Fallow or / Summer Resistant Crop

FIGURE 5-3. Examples of management practice sequences for a root knot nematode infested garden site or field. The site should be divided into at least three subsections to facilitate various treatments and vegetable growing sequences. Choose from the schemes shown here. See text for discussion of methods.

Figure 5-3 along with the discussions below and devise a strategy that fits your particular circumstances. Any good program for nematode management requires long-term planning. Such a program of rotating management practices will have to be maintained as long as susceptible crops are to be grown in the soil, because no currently available methods can permanently eradicate a root knot nematode problem.

Crop rotation. Rotation to a crop where nematodes do not reproduce is an especially good management strategy for cyst nematodes, which have only a few crop hosts. For instance, the sugarbeet nematode, *Heterodera*

schachtii, parasitizes only cole crops, other crucifers such as kale, turnips and mustards, and beets, spinach, rutabaga, and related crops and weeds; the cabbage cyst nematode, *H. cruciferae,* is limited to cole crops and other crucifer crops and weeds. Three to 5 years are required between host crops in an infested field. Not only must fields be kept free of crops that host the nematode, related weeds must be kept out of the field as well. To determine if cyst nematode populations have been reduced to levels that will not cause economic damage, you must take a series of soil samples throughout the rotation program. See UC ANR Publication 3307, *Integrated Pest Management for Cole Crops and Lettuce,*

for more details on this rotation program for commercially grown cole crops.

Rotation is not always a practical solution for root knot nematodes in small farms because so many vegetable crops and weeds are hosts; however, with careful long-term planning, rotation in combination with fallowing or solarization can be used to manage most problems. Chemical treatment is also an option. Choosing suitable rotations can be confusing; even crops, such as corn and onions, that are not seriously damaged by root knot nematodes may still allow them to build up in the soil; other crop varieties are resistant to one or more species of root knot nematode but allow buildup of others.

Annual crops that are useful for reducing root knot populations include small grains, such as wheat and barley, sudangrass, and resistant tomato and bean varieties. Sweet potato varieties with nematode resistance can also be used to lower nematode populations; however, they are still damaged in heavily infested fields. It is best to plant sweet potatoes in lightly infested fields or after a fallow season or a season of a truly tolerant crop such as resistant tomato varieties. When planning a rotational program, it is essential to know the species of nematode at your planting site and the precise resistance characteristics of your proposed rotational crop.

Fallow. Nematode populations may be reduced by leaving the soil bare for one season or more, a practice known as fallowing. The area must be kept absolutely weed free to keep nematodes from surviving on alternative hosts. Fallowing the soil for one year will lower root knot nematode populations enough to successfully grow a susceptible annual crop; 2 years of fallow will lower populations still further. Three to 5 years would be required to lower cyst nematodes to acceptable levels. Fallowing is most effective if the field is kept wet, especially in the summer, to induce nematode egg hatch, and weeds are controlled. Because weeds are very difficult to control in irrigated areas, many users choose to keep fallow fields dry; however, they may not obtain adequate nematode control. Once the crop is grown, fallowing or other control methods will have to be repeated when you again

observe nematode damage on crops, which may occur even after one season. A practical way to carry out a fallowing program is to split the garden or infested field into thirds and fallow one third every 1 or 2 years on a rotating schedule. (See Figure 5-3.)

Resistant or tolerant varieties and rootstocks. One of the simplest ways to manage nematodes is to use vegetable varieties or rootstocks that are not susceptible to damage by the problem nematode. Various levels of tolerance and resistance are available. For instance, stone fruit and almond trees can be protected from root knot nematode damage with Nemaguard rootstock, and tomato varieties with VFN after their name are resistant to root knot nematodes. Root knot resistant grape varieties include Harmony and Freedom. Troyer and Trifoliate rootstocks with resistance to citrus nematode are available for citrus trees. Check with your nurseryman for resistant varieties suitable for growing in your area. Table 5-1 lists some resistant varieties. Unfortunately, for many susceptible crops, no resistance is available.

Solarization and other heat treatments. In gardens and small farms, solarization can be used to temporarily reduce nematode populations in the top 12 inches of soil, allowing successful production of shallow-rooted crops. It will not provide long-term protection for susceptible fruit trees or vines. For the solarization process to be effective, the soil must be moistened and then clear plastic tarps must be left on

for 4 to 6 weeks during the hottest part of the summer; effectiveness may be limited in the cooler coastal areas where summer temperatures commonly remain below 80°F. Combining soil solarization with crop residues or animal manures may increase effectiveness. For example, the southern root knot nematode, which was incompletely controlled in lettuce by either solarization or composted chicken manure, was completely controlled by combining the two with a big yield increase. Soil solarization is described in more detail in Chapter 2.

Root knot nematodes, including egg stages, are readily killed when soil temperature exceeds 125°F for 30 minutes or 130°F for 5 minutes. There have been tests with hot water at 155°F added to deep trenches in summer time but the surface 2 feet of soil does not reach 125°F unless 180°F water is applied to deeply prepared soil that is already 100°F. A brief solarization can produce 100°F soil. Some home gardeners may have access to equipment, such as solar hot water heaters or steam cleaners, that could generate and transfer the needed additional heat deep into the soil. When carrying out heat treatments, use a soil thermometer to be sure temperatures are reaching the required levels 2 feet below the soil surface. Substantial trenching and digging may be required.

Pesticides and other nematicidal agents. There are currently no nematicides available for use by nonlicensed home gardeners in their backyard vegetable gardens. Although some products are available for application by licensed professionals—especially on commercial farms—their availability and regulations regarding their use change frequently. Normally they should not be necessary in home gardens.

Gardeners and limited-scale growers need to determine through the agricultural commissioner's Office whether use of nematicides and fumigants is permitted within their county or city location, and then they should contact a licensed applicator.

Many other nonpesticidal products have been reported to be nematicidal and many actually are; however, the difficulty with nematode control is in getting the active ingredient deep enough into the soil to reach the target nematode. Materials that can be nematicidal include manures, various composts, and other organic materials, ammonium forms of nitrogen, sugar, and certain salts. Some of these materials are toxic to crop plants as well as nematodes. However, the major problem in every case is moving the material to the site of the pest several inches or feet below the soil surface and still avoid damage to established plants. Unless such methods are developed, use of these materials cannot be recommended.

Marigolds and other nematode-suppressive plants.

For many years it has been suggested that certain types of marigolds suppress some kinds of root knot nematode, and many gardeners have tried this technique—mostly with poor results. Apparently there can be some reduction in nematode popula-

tions with marigold treatments, but specific, consistent treatment procedures have not yet been developed. New research in this area is currently underway at the University of California. Solarization, fallowing, or rotation with resistant tomato varieties or sweet potatoes is a more reliable control method and easier to carry out. They also have a suppressive effect on a broader range of pests.

The marigold species most commonly reported to suppress nematodes in the soil in which it is growing is the French marigold, *Tagetes patula.* Common varieties of this species include Nemagold, Petite Blanc, and Queen Sophia. Other species of marigolds may or may not be effective. Although French marigolds may inhibit most species of root knot nematode under some conditions, they are a very good host for the northern root knot nematode, *Meloidogyne hapla,* and will actually increase the numbers of this pest in the soil. *M. hapla* is a common species of root knot nematode and can only be positively identified by a trained nematologist in the laboratory.

Intercropping French marigolds with annual crops or under trees or vines will not reduce nematode damage on susceptible crops. Marigolds must be grown as a rotation crop and in a solid planting for any effect to be seen. As with most other control measures, nematode populations are likely to increase fairly rapidly as soon as susceptible crops or weeds are grown.

A few other crops in addition to marigolds have been reported

to abate nematodes. These include asparagus, garlic, and onion. Very little research data are available, and the results that do exist are even less consistent than with marigolds.

Soil amendments. Various organic substances can be added to the soil to decrease the impact of nematodes on crop plants. These amendments, which include peat, manure, green chop and composts, may contain substances deleterious to nematodes as discussed above, but are probably most useful for their ability to increase the water-holding capacity of soil, especially in sandy soils. Plants that become water-stressed are damaged more in the presence of root knot nematode. For instance, root knot nematode damage can be reduced to one half by a more frequent irrigation schedule. The same effect can be accomplished by increasing soil water-holding capacity and nutrient availability. There will be just as many or more nematodes in the soil, but they will cause less damage.

Orchard floor management. Cover crops in orchards and vineyards can have a significant impact on nematode problems. Since they normally add organic content and increase water penetration, cover crops can improve root growth, overall plant health, and tree or vine tolerance to nematode damage. However, most legume cover crops are good hosts for nematodes and will increase populations over time. As a general rule, legumes should not be used as cover crops at nematode-infested sites when soil temperatures exceed 60°F, especially in the warm areas of the San Joaquin Valley south of Stockton. Grasses are better cover crops for this region; however, they deplete nitrogen supplies and special efforts must be taken to restore nitrogen.

Two recently developed varieties of vetch—Cahaba White Vetch and Nova Vetch—do not host root knot nematodes. Nova has the advantage of being resistant to ring nematodes as well. These legumes have the advantage of preventing low nematode populations from increasing while not depleting nitrogen supplies. These varieties are available from various seed companies, but may be a bit difficult to obtain. Other varieties of vetch are not recommended because they increase root knot populations. Research is currently underway to develop a cover crop rotation program that shifts plant types every few years.

Planting and harvest dates.
Root knot nematodes cause the most damage during the warm summer months. Juveniles of most species cannot penetrate roots at soil temperatures below 64°F. (*Meloidogyne hapla* is adapted to cooler temperatures and can infect at 50°F.) Therefore, damage to susceptible fall planted crops such as carrots can some-times be reduced by waiting until soil temperatures drop below these activity thresholds. Likewise, early spring plantings may result in healthier plants than later plantings in infested sites if seedling roots get an opportunity to develop substantially before nematodes are able to invade them. Seedlings planted in cool, wet soil are more susceptible to seedling diseases, so this strategy should be used primarily at sites where you are certain that a damaging level of nematodes is present.

Early crop termination can help to limit nematode population buildup. Remove and destroy annual vegetable and ornamental plants in infested soil as soon as harvest or desired bloom period is over, so that nematodes are not allowed to continue feeding and reproducing on root systems. For example, a tomato plant with one thousand root knot females on July 1 can result in several hundred thousand root knot eggs by August 1.

Biological control.
Many organisms occurring naturally in the soil attack or kill pest nematodes. Some of the more common predators and parasites include fungi, bacteria, other nematodes, and soil-dwelling insects such as collembola. However, these biological control agents cannot be relied on for control where nematodes have developed to damaging populations. Little is known about how to manipulate them to provide control. To date, no one has consistently reduced nematode damage through addition of biological agents.

Sanitation.
Relative to most insects and many plant pathogens, nematodes are not very mobile. An undisturbed infestation will not normally spread more than 3 to 6 feet in a season, and nematodes are usually introduced into new fields or areas in infested soil, irrigation water, or plants. In theory, procedures that limit the spread of nematodes from infested to noninfested areas can be important; however, in reality, most nematodes are already quite widely distributed. One of the few examples of the use of an exclusion strategy in California is in the management of cyst nematodes in large commercial operations where equipment or irrigation water may be used on several fields. In gardens and small farms, the main sanitation practice of importance is to prevent introduction of nematodes on planting material. Where soil has been solarized or fumigated to remove nematodes, take care to prevent reinfestation from movement of soil or water from infested areas of the garden or farm.

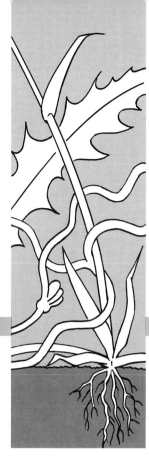

CHAPTER SIX

Weeds

WEEDS COMPETE WITH CROP plants for sunlight, water, and nutrients, and their growth must be limited to obtain a reasonable yield of any food crop. Weeds may also interfere with harvesting or other field operations or pose fire hazards. Generally, a weed is defined as "any plant growing where it is not wanted." Many weedy species can provide benefits in the garden or farm under certain circumstances such as providing food and shelter for insect predators and parasites or keeping dust down. On the other hand, crop plants act as weeds when they are growing where they are not wanted.

In any situation, maintaining a healthy, vigorously growing crop is one of the most important factors in reducing losses due to weed competition. The best weed management program for your garden or farm will depend on the crops grown, size of your operation, and the weed species that have caused problems in the

past. An integrated program using a variety of different methods will give the best long-term control of the range of weeds likely to occur.

Weed management options for orchard crops and vegetables are quite different. Weeds compete with fruit and nut trees principally for water and nutrients; where these two requirements are not in a critically short supply, some weed growth can be tolerated and is often beneficial. Thus, in orchards, weeds can be grown as ground covers or cover crops, mowed, or allowed to grow in strips between trees during much of the year. On the other hand, in vegetable plantings, weeds can quickly shade out young crop plants as well as rob the soil of nutrients and water, so weed control is essential—especially when these crops are young.

The length of time that weeds must be intensely controlled in vegetables varies from crop to crop. For instance, crops like corn, potatoes, pole tomatoes, and brussels sprouts grow tall and after 3 weeks to a month are able to shade out weeds growing at their base. Beans should be kept weed free for the first 5 weeks after seeding. On the other hand, lettuce and spinach, strawberries, and celery produce less shade and require more weed control later in their growth. Onions, garlic, and carrots give so little shade that they require intensive season-long weed control. The principal methods for weed control in vegetables are cultivation, handweeding, mulching, solarization, and—for larger plantings of one crop—herbicides.

germinate whenever water is present. Common winter annual weeds causing problems in California gardens and small farms include annual bluegrass, burclover, burning nettle, chickweeds, common groundsel, henbit, fiddleneck, filaree, little mallow (cheeseweed), mustards, shepherdspurse, sowthistle, wild oats, and wild barley. Summer annuals germinate in the spring or early summer, and flower and produce seed in the fall before dying in the winter. Major species include

barnyardgrass (watergrass), crabgrasses, common lambsquarters, fleabane, pigweeds, puncturevine, spotted spurge, purslane, sprangletop, and nightshades. In California's mild climate, certain annuals may behave as biennials or short-lived perennials—for example, horseweed, little mallow, and sweet clovers. Biennials complete their life cycle in two growing seasons, producing vegetative parts in the first growing season and flowers and seeds in the second. Common biennial

Types of Weeds

Weed species can be grouped according to their life cycle as annuals, perennials or biennials (Figure 6-1.). Annual plants complete their cycle in one growing season. Winter annuals normally germinate in the fall, grow during the winter, and flower and produce seed in the spring before dying in early summer; however, in mild coastal areas, winter annuals may

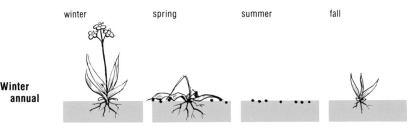

Winter annual

Winter annuals germinate in the fall, mature in the winter, and die early summer. The seeds remain dormant until the fall.

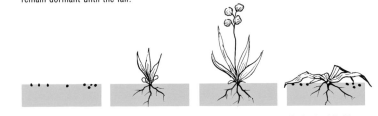

Summer annual

Summer annuals germinate in the spring, mature in the summer, and die in the fall. The seeds remain dormant until the spring.

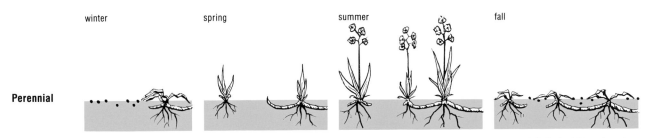

Perennial

Herbaceous perennials grow new plants from seeds or vegetative parts, such as rhizomes, bulbs, tubers or root stocks, in the spring.

They mature in the summer and die in the fall; seeds and underground parts overwinter.

FIGURE 6-1. Life cycles of winter annual, summer annual, and perennial weeds.

weeds in California include bristly oxtongue, prickly lettuce, and milk thistle.

Perennials, which can live 3 years or longer, may be either herbaceous or woody plants. Herbaceous perennials often die back during the winter but regrow during the spring or early summer from underground rhizomes, bulbs, tubers, or crowns on tap roots. Commonly found herbaceous perennial weeds include bermudagrass, dallisgrass, field bindweed, johnsongrass, oxalis, and nutsedges. Because of their underground resources, perennials are much more difficult to control than annual weeds.

Choosing appropriate management methods requires that you know the identity of the weeds in your garden. If you choose to use herbicides, proper identification is essential. A few of the most troublesome garden weed species are pictured and described at the end of this chapter. Additional help in identifying plants can be obtained from University of California Cooperative Extension offices, your County Agricultural Commissioner's office or from nurseries or botanical gardens. University of California Agriculture and Natural Resources Publication 4030, *Growers Weed Identification Handbook*, and several other useful references for identifying weeds are listed in the References at the end of this book.

Weed Management Methods

EXCLUSION

One of the most important components of a weed management program is keeping weed seeds and rhizomes, stolons and tubers of perennial weeds out of your cropped area. Never let weeds go to seed in your garden or in areas surrounding your garden. Most weeds are prodigious seed producers; for example, one pigweed plant can produce over 250,000 seeds, one black nightshade over 800,000 and one barnyard grass plant over 1 million seeds. Remember that many weed seeds and other propagules have the ability to remain dormant in the soil for a number of years. If you neglect weed control one year, it may take several years to get weed populations down to earlier levels. On the other hand, persistence with weed control efforts ensures less severe and easier to manage weed problems as the years go by.

Seeds and other plant propagules move into the field or orchard on cultivation equipment, with water, on clothing, and in compost, mulches, and manure. Make every effort to limit these avenues of weed entry. Never use fresh manure; compost it or fumigate it before application. Be sure your composting system is hot enough to kill weed seeds (see Chapter 2), and check plant-derived or organic mulches for weed seeds and propagules; even soil sold as "clean topsoil" in garden stores may contain weed seeds. In the home garden, bird seed can be an important source of weeds.

ROTATION

Rotation has many benefits, particularly on larger plantings of vegetables, and these are detailed in Chapter 2. You can improve your weed control program by planting crops with different types of weed control methods in subsequent seasons or years. For instance, corn is a good crop to plant in areas of your garden where you have had severe weed problems. Corn planted in rows can be easily cultivated through most of the season and its tall growth shades out most weeds. Extremely severe infestations can often be eliminated by planting corn in the same area for 2 years in a row and keeping weeds under control with cultivation. Potatoes are also a good rotation crop because the tubers must be dug up and the turning of the soil discourages some weed species. (Potatoes are not, however, a good rotation crop for areas heavily infested with nutsedge, because nutsedge rhizomes will pierce tubers and allow rot to occur.) Generally, crops that provide little shade and are difficult to cultivate, such as lettuce, radishes, onions and carrots, should be followed by crops that are easily cultivated or produce a lot of shade; in addition to corn and potatoes, broccoli, cauliflower, brussels sprouts, and tomatoes are good candidates. On larger farms, rotations with alfalfa or rice also assist in weed control.

In commercial vegetable production, crops can be rotated to take advantage of the different herbicide materials available in each crop.

SOLARIZATION

In many areas of California, solarization of soil can provide an excellent way of eliminating most common weed species from cultivated areas for a year or longer. To solarize your soil, you must cover the area with a sheet of clear plastic for 4 to 6 weeks during the hottest part of the year. The plastic traps heat and, especially in the presence of moisture, raises the temperature of the upper layer of the soil to levels that kill many weeds and weed seeds as well as other pest organisms. Solarization is less effective in the cooler coastal areas, where summer temperatures usually remain below 80° F.

Late summer and fall planted crops, such as onions, garlic, carrots, lettuce, and radishes, that pose special difficulties for weed control are particularly good candidates for soil solarization. The procedure is also especially recommended for areas severely infested with weeds such as barnyardgrass, annual bluegrass, nightshades, cheeseweed, common groundsel, pigweeds, mustards, and other winter annual weeds. With special attention to proper conditions, solarization can be used to eradicate bermudagrass. However, solarization will not completely prevent regrowth of field bindweed, johnsongrass, or nutsedge from underground stolons, rhizomes, or tubers, although it will control seeds and reduce established populations. Two summer annual species—purslane and crabgrass—are only partially controlled; but most other annual species, especially winter annuals, are well controlled for more than a year after treatment. See Chapter 2 for instructions on how to solarize and species controlled.

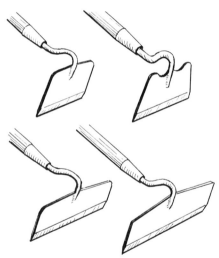

Garden hoe or chopping hoe

Garden hoes come in just about every size and shape. Use a chopping motion to cut weeds off at ground level. Keep hoe blades sharpened for most effective results. Use this type of hoe for removal of weed seedlings and small annual weeds and digging furrows and hilling up and around growing crop plants.

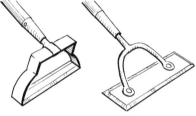

Push-pull hoe

(also called scuffle hoe or hula hoe)

Push-pull hoes are good for fast removal of weed seedlings and annual weeds on level ground. Instead of chopping, you push the hoe forward and then pull back. Weeds are usually destroyed with both strokes.

Hand or push cultivator

Hand cultivators (or wheel-hoe cultivators) are efficient, non-motorized cultivators for weeding between rows in larger gardens and small farms.

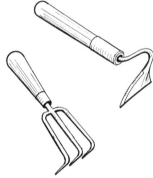

Hand tine

These hand held cultivators are good for loosening soil around larger weeds and cultivating weed seedlings in small areas.

Weeding hoe

Triangular point on one side of the weeding hoe cuts weeds; the two-pronged side is used to pull weeds out of the soil.

FIGURE 6-2. Implements for hand hoeing.

HANDWEEDING AND HOEING

Handweeding is the oldest form of weed control, and it is still *the primary method* for controlling weeds in gardens and small farms. Many tools are available to make the job easier. Garden and hardware stores display a variety of hoes of various sizes and shapes. Figure 6-2 shows some of the most useful types. Both broadleaf and grass annuals can be controlled with the hoe, but grass species must be cut off below the soil surface to prevent new sprouts from growing out of the crown. Be careful not to hoe so deeply that you injure the crop's lateral roots (Figure 6-3). To effectively control established perennial weeds, you must dig up and destroy all roots and underground stems and tubers that may grow into new plants. Trowels and other digging tools are useful for removing underground parts of plants. However, dandelion weeders, although popular, provide only temporary suppression of dandelion plants because dandelions are able to regenerate from taproots cut off as deep as 4 or 5 inches below the soil surface.

It is easier to remove weeds when they are in the seedling stage before they have developed a substantial root system, so weed regularly—once a week or so—especially when crop plants are young. Once weeds are large, hoes and other cultivating tools are not very effective and weeds must be removed by hand. Always remove weeds before they flower and begin producing seeds that can further infest your garden. If

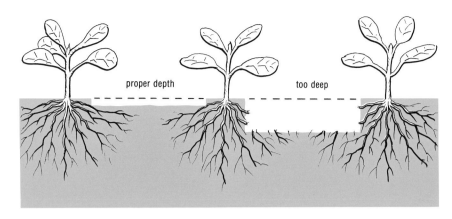

FIGURE 6-3. Deep cultivation destroys necessary roots. Shallow cultivation is preferable.

you leave hoed weeds in furrows or elsewhere in your garden or orchard, do not irrigate until they thoroughly dry; some species, such as purslane, can reroot.

For some crops, such as corn, potatoes, and tomatoes, you can control weeds in the seed row and build a furrow at the same time after crops get to be a few inches high. Use a hoe to dig a furrow between rows and place the removed soil around the seedling crop to smother weeds growing in the seed row. If you use the furrows to irrigate and are careful to keep the water from overflowing or absorbing onto the tops of the beds, the beds should remain mostly weed free. This method, which was developed by native Americans, is illustrated in Figure 6-4. It can be used for any vegetable crop that has a stem that grows at least an inch or two above the soil surface and is planted in rows; leafy vegetables, such as lettuce and spinach, and crops without stems, such as carrots, onions and garlic, could be smothered by the added soil.

MACHINE TILLAGE

Tillage may have many purposes but a primary benefit is the control of weeds. Most cultivations are made just for weed control; others are made to turn under crop debris, incorporate fertilizer, improve water penetration or otherwise enhance growing conditions for the crop; however, all cultivations destroy some weeds and by improving crop growth, such operations increase the crop's ability to compete with weeds, thereby assisting in weed management.

Tillage readily controls annual weeds, biennial weeds without a taproot, and seedlings of perennial weeds. As with handweeding, the younger the plants, the easier they are to control. Mature perennial weeds can sometimes be controlled by repeated cultivations under dry soil conditions that cause the plant to use up the reserves stored in its roots or rhizomes through continuous resprouting. However, cultivation can actually increase problems with certain perennial weeds

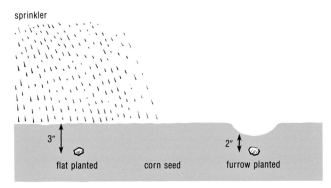

1. Plant seeds on level surface for sprinkler irrigation or in shallow furrows.

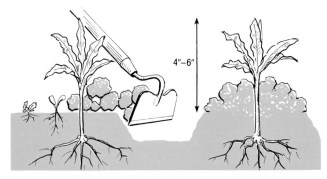

2. After plants have grown 4 to 6 inches high or have about 5 true leaves, use a hoe to dig a furrow between rows. Toss soil around crop plants to smother weeds growing around their bases.

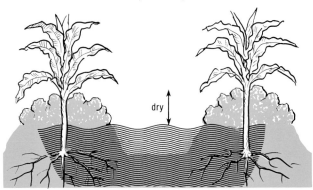

3. Irrigate using newly formed furrow, keeping bed surface dry so new weeds cannot germinate there.

4. Remove weeds growing in the furrow with a hoe.

FIGURE 6-4. Use a hoe to dig a furrow and smother weeds around crop plants. Smothering weeds works for most vegetables that have vertical stems that grow up a few inches above the soil surface. Crop plants must be planted in rows.

such as johnsongrass, bermudagrass or field bindweed by chopping up underground stems and distributing them throughout the field. Also nutlets, tubers or bublets of perennial weeds, such as nutsedge are easily distributed by cultivation.

Rototillers and hand cultivators provide all the tillage flexibility you will need in most gardens and some very small farms. Many brands of rototillers are available; each has different features and attachments. Power spaders are a new tillage tool that may reduce soil compaction associated with rototillers. On larger farms, tractor-pulled implements are required for efficient tillage.

Various shaped disks, hoes, and other cultivating blades can be arranged on the rototiller or on a tool bar on a tractor to vary the way the soil is moved. Prior to planting, the field or garden plot may be plowed or disked one or more times to turn under crop debris and weeds, then gone over with a harrow attachment to improve soil texture followed by a rake or roller to level the soil. It is a good idea to preirrigate the area to allow weed seeds to germinate and then cultivate lightly about 2 weeks later to remove weed seedlings. Plant crop seeds or seedlings immediately.

After planting, various configurations of tillage tools can be used. Not all the attachments dis-

cussed below are available for rototillers. For deep-planted, large-seeded crops, a rotary hoe can be used to control rapidly germinating surface weeds soon after seeds are planted and before the crop begins to poke through the soil surface. It also breaks the soil crust and helps with seedling emergence where crusting is a problem. Row cultivators can be fitted with various attachments such as sweeps, shovels, knives, or disk hillers arranged to cut weeds, cultivate the area between rows, and move soil into the seed row to smother weed seedlings; as the crop gets larger, sweeps can be adjusted to move up to an inch of soil into the seed row for weed control depending on the crop. This system is especially good for nutsedge control. Rolling cultivators are versatile implements for cultivating between rows. They can be adjusted to move the soil slightly when the crop is small or more aggressively as the crop gets larger. Row weeders, when set just below the soil line, pinch the soil in the seed row causing it to buckle and dislodge weed seedlings; however, the crop plants must be large enough that they are not dislodged as well.

Whether you are using a plow, a rototiller, a push cultivator or a hand hoe, the most important requirement for precision cultivation is to accurately set your equipment for each operation. Beds must be uniformly aligned and straight so you can effectively use the equipment without damaging the crop. Be sure rows are far enough apart to accommodate the cultivation without injuring roots of the crop plants, even if you are only using a hand hoe.

Timing of cultivation in relation to other management operations is important. Allow enough time after an irrigation to allow weeds to germinate before cultivating, but never irrigate immediately after a cultivation because some weeds may reroot. Although tillage has benefits for weed control, too many trips through the field with the tractor —especially when the soil is wet— can cause soil compaction that can have detrimental effects on future crops. Carry out as many operations as possible each time you take the tractor through the field.

Once the crop begins to grow over the row, cultivation with larger pieces of equipment can no longer be carried out without damaging the crop. However, by this time, the crop can shade out most weed seedlings and compete well for water and nutrients, so little further weed control is usually necessary. Handhoeing may be required in some fields and some crops.

MULCHING

Mulching (Figure 6-5) can be the easiest and one of the most effective ways to suppress weeds in many small farm and garden situations. A mulch is simply a layer of material covering the soil surface to exclude sunlight. Not only do mulches provide good weed control, they also help conserve soil moisture, prevent soil erosion, and can be used to improve water penetration or regulate soil temperature around crop plants. One drawback with mulches is their propensity to provide hiding places for snails, slugs, earwigs, sowbugs, and certain other invertebrate pests. Mice, moles and gophers may prefer mulched areas because of the protection and food they may provide. Many materials, both manufactured and natural organic and inorganic, have been used as mulches. Some of the most common and effective are listed in Table 6-1.

Before any mulch is applied, be sure the soil is weed free. Although very effective against newly germinating annual weeds, mulch often fails to control established perennial weeds such as johnsongrass, bermudagrass, and field bindweed. Remove these species manually as soon as they appear.

Most plant-based mulches are biodegradable and are usually dug into the soil after a period of time to add organic matter and improve soil texture. Generally these mulches cool the soil in the spring; this makes them ideal during hot summer weather but can slow growth of spring seedlings and thus should be applied after seedlings are well established. Take special care with warm weather crops such as eggplants and bell peppers; mulching too early can greatly depress their growth. Often mulches are added gradually, an inch at a time, during the spring as the soil warms up and the crop grows taller. Six or more inches may be added where weed problems are severe. In fall and winter, mulches will warm the soil.

FIGURE 6-5. **Mulches placed around crop plants are one of the easiest and most effective ways to control weeds.**

TABLE 6-1.

Mulching Materials.

MATERIAL	COMMENTS
Organic materials (general)	Can conserve moisture, prevent surface crusting, improve water penetration, harbor insect pests.
Bark chips and ground bark	Can harbor insects and termites; often placed over plastic as decorative material.
Compost	Excellent source of organic matter; may harbor certain weed seeds or plant pathogens if not properly prepared.
Grass clippings	Readily available, can reapply over time, may contain weed seeds or bermudagrass rhizomes, may mat and reduce water penetration if not dried first.
Hay and straw	Allows good water penetration; may contain grain seed.
Rice hulls	Has benefits as a soil amendment; slow to degrade; frequently contains weed seeds unless composted or rolled to crush seeds.
Leaf mold	Can add needed acidity to alkaline soils; must be carefully prepared or purchased.
Newspapers (shredded)	Readily available, low cost, no weed seeds; attracts earwigs, sowbugs; some concern about toxins in inks; not stable in windy locations.
Peat moss	Increases water-holding capacity but resistant to wetting when dry; adds acidity to alkaline soil; expensive.
Pine needles	Adds acidity; pine resins may be toxic to some plants.
Sawdust	A fine, short-term soil amendment; will mat and inhibit water penetration; robs soil of nitrogen but composting will eliminate some problems; need to add additional nitrogen.
Wood chips	Robs soil of nitrogen; less depletion if rotted prior to application; need to add additional nitrogen.
Pressed heavy fibrous paper for mulching (e.g., Hortipaper)	Good water and air penetration, easy application; must be purchased.
Nonorganic mulches	Do not provide organic matter for soil; often unattractive; most require special irrigation procedures; must be removed from field; conserve moisture.
Black plastic	Effective; need to add drip irrigation or furrows; warms up soil in spring slightly; unattractive.

continued

TABLE 6-1.

Mulching Materials, continued.

MATERIAL	COMMENTS
Clear plastic	Same as black plastic except does not control weeds beneath unless solarization procedures followed.
Aluminum foil	Disorients aphids within twelve inches, expensive, may reflect too much heat in summer.
Nonwoven polypropylene fabric	Allows air and water penetration.
Photodegradable plastic film	May not need to be removed from garden or field; degrades during the life of the crop, although degradation may not be complete with some products.

Organic mulching materials that are not at least partially decomposed can rob soil of nitrogen when they are mixed into soil after mulching is finished, so add ammonium nitrate or similar fertilizer at 1 pt/100 row feet when materials such as raw sawdust, shredded autumn leaves, straw, or wood chips are dug into the soil. Many materials, such as leaf mold and pine needles, can turn the soil acidic; this is not a major concern in most of California where soils tend to be more alkaline. Certain fine mulches such as raw sawdust, peat moss and grass clippings can mat and cake and prevent water penetration. Reduction of water penetration can also be a problem with unshredded leaves because their flat surfaces will allow the water to roll off to the edge of the mulched area.

Compost makes an ideal mulch. Three to 4 inches deposited on the soil surface will eliminate most annual weed growth, allow good water penetration, and provide a good soil conditioner when the mulch is dug into the soil. However, it is important that the compost be properly decomposed. Weed seeds, plant pathogens, and nematodes can survive if the compost does not reach high enough temperatures. (See section on composting in Chapter 2.)

The most common manufactured mulching material is black plastic. Besides providing good weed control, black plastic can warm the soil at shallow depths in the spring, which can help warm season crops get an early start. Clear plastic will increase soil temperature much more than black plastic, but weeds will germinate and readily grow under it. When temperatures rise in areas with hot summers, cut out or cover parts of the plastic directly over the roots of small plants to prevent overheating. Plastic mulch is especially good for growing vine crops such as melons and strawberries because it keeps the fruit clean and these crops are hard to cultivate with machines or hoes. Getting adequate water under the plastic to the crop can be a problem; plastic works best when a drip irrigation system is installed beneath but furrow irrigation will adequately wet the soil beneath plastic mulch if the edges are not tucked too deeply at the bed shoulders. Transplants or seeds are planted through holes cut in the plastic after it is installed. Plastic becomes brittle and deteriorates after a season or two. Be sure to remove and dispose of it before it becomes so brittle that it cracks into many pieces. Problems with nutrient deficiencies and plant pathogens associated with lack of air circulation are occasionally reported with plastic mulches. But these problems can be prevented with careful attention to the nutrient requirements and proper tillage between crops.

Several new types of mulching materials have recently become available commercially. These include products made of pressed heavy fibrous paper and nonwoven polypropylene fabric. Both materials allow much better penetration of air and water than conventional polyethylene plastics or woven polypropylene fabrics that are sometimes used. Another innovation is photodegradable plastic film, which becomes brittle and degrades as the crop matures.

BIOLOGICAL CONTROL

Biological control of weeds by plant pathogens and insects is going on all the time in orchards, farms, and gardens and provides some degree of control. For instance, every year a substantial percentage of weed seeds in the soil are destroyed by pathogens and invertebrates. Check the weeds in your garden and you will usually see evidence of insect feeding or disease. Generally, however, these naturally occurring biological control agents are not effective enough to reduce weeds to nondamaging levels.

One practical application of biological control of weeds is the use of grazing animals. Geese, chickens, pigs, goats, and sheep have been used with varying degrees of success to control weeds on small farms in California. Since chickens and sheep and other larger animals will eat tender vegetables, their use is mostly restricted to orchards and noncrop situations. Larger animals cannot be used in young orchards or older orchards where branches are low enough for them to reach. Goats should not be used at all because they damage the trunk bark of trees.

Geese (Figure 6-6) are probably the most widely used animal for weed control and are the easiest to handle. They prefer seedlings or young plants of grassy species and nutsedges and have been used in corn, vegetable gardens, strawberries, and cotton as well as in vineyards and orchards. They must be contained in a fenced area; a temporary fence of chicken wire and laths about 3 feet tall works well.

Protect them from dogs and coyotes and provide clean water, shade, and a small amount of supplemental feed. Remove geese from the area if you use pesticides, and do not replace them until the reentry period on the pesticide label has passed. Although there are no formal recommendations for food crops, in cotton, geese are stocked at a rate of 3 to 5 geese per acre after cotton has emerged but before it reaches the 4-leaf stage. The stand of weeds must be adequate and young and tender enough to support the geese, or they will eat the crop seedlings. In orchards geese are stocked at a rate of about 12 geese per acre because there is more weed growth, and trees will not be damaged by the geese.

HERBICIDES

Herbicides are valuable weed control tools in commercial agriculture, but their use in home gardens should be limited primarily to one-time spot treatments of problem weeds. There are several good reasons for avoiding use of herbicides in the home vegetable garden:

Toxicity to crops. Herbicides are formulated to kill certain species of plants; however, many weed species are closely related to crop plants, so herbicides often injure certain crops as well. In a home vegetable garden, many different crops will be grown close together and rotated on the same soil. It would be difficult, if not impossible, to apply a herbicide in this situation without injuring a neighboring or subsequent crop.

Difficulty in adjusting rates. Herbicides must be applied in precise amounts or they will injure crop plants. It may be impractical to measure out the small quantities needed and apply them correctly to the small areas of the garden sown to each different species of vegetable.

FIGURE 6-6. Weeder geese have been used effectively to manage grasses and sedges in cotton and other crops.

Residues. Residues of some herbicides may remain in the soil and affect the growth of a following crop. Choice of subsequent crops must be coordinated with herbicides used to prevent injury. Home gardens are not large enough or uniform enough to reliably avoid these residue problems.

Ease of handweeding and hoeing. It takes much less time and effort in the small vegetable garden to hand pull or hoe young weed seedlings than it does to read several chemical labels, decide on the proper chemical, check and calibrate the sprayer, measure the needed amount of chemical, make the application, and then wash the sprayer.

Types of herbicides. Herbicides can be classified according to when they are applied relative to the growth of weeds. *Preplant soil fumigants* are applied to prepared soil before the crop is planted. Soil fumigants are broad spectrum and kill most plants in an area. A substantial waiting period is required before crops can be planted in treated soil. These materials can only be applied by licensed applicators and should not generally be needed in gardens. *Preemergence* herbicides kill weed seeds or germinating weeds but generally will not kill established weeds. They may be applied before or after the crop is planted depending on the material and the situation. *Postemergence* herbicides are applied to the foliage of established weeds. They also may be applied before or after the crop is planted. Materials applied after the crop has been planted must be *selective;* that is, they must kill only the undesirable weed species while leaving the crop uninjured. Because no herbicides are so selective that they kill all weeds while leaving crops unaffected, most are effective only against certain groups of weeds. Some postemergence materials are *nonselective* and kill or injure all plants with which they come in contact. These nonselective materials must be applied before the crop is planted or they must be applied in such a way that they do not contact susceptible plants; special care must be taken not to allow them to get on foliage of desirable plants adjacent to the treatment area. Check with your county Cooperative Extension office for specific recommendations for herbicides for use on small farms.

Never use a herbicide without reading the label carefully. Labels give directions for proper use, rates, precautions and warnings and indicate how long residues may remain in the soil. The label also tells you what crops the material can be used on and what general types of weed species it is effective against. Never use a herbicide on a crop not listed on the label because it is likely to severely injure the crop.

Glyphosate. Glyphosate (Roundup) may be one exception to the rule about avoiding herbicides in the vegetable garden. It is a broad spectrum, foliar-applied, postemergence herbicide that kills most plants including many hard-to-control perennial weeds such as bermudagrass, field bindweed, dallisgrass, and johnsongrass. Where these species are causing severe problems, they can be managed with one or more treatments with glyphosate followed by a vigorous program of handweeding to assure that seedlings of these species do not become established again. Glyphosate does not control equisetum (scouring rush), or easily control bamboo, English or Algerian ivy, strawberry or white clovers. Control of filaree, cheeseweed, and nutsedge requires special methods or repeat treatments. Although glyphosate readily controls almost all annual weeds, its use against annual weeds in gardens is probably not justified because these weeds can be easily controlled with much safer methods such as cultivation.

Glyphosate must be applied with great care because of its potential hazard to surrounding plants. Even a few drops of the herbicide can severely injure young trees, or kill lawn or herbaceous plants if their leaves come in contact with the spray.

Although homeowners can buy ready-to-use squirt bottles of glyphosate, these formulations are not as effective against established perennial weeds as solutions you mix up yourself. Squirt bottle formulations are usually formulated at rates more suitable for controlling annual weeds. The correct rate for the species you wish to control will be listed on the label of the glyphosate product.

Various types of equipment can be used to apply glyphosate. If you plan to use it frequently, a hand-held controlled droplet applicator (Figure 6-7) especially designed for translocated herbicides is a good investment. Wick or wiper applicators are also useful; for very small spot treatments

FIGURE 6-7. A hand-held controlled droplet applicator for glyphosate application.

the solution can be applied with a brush. If you use a small pump sprayer, use a low pressure one and place a funnel over the nozzle to avoid drift. Do not use an insecticide nozzle; choose a flat fan or cone nozzle. Never apply glyphosate through a hose end sprayer; you will get too dilute a solution and poor distribution. Complete coverage of leaves with glyphosate is not important because it translocates throughout the plant.

Perrennial weeds should be flowering or in early fruiting stages for best control. Glyphosate is less effective when target weeds are water stressed or dusty. When treating larger areas such as orchards that are infested with perennials with many rhizomes, such as bermudagrass or johnsongrass, consider disking the perennials first to cut the rhizomes into smaller pieces, irrigate the area so the rhizomes will sprout, and then treat all the sprouts with glyphosate when

they approach flowering stage. Because the weeds will have less food reserves in each short rhizome segment than in longer ones, this procedure will make these weeds easier to control. More than one treatment may be required, especially in an orchard situation.

Weed Management when Planting and Transplanting Vegetables

Choose your planting site carefully to give your vegetables the edge over their weed competitors. Select a sunny location that has not had a persistent problem with perennial weeds such as bermudagrass, nutsedge, or field bindweed. Medium-textured, soft, workable soils that drain well will allow growth of more vigrous crops, so choose these over areas with clay or light sandy soils when possible.

Be sure the garden is weed free before you plant. Cultivate, weed, and rake the plot, irrigate it to let weeds germinate and then cultivate or hoe the garden again and plant immediately (Figure 6-8). Be sure the garden is level to allow good distribution and drainage of water. Furrow irrigation can help reduce weed problems if tops of beds can be kept dry while still properly irrigating the crop. This is easier to do with transplants planted on raised beds or if furrows are made after the crop has begun to grow (Figure 6-4).

Timing of planting can have a dramatic impact on weed problems in vegetable crops. Choose planting times and temperatures appropriate for the crop and variety grown. Getting your crop growing vigorously before weeds germinate will give the crop a head start and weeds may never catch up. However, it does no good to plant early if conditions are too cold or too hot to allow vigorous crop growth. Crop plants become more susceptible to seedling diseases and grow less vigorously than weeds.

One way to give crop plants a competitive edge over weeds is to transplant them into a weed-free field after they have developed several leaves in the greenhouse or seedbed. These older seedlings can shade out newly germinating weeds and have a better developed root system so they compete better for water and nutrients. Growing seedlings in the greenhouse also allows you to plant earlier in the spring, often before summer weeds like barnyardgrass have started to germinate. It is easier to cultivate around transplanted vegetables than the tiny seedlings grown from seed because you can see them better and they are less likely to be injured or buried. Most vegetables can be successfully transplanted; only a few, such as carrots and beets, do not tolerate transplanting. Plants grown from bulbs or tubers such as garlic, shallots, and potatoes are generally not transplanted. Transplants are preferred for broccoli, brussels sprouts, cabbage, cauliflower, celery, eggplant, herbs, onion, parsley, peppers, and tomatoes. Lettuce, leeks, melons, spinach, and cucumbers can also be transplanted, although they are more commonly grown from seed.

1. First cultivate and rake the area to remove weeds and old crop debris.

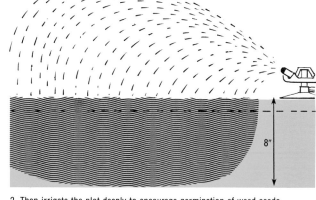

2. Then irrigate the plot deeply to encourage germination of weed seeds.

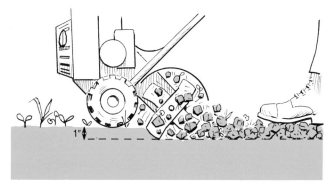

3. In a week or two, after a substantial number of weeds have germinated, cultivate the area again to kill the weed seedlings. You may use a hoe, rototiller, or other implement.

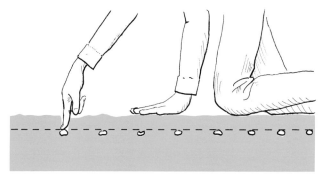

4. Plant crop seeds or transplants immediately.

FIGURE 6-8. Preparing a weed-free planting bed.

Managing Weeds Around Backyard Fruit Trees

Keep a 3- to 5-foot diameter area around individually planted fruit and nut trees weed-free to allow maximum use of water and nutrients by the tree. Take special care to eliminate lawn and other perennial grass plants from directly beneath the tree; these grasses are very competitive and tree trunks are often damaged when the lawn-mower is brought close to the tree. Mulching provides the easiest way to keep the area beneath the tree weed-free. Any mulch that allows good penetration of water and oxygen to the roots of the tree can be used. Rock mulches work well and they do not have to be replenished. Do not allow organic mulches to pile up against the trunk because they may retain water and promote rot. Black plastic is not recommended because it inhibits water penetration, exchange of oxygen, and is sometimes associated with root rot.

Managing Weeds in Orchards

Plants growing on the orchard floor can have benefits and drawbacks for fruit production. Their root systems, especially those of annual grasses, can penetrate plow pans and improve water penetration in many soils. Plants on the orchard floor provide a firmer soil surface for better year round access into the orchard and help reduce soil erosion as well as dust, which can promote mite outbreaks. In addition, certain plants may provide refuge, shelter or food for beneficial insects and mites as well as pest species.

If not properly managed, however, weeds can create numerous problems. They compete with orchard trees for water, nutrients, and sunlight—especially in newly planted orchards. Young orchards with severe weed infestations may take longer to come into profitable production. Competition is not as severe after 3 or 4 years unless water is in short supply, but deep-rooted perennials may continue to lower productivity. Weeds also provide conditions favorable for buildup of other pests, especially certain insects, by providing alternate food, shelter, or overwintering sites. Weeds growing around the base of trees are a special problem because they tend to retain moisture, increasing the potential for crown rots, and attract or shelter rodents.

Weeds pose several physical problems. When weeds dry out in the summer, they create a fire hazard. Weedy orchards are slightly cooler in the spring than orchards with bare ground, increasing the potential for frost damage. Taller weeds cool the orchard more than mowed weeds. At harvest time, weeds can interfere with mechanical operations, especially in nut crops where nuts are shaken off the tree and swept up from the orchard floor.

The three main methods for controlling weeds in orchards are tilling, mowing, and using herbicides. Although any method can be used alone, a combination of methods usually provides the best control. Growing a cover crop may have advantages during certain times of the year and intercropping may have benefits in young orchards. Intercropping is discussed in Chapter 2.

No matter what management strategy you choose, make every effort to prevent introduction of undesirable weeds, particularly perennial species, and eliminate conditions that may enhance development of weed problems. Never let undesirable species go to seed in the orchard. To avoid spreading weeds, keep irrigation ditches weed free and keep equipment clean and free of seeds and underground portions of perennial weeds. Drain areas with standing water; they provide ideal conditions for dallisgrass, sprangletop, nutsedge, and curly dock. By letting the top 2 to 4 inches of soil dry completely between irrigations, you can discourage weed growth, especially the establishment of seedlings.

Biological control of weeds in orchards is limited. Puncturevine stem and seed weevils can help reduce the number of puncturevines in unsprayed orchards, but they are more successful in controlling the weed in undisturbed locations such as roadsides. Purslane sawfly larvae feed inside leaves of purslane and reduce plant vigor and numbers of seeds, but removal of these species will merely allow the survival of other weed species that take their place. Use of grazing animals, especially geese, is a form of biological control of weeds that can be effectively used in some situations.

Tillage. Total or clean cultivation (tillage) keeps the orchard free of weed growth during most of the year. Often a newly planted orchard is tilled for the first year or two to eliminate established weeds and then converted to mowing or strip weed control. Cultivation is often the most practical way to manage orchards that are irrigated with furrows or contour checks, which must be removed and replaced periodically. Tillage is a good idea in areas where frost is a problem, because a clean, smooth, moist floor allows the orchard to stay warmer than a weed cover. Where irrigation water is limited, total cultivation is one way to reduce water use by weeds.

To carry out a cultivation program, disk or harrow the orchard floor once during late winter or early spring to turn under the winter growth of weeds and four or more times during the summer, depending on the number of irrigations and amount of weed growth. You will probably have to weed around the trunks by hand several times a year, especially if your orchard is young. In apple and peach orchards, disking can only continue until trees are propped.

Total cultivation requires disking in both directions with heavy equipment that tends to compact the soil, decreasing water penetration. In addition, disking may injure the lower trunks of trees, making them susceptible to crown diseases. Disking also cuts feeder roots in the top 6 inches of soil, where nutrients, oxygen, and water are often most abundant, so do not disk any deeper than you need to. Do not disk citrus groves because citrus feeder roots are so shallow that water and nutrient uptake will be substan-

tially impaired.

Tillage can increase problems with perennial weeds by spreading underground propagules. However, when used in combination with a herbicide treatment, good control of perennials can be achieved. Either treat with a translocated herbicide that is effective against the problem weed and then cultivate, or disk the perennials to cut rhizomes into smaller pieces, irrigate to sprout rhizomes, and treat with glyphosate.

Mowing. A ground cover or cover crop may be maintained on the orchard floor for all or part of the year and kept under control by frequent mowing. Total mowing offers several advantages over total tillage. Mowing uses lighter equipment that is less expensive and easier to operate, it reduces soil compaction, and it makes the orchard accessible during wet conditions. Roots of cover plants help maintain good water penetration. Tree roots in the upper inches of soil are not damaged, and the ground cover reduces dust.

Winter ground cover is usually a volunteer cover provided by annual plants. If the volunteer ground cover is not adequate, it may be necessary to plant an annual fall-seeded cover crop, such as wheat, cereal rye, or barley at 50 to 80 pounds per acre or a reseeding grass or legume, to start the program. Be aware that cereal ground covers will increase your orchard's need for nitrogen and legumes can substantially increase nematode and certain insect pest problems. For further information on plants to use in cover crops, see *Cover*

Crops in California Orchards and Vineyards, listed in References.

To maintain your ground cover, begin mowing in February before trees become susceptible to frost damage. Continue mowing as necessary until harvest. Generally plants should be mowed before they reach 6 to 8 inches (12 to 20 cm) or before seed heads form on tall growing weeds. Keeping the cover this height will allow desirable, low-growing winter annuals, such as redmaids, chickweed, or annual bluegrass, to reseed themselves. If you have planted a reseeding cover crop, you will have to stop mowing for a period in the spring to allow the crop to go to seed. To avoid injuring the tree trunks with mowing equipment, weed around trunks by hand or treat with herbicides.

Crops such as walnuts and almonds require a clean orchard floor at harvest time to allow nuts to be swept up by harvesting equipment. Mow the orchard frequently enough throughout the season so clippings will have had sufficient time to decompose prior to harvest. Flail the ground cover after the last irrigation before harvest, and set your equipment to cut as close to the ground as possible without disturbing the soil surface.

There are a few drawbacks with a mowed ground cover on the orchard floor. While the cooling effect of the ground cover can be advantageous in areas with hot summers, it can increase frost hazards somewhat in the spring. You can minimize the problem by keeping the ground cover closely mowed in late winter and early spring if weather permits entry of the equipment. Maintaining a ground cover requires more water and nutrients than a clean-cultivated orchard. Under mowing, perennials, such as bermudagrass, field bindweed, and dandelions, tend to become dominant in the weed population because they grow longer and compete better for nutrients and water than annuals do when mowed. Nutsedge is also favored by mowing in the lighter soils of the San Joaquin Valley. Finally, total mowing does not allow the use of contour checks for irrigation.

Strip weed control. Keeping weeds away from trunks without injuring trees is difficult and expensive if you rely on only mowing or cultivation for weed control. These shortcomings are avoided in most orchards by maintaining a 2- to 6-foot-wide strip within the tree row weed free and mowing or disking weeds between the rows (Figure 6-9). Herbicides are most often used to control weeds in the weed-free strips but mulches, including straw and newspaper, can also be used, as long as the area directly around the trunk is kept free of moisture-retaining mulching materials. When using mulches, handweed the area directly around the trunk.

Use strip weed control in both young and established orchards and with drip, sprinkler, or flood irrigation with permanent berms established down the tree row. It saves time and energy because mowing and disking operations only have to be done in one direction, reducing travel through the orchard by 40 to 50%. By keeping the tree row free of weeds, there is less competition for nutrients and water, and tree trunks are kept dry. Strip weed control with a cover crop also offers all the advantages listed under mowing.

When practicing strip weed control, watch for perennials in the mowed strip, on treated berms, and at the orchard's borders. Consider spot treating perennials with a postemergence herbicide. Once perennials become established, they grow vigorously in the absence of competition from annual weeds in the herbicide-treated strips.

Basal weed control. Basal weed control involves controlling the weeds in a 4- to 8-foot perimeter around the base of the trees with herbicides or a combination of mulching and hand-weeding and mowing or disking the rest of the orchard. Basal weed control requires less herbicides or mulching but more mowing or disking than strip weed control. Like strip weed control, it eliminates the need for mowing or tilling close to the trees, greatly reducing likelihood of trunk injury. This method is not suitable for orchards with permanent berms or checks because it requires two-way mowing or tilling.

Grazing animals. Various animals can be used to control weeds in orchards. They must be properly fenced and moved from area to area for best results. Geese are excellent grazers and feed mostly on grasses and nutsedges. Release geese when they are about 6 weeks old. No stocking rates have been experimentally established, but 12 geese per acre of orchard are often used. In very small orchards, chickens can be used; but they must be protected from predators and provided

FIGURE 6-9. Weeds in orchards are frequently managed by mowing center strips between tree rows and keeping the area directly around the trunk weed free with herbicides or mulches.

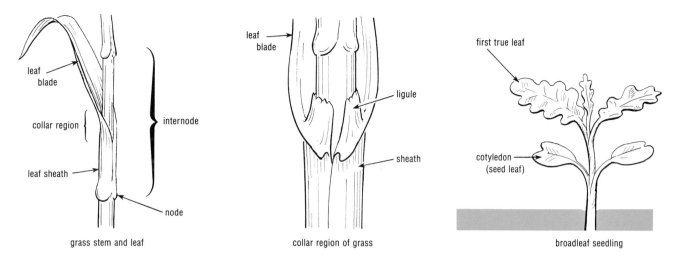

leaf blade

collar region {

leaf sheath

internode

node

grass stem and leaf

leaf blade

ligule

sheath

collar region of grass

first true leaf

cotyledon (seed leaf)

broadleaf seedling

FIGURE 6-10. Terms for vegetative parts of weeds used in identification.

with nesting areas in a chicken-house. Larger animals such as sheep, goats, pigs and cattle are sometimes allowed to graze in orchards; but they will damage trees in leaf if branches are low enough for them to reach. Goats will damage bark on tree trunks and should not be used for weed control in agricultural situations.

SPECIAL PROBLEM WEEDS

Dozens of species of plants are likely to invade your garden or small farm and cause problems as weeds over the years. Most are fairly easy to control with a diligent program of handweeding, cultivation, mulching and sanitation. The most difficult to control are perennial species that produce underground stems, rhizomes or tubers from which new plants can sprout and which may easily be spread during cultivation. Certain summer annual weeds are also troublesome. The worst pests are those that grow rapidly and quickly shade out crop plants

and those that have seeds that survive for many years in the soil. Winter annual species are usually not as difficult to control because they grow more slowly and can be removed with regular weeding before they set seed.

Eight of the most common and difficult to manage species are described and pictured here. For photographs and descriptions of other species, see the publications listed in the References, especially the *Growers Weed Identification Handbook*, UC ANR Publication 4030, which contains color photos of seedling as well as mature plants of hundreds of California weed species. Figures 6-10 and 6-11 illustrate terms commonly used in identifying and describing weed species.

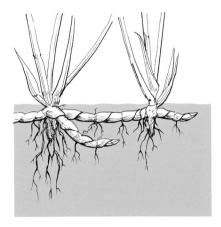

Johnsongrass rhizomes

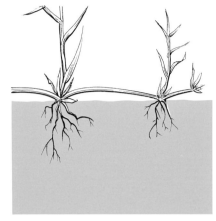

Bermudagrass stolon

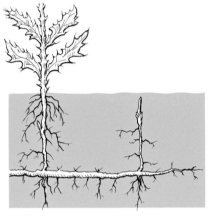

Creeping root system
of Canada thistle

Wild garlic bulb

FIGURE 6-11. **Vegetative reproductive structures of perennial plants.**

Nutsedge tubers

Pigweeds
Amaranthus spp.

Pigweeds are common invaders of gardens and orchards. They are annual plants that germinate from seeds from late winter through summer. Several species occur. The most common is redroot pigweed, which grows so large so quickly that it rapidly shades out vegetables. Another common pigweed, prostrate pigweed, has a tough, fleshy root that often remains after plants are pulled from the soil and resprouts to produce a new plant. Pigweeds can be confused with lambs-quarters, *Chenopodium album*, but lack the mealy texture on their leaves common to lambsquarters and other weeds in the goosefoot family.

Description. Seedlings of all common pigweeds are similar. Seed leaves are long and narrow and are often red underneath. First true leaves have a shallow notch at the tip that distinguishes them from seedlings of nightshades, lambsquarters, and many other species.

Leaves of redroot pigweed, *Amaranthus retroflexus*, are large and rough to the touch. Leaf blades may be several inches long, with long petioles and prominent veins on the lower surface. Flowers are arranged in dense spikes at the tops of the main stem and branches and in smaller clusters in the leaf axils. Plants may grow up to 7 feet tall when conditions are ideal.

Prostrate pigweed, *Amaranthus blitoides*, forms dense mats that may be 3 feet (90 cm) or more across. The inconspicuous flowers are in small clusters in axils. Leaves of prostrate pigweed have distinctive, light-colored edges and a tiny bristle at the tip.

Mature plants of tumble pigweed, *Amaranthus albus*, may be from 6 to 48 inches (15 to 100 cm) tall. Leaf length seldom exceeds 2 inches. Stems are smooth and white. Flowers are in small clusters in leaf axils; there are no dense spikes at the tops of stems. Stems of older plants break in the wind, and plants can be blown around, scattering seeds, and are often called tumbleweed.

Management. Because pigweed flowers are green and inconspicuous, it is not always apparent to the casual observer that they are flowering. However, learn to recognize this stage immediately. Removing plants before they flower is essential to prevent even worse infestations in subsequent crops. A single pigweed plant may produce 250,000 seeds.

Pigweeds are easy to control if seedlings are handweeded or cultivated. Larger plants are harder to remove because of their deep taproot. Solarizing the planted area will control pigweeds and most other annual weeds for one year or more.

Pigweed seedlings have long narrow seed leaves. First true leaves have a shallow notch at the tip.

Leaves of prostrate pigweed have distinctive light colored edges and a tiny bristle at the tip.

Redroot pig weed has dense spikes of flowers and may grow to 7 feet tall.

Prostrate pigweed forms dense mats. Note the inconspicuous flowers.

Tumble pigweed plants have flowers in leaf axils rather than in spikes and have smaller leaves than redroot pigweed.

Purslane
Portulaca oleracea

Purslane thrives under dry conditions but also competes well in irrigated situations. Plants prefer loose, nutrient-rich, sandy soil. Cultural practices such as high fertilizer rates, good moisture levels, and shallow seeding of crops aggravate purslane problems. The weed is an annual that grows rapidly in spring and summer.

Description. The mature purslane plant may form a mat or grow up to a foot tall. The plant branches at the base and along the stems. Leaves are very succulent and often tinged red. Small yellow flowers are born singly or in clusters of two or three in stem axils or at tips of stems. Flowers usually open only on sunny mornings. Purslane seeds are very tiny and produced in abundance.

Management. Handweeding or cultivating small plants provides good control; however, medium to large purslane plants can withstand several days of desiccation and often resprout when cultivation is followed by irrigation; always remove purslane plants from the garden area after handweeding. Purslane is one of the few annual weeds that is difficult to control with solarization; however, it can be reduced to tolerable levels with solarization in the San Joaquin Valley.

Purslane plants have thick leaves and small yellow flowers.

Sowthistles
Sonchus spp.

Sowthistles are among the most common weeds in farms and gardens in spring. The two most common species are annual sowthistle, *Sonchus oleraceus*, and spiny sowthistle, *Sonchus asper*, which can be easily distinguished by the prickly edges of the leaves on spiny sowthistle. These weeds can be seen at any time of the year in California's mild climates, but commonly germinate from late fall to early spring with the highest numbers of mature plants present in spring and early summer.

Description. The sowthistle seedling has markedly stalked, almost spoon-shaped seed leaves, rounded at the tip. The seed leaves often have a grayish powdery bloom, but later leaves have only a few hairs. Later leaves have prickles or teeth along the margin with a winged stalk on the third or fourth and later true leaves.

Sowthistle plants can grow several feet high. Flowers are yellow and mature into fluffy white seed heads. Stems are hollow and exude a milky juice when cut open. Leaves are pointed with toothed margins; leaves usually clasp the stems with a pair of clawlike lobes.

Management. Although these sowthistles are annuals, they develop strong taproots that are often difficult to remove as the plants grow larger. Seedlings are easy to cultivate and should be weeded out as soon as they are seen. Sowthistles are often covered with aphids in the spring, providing alternate hosts for natural enemies of aphids before they build up in crops; this benefical aspect does not justify leaving them in the garden, however.

Sowthistles have yellow flowers that mature into fluffy, white seed heads.

Sowthistle seedlings have spoon-shaped seed leaves. Later leaves have teeth along the margin.

Barnyardgrass
Echinochloa crus-galli

Barnyardgrass is a summer annual grass that germinates from seeds from late winter or early spring throughout the summer. Barnyardgrass seeds are often brought onto the farm with surface irrigation water or rice hull mulches. The weed grows best in bare spots where other plants are not growing well. Where moisture is available, barnyardgrass seedlings quickly grow an extensive root system. Plants root at lower nodes to form large clumps, often several feet tall.

Description. To identify barnyardgrass, pull back a leaf and look at the collar region. Barnyardgrass is the only common summer annual grass (aside from the closely related striped junglegrass, *Echinochloa colonum*) that lacks a ligule. Stems of young barnyardgrass plants often grow outward along the ground before turning upward and commonly root at inner nodes. Stems are flattened near the base.

Mature barnyardgrass can be up to 6 feet (2 m) tall in moist situations but may be only 6 inches (15 cm) in dry places. Flower heads are extremely variable, but they often droop slightly with lower flower branches further apart than upper ones.

Stems of young barnyardgrass plants often grow outward along the ground before turning upward. Stems are flattened near the base.

Barnyardgrass is the only common summer grass that has no hairs or membranes at the collar region.

Barnyardgrass flower heads often droop slightly.

Management. Plants must be dislodged when they are small and have not yet formed secondary roots; otherwise, the weed will be extremely difficult to control with normal cultivation or hoeing procedures. Barnyardgrass can be controlled for a year or more by solarizing the planting area.

Large Crabgrass
Digitaria sanguinalis

Large crabgrass is one of the most common weeds in gardens, small farms, and orchards. It is a particular problem in vining cucurbit crops. It is a summer annual weed, but new seedlings may begin to appear as early as February in warmer parts of California. Seeds continue to germinate through the summer. Crabgrass grows best under conditions of regular irrigation or rainfall and warm temperatures. The plant forms roots on the swollen lower nodes of stems and may form a large mat. Most spread, however, is from dispersal of seed.

A related species, smooth crabgrass, *Digitaria ischaemum*, also occurs in California. Leaves of smooth crabgrass are more slender and lack the stiff white hairs characteristic of large crabgrass. Management practices for both species are similar.

Description. Large crabgrass plants usually have many stems growing out of a branched base. Plants may spread by rooting at the swollen lower joints of stems. Leafblades are ¼ to ⅓ inch wide and 2 to 5 inches long with edges covered with long, stiff white hairs.

Crabgrasses superficially resemble bermudagrass but can be distinguished by examining the collar region and the way the branches of the flower structure diverge. The crabgrass ligule is stiff and papery. There may be stiff, white hairs on the edges of the leaf, but there are none on the ligule as on bermudagrass. The crabgrass flowers are held in 3 to 7 slender fingerlike racemes on the end of the flowering stem. They superficially resemble the flowers of bermudagrass, but the branches arise slightly apart whereas those of bermudagrass grow out of a common base.

The large crabgrass seedling has rather wide (¼ to ⅓ inch) leaves that are covered with coarse hairs. Leaves unroll as they grow out of the center. Check the ligule on the collar region for crabgrass characters.

Management. Crabgrass can be managed with a regular program of handweeding and cultivation or mulching, but vigilance is required. Large plants with extensive secondary root systems are difficult to remove. Crabgrass seed is only partially controlled with solarization.

Crabgrass flowers occur in three to seven slender racemes on the end of the flowering stem.

The crabgrass seedling has rather wide, hairy leaves.

Bermudagrass
Cynodon dactylon

Bermudagrass is a perennial grass that is frequently used for lawns but also is a troublesome weed in many gardens and orchards. Infestations are very difficult to eliminate because of the aggressive habit of the weed. The plant grows rapidly when temperatures are warm and moisture abundant, sending out runners above ground (stolons) and below the soil surface (rhizomes). Chopping plants with a hoe or a disk just aggravates the problem, because each piece of rhizome or stolon can root and produce a new plant. On the other hand, transferring soil-containing parts of stems or seeds will infest other parts of your garden.

Description. The mature bermudagrass plant forms dense mats with spreading and branching stolons that root at the nodes. Erect stems arise from the axils of the short leaves on the stolons; the erect stems have longer leaves than the stolons. If you dig beneath the soil surface, you will find the underground stems or rhizomes.

The bermudagrass flower head consists of 3 to 7 slender spikes radiating from one point at the tip of the stem; this distinguishes it from crabgrass which has spikes rising from several points. The collar region of bermudagrass has a fringe of short, white hairs and often has tufts of long hairs on the margins.

Management. Bermudagrass can be controlled by consistently

The flowering spikes of bermudagrass radiate from a single point at the tip of the stem.

The collar region of bermudagrass has a fringe of short, white hairs.

Mature bermudagrass often grows in dense mats.

removing plants as they emerge and by keeping additional seeds or stems from entering the garden. The root system tolerates salt and drought, but is killed when exposed to the sun. Although mulches will reduce growth, bermudagrass can grow through most of them. Solarization will control an infestation in the warmer regions of California such as the Sacramento and San Joaquin valleys if proper procedures are carefully followed: the plastic must be kept close to the soil, the soil kept moist, no holes are pierced in plastic for at least 6 weeks, and the solarization process is carried out during the hottest part of the summer. Avoid cultivating bermudagrass before solarization; it may bury rhizomes deeper under the soil.

Although bermudagrass can be controlled through very persistent use of cultural practices, a severe infestation of bermudagrass may be one situation where use of a herbicide is justified in a garden. Glyphosate will control the weed if applied when the bermudagrass is blooming or has just formed seed heads and while the plant has plenty of moisture. Follow the instructions for application presented earlier in the chapter, taking extreme care to keep glyphosate from coming in contact with any other plants. Once you have removed bermudagrass from the garden, be sure to avoid reintroduction with infested soil, sod, compost, mulches or tools. Pull new plants as soon as you see them.

Nutsedges
Cyperus spp.

Nutsedges, also incorrectly called nutgrasses, are perennial weeds that superficially resemble grasses. The most common species in most areas is yellow nutsedge, *Cyperus esculentus*, but purple nutsedge, *C. rotundus*, also occurs in some areas. The two species can be distinguished by examining their tubers. Although they produce seeds, nutsedges grow mainly from tubers or "nutlets" formed on rhizomes, mostly in the upper foot of soil. Tubers are commonly spread by movement of soil or cultivating equipment. Infestations often begin in areas where water collects or that are poorly drained.

Description. Young nutsedge plants resemble grasses, but the leaves are thicker and stiffer than most grasses. Nutsedge leaves are V-shaped in cross section and are arranged in sets of three at the base, whereas grass leaves are opposite in sets of two. Nutsedge stems are triangular in cross section.

Individual nutsedge plants spread to form a dense clump. Flowering stems are triangular in cross section and there are three long, leaflike bracts at the base of each flower head.

Tubers of yellow nutsedge are produced singly and range in size from ⅛ to ¾ inch (3 to 18 mm) in diameter. They are smooth when mature but have loose scales when smaller. Tubers of purple nutsedge are produced in chains, several on a single rhizome. They are oblong and retain

Leaves of young nutsedge grow out of the base in sets of three rather than two as in grasses.

Nutsedge plants spread to form a dense clump.

Nutsedge plants produce tubers under the soil, which are an important means of propagation.

their scales when mature. Yellow nutsedge tubers have a pleasant almondlike taste; purple nutsedge tubers are bitter.

Management. Nutsedge can be controlled if cultivated before plants have 5 to 6 leaves, the stage when they begin to produce tubers. Once tubers form however, an infestation is extremely difficult to manage. Tillage is not effective against yellow nutsedge because the tubers can survive up to 4 years in dry soil. Cultivation will only spread the infestation. Purple nutsedge can be controlled with repeated summer tillage of dry soil on fallow land since its tubers are killed by drying, but tillage on moist soil or soil with a high water table or large clods will not be very effective.

Nutsedge is not effectively controlled with mulching, solarization or the herbicide glyphosate although all these treatments will reduce populations somewhat. Some herbicides are available for control of nutsedge on commercial farms. Check with your local Cooperative Extension office for more information. Landscape fab-rics made from polypropylene polymers will suppress growth. Black plastic will not.

The best approach in gardens and organic farms is to keep removing small nutsedge plants as soon as you see them. If pursued diligently over the years the nutsedge population will slowly dwindle. Yellow nutsedge plants do not compete well in the shade so planting crops, such as corn that grow quickly and provide a lot of shade will help if you continue to cultivate to remove nutsedge plants before they reach the 5 to 6 leaf stage. In commercial operations a rotation with a dense stand of alfalfa for a few years may help suppress a severe infestation.

Field Bindweed
Convolvulus arvensis

Field bindweed, also called perennial morningglory, is a widespread perennial weed that spreads from an extensive rootstock as well as from seed. The root system may be 10 feet (3 m) or more deep. The underground stems or rhizomes, may be several feet long and can grow into a series of new plants if chopped up by cultivation when the soil is moist. Seeds remain viable for years. Field bindweed is favored by heavy soils; the weed is seldom serious in sandy soils.

Description. Seed leaves of field bindweed are nearly square with a shallow notch at the tip while early true leaves are spade shaped. Petioles are flattened and grooved on the upper surface. Plants sprouting from rhizomes lack seed leaves. Trailing stems of field bindweed may be several feet long. Trumpet-shaped white to purplish white flowers close each afternoon and reopen the following day.

Management. Field bindweed can be easily controlled with cultivation when it is a seedling. However once it develops 5 true leaves it begins to develop its extensive root system and becomes extremely difficult to control. Maintain a regular practice of surveying for bindweed seedlings and removing them from your garden or farm.

On large acreages, tillage alone will control but not eliminate field bindweed if frequently repeated in dry soil. Tillage is most effective when shoots are allowed to grow for about 10 days after emergence from rhizomes since they continue to draw energy from the rootstock before they can produce new food resources to be stored in the rhizomes. Sixteen or more cultivations may be needed over a period of several years. However, a single deep cultivation of dry soil can set back field bindweed enough that it will not interfere with an annual crop. In commercial fields where deep tillage is needed for another purpose such as to break up a compacted soil layer this may be a feasible option. Use wide sweeps to cut roots and rhizomes about 16 to 18 inches below the surface in dry soil.

Growing alfalfa for 3 or 4 years or rice for 2 years has reduced field bindweed enough in some cases to allow growing an annual crop for one or two seasons. A vigorous alfalfa stand shades out the bindweed, and the flooding in rice culture rots the rootstock.

Solarization will not eliminate established populations of field bindweed, and the weed will grow through most mulches. However, the weed can be controlled with applications of the herbicide glyphosate. Glyphosate is most effective on vigorously growing bindweed that has a few flowers, but has not reached full bloom. When absorbed by vigorous foliage, the glyphosate is carried to the roots and rhizomes, destroying part of the rootstock. Plants must be kept well irrigated and dust free for maximum effect. Multiple treatments are usually necessary to destroy the entire rootstock. Glyphosate is extremely toxic to crop plants, so avoid all drift and possible contact. Follow instructions presented earlier in this chapter, and read the label carefully. For commercial crops, other herbicides are available to control field bindweed.

The trumpet-shaped flowers of field bindweed open in the morning and close in the afternoon.

This field bindweed seedling shows the notched seed leaves and the spade-shaped true leaves.

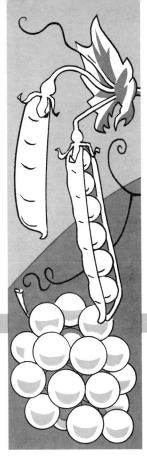

CHAPTER SEVEN

Crop Tables

THIS CHAPTER CONTAINS LISTINGS of common pests on vegetables and fruits in California. Use these crop tables to determine what pests most frequently occur on the crops you grow. In most cases, they refer you to other parts of this book where you can get more information on identification, biology and management. For those pests not covered elsewhere in the book, a brief description of management practices is provided. For vertebrate pests, see University of California Agriculture and Natural Resources Publication 21385, *Wildlife Pest Control around Gardens and Homes*. Seedling pests of vegetables are listed in the first table.

GENERAL PESTS ATTACKING SEEDLINGS OF MANY VEGETABLE CROPS

WHAT THE PROBLEM LOOKS LIKE	PROBABLE CAUSE	COMMENTS
Seeds fail to germinate and/or emerge.	**Seedcorn maggots** **Garden symphylan** **Wireworms** **Seedrot and damping-off diseases** **Birds** **Seeds planted too deep or shallow**	See page 126. See page 131. See page 84. See page 139. See Pub. 21385.
Newly emerged seedling dies or topples over; bottom of stem shriveled.	**Damping-off disease** **Heat damage in summer**	See page 139.
Seedling stems eaten off at soil line.	**Cutworms**	See page 77.
Seedling leaves and stem eaten off.	**Earwigs** **Snails and slugs** **Sowbugs and pillbugs** **Caterpillars** **Darkling beetles** **Vegetable weevils** **Rabbits**	See page 128. See page 132. See page 129. See page 67–78. See page 84. See page 87. See pub. 21385.
Seedling roots eaten off.	**Wireworms** **Maggots** **Gophers**	See page 84. See page 126. See pub. 21385
Small round pits or holes chewed in leaves.	**Flea beetles**	See page 82.
Yellow spots or streaks on leaves, may be puckered and have tiny black spots of feces.	**Thrips**	See page 125.
Threadlike, twisting white lines in leaves.	**Leafminers**	See page 124.
Small, sedentary pear-shaped insects on leaves.	**Aphids**	See page 96–102.
Plants completely removed.	**Birds or gophers**	See Pub. 21385.
Thin and spindly plants.	**Weed competition**	Keep vegetables weed free for first 6 weeks. See page 183.

ARTICHOKES

WHAT THE PROBLEM LOOKS LIKE	PROBABLE CAUSE	COMMENTS
Holes and discoloration of choke bracts, new stems and leaves; holes bored in stem and buds.	**Artichoke plume moth.** Look for caterpillars in chokes.	Soak artichoke stumps before planting or replanting in a suspension of *Steinernema carpocapsae* nematodes (150 nematodes per ml water) for 10 seconds. Hold, covered with a plastic sheet, for 48 hours before planting. Cut all plants down to ground level once a year. Chop and cover cuttings with at least 6 inches of soil. Remove all thistles and related plants. Pick off and destroy infested buds. Sprays of *Bacillus thuringiensis* and predaceous nematodes may help reduce numbers. Pheromones are available for monitoring.
Sticky exudate on chokes, may be covered with black, sooty mold.	**Aphids**	See pages 96–102.
Blackening of choke surfaces, jagged holes in leaves and stems.	**Snails and slugs**	See page 132.
Curled leaves, dwarfed plant, reduced choke production. Chokes may be small and misshapen.	**Curly dwarf virus**	Remove and destroy infected plants immediately. Use noninfected stock for new plants. Remove all milk thistle as it is an alternate host.
Grey or brown fungus growth.	**Botrytis fungus**	Most common during rainy weather. Remove infected plant parts. When storing chokes, remove infected chokes immediately. See page 146.

For problems on seedlings please see page 213.

ARTICHOKES

ASPARAGUS

WHAT THE PROBLEM LOOKS LIKE	PROBABLE CAUSE	COMMENTS
Newly emerging spears chewed; look for silvery slime trails.	**Snails and slugs**	See page 132.
Growing tips chewed, and scarred with brown blemishes. May be many elongated, black eggs attached by one end to spears. Black stains on spears.	**Asparagus beetle** (*Crioceris asparagi* or *C. duodecimpunctata*). Larva is a dark green grey grub about 9 mm long when full grown. Adult is blue black beetle with red phrothorax and yellow spots and red borders on wings or orange with black spots.	Prompt removal of spears. Handpicking. Wash eggs, larvae and beetles off with a strong stream of water.
Brown rusty spots on spears and fern branches.	**Rust.** Caused by the fungus *Puccinia asparagi*.	Most prevalent in more humid areas. Sulfur and other fungicides helpful in control. Cut down diseased ferns at crown and destroy. Resistant varieties available.
Weak, spindly spears, cross sections of shoots and crowns show discoloration.	**Fusarium wilt.** Caused by the fungus *Fusarium oxysporum* or *F. moniliforme*.	Disease is introduced on transplants. Once in soil, little can be done to prevent it from infecting other plants. Plant from clean seed only.
Spears thin and weak, or spears do not come up at all.	Plants have been weakened by **poor cultural practices, weed competition, pests, frost** or **drought,** or **harvesting** too heavily.	Do not harvest to allow plants to recover. Control damage by pests and weeds. Mulch soil to protect plants from freezing.
Very fine whitish or yellowish stippling on on shoots.	**Spider mites**	See page 116. Sulfur is effective.
Plants stunted and rosetted; aphids on young ferns, especially in summer	**European asparagus aphid** (*Brachycorynella asparagi*). A new pest that invaded California in the1980s.	Incorporate ferns into soil in fall to destroy eggs. See also pages 96–102.
Spears bent, drought stricken, white or light green.	**Phytophthora crown and spear rot.**	Most common in wet years. See page 162.
Small holes eaten out of new spears; spears may curl.	**Cutworms**	See page 77.

For problems on seedlings please see page 213.

BEANS

WHAT THE PROBLEM LOOKS LIKE	PROBABLE CAUSE	COMMENTS
Seeds rot and/or plants do not come up, or seedlings collapse soon after they come up.	**Damping-off fungi** which rot seeds or seedling; or **seedcorn maggot**, legless off-white grubs without distinct head which tunnel into and destroy seed or germinating seedlings.	See pages 139 and 126.
Leaves yellow, plants weak, wilting, or dying. Sunken red oval spots at base and beneath the soil of seedling stem.	**Rhizoctonia root** or **stem rot.** Mostly a problem on seedlings.	Favored by warm soil temperatures. Plant shallowly. See page 139.
Plants wilt and turn yellow. Roots and below ground stems have brick red spots that later turn brown and decay.	**Fusarium root rot**	See page 160.
Very fine whitish to yellowish stippling on upper leaf surface; leaves look burned when heavily infested. Fine webbing on under-surface of leaves gives gray appearance to portions of underleaf surface.	**Spider mites**	See page 116.
Curled, deformed leaves. Leaves may have shiny appearance from honeydew or black-ened look from sooty mold. Plants may be stunted.	**Aphids**	See page 96–102.
White stippling on upper surface of leaves. Severely affected leaves show tip or margi-nal burn, or appear completely dry. Under-surface of leaves often shows small white cast skins of insects and tiny dark varnish-like spots.	**Leafhoppers**	See page 94.
Leaves turn yellow or appear dry and slightly curled. Some leaves may look shiny from honeydew or blackened from sooty mold. Clouds of tiny white insects fly up when plant is disturbed.	**Whiteflies**	Whiteflies in moderate level are a nuisance problem only and do not reduce yields measurably. See page 114.
Stunted seedling plants with distorted leaves. Portions of leaves may be yellowed. Problem usually does not develop until adjacent grasslands dry up from lack of soil moisture.	**Thrips.** Tiny, slender insects. Adults are tan to black and look like slivers of wood. Young are yellow. Both stages feed in growing points of plants.	Plant as early as possible, before grass-lands dry up. Larger plants much less affected than seedling plants. or Ignore the problem; unless infestation is uncommonly heavy, plants will outgrow the problem and recover. See page 125.
Holes or skeletonized areas on leaves, flowers and pollen eaten.	**Cucumber beetles.** Greenish yellow beetles about ¼ inch long with black spots and a black head. Also could be flea beetles if holes are tiny (see page 82).	Moderate damage can be tolerated. See page 85.

WHAT THE PROBLEM LOOKS LIKE	PROBABLE CAUSE	COMMENTS
Buds and flowers drop off. Maturing beans pitted and blemished.	**Lygus bugs**	Lygus often migrate into beans and other crops when nearby alfalfa is mowed. A few lygus bugs can be tolerated. See page 91.
Blossoms drop off.	**Hot weather** (over 90°F); **low soil moisture**; or **smog** during blossoming period.	Be sure soil does not dry out too much between irrigations. Plant early to avoid hot weather during flowering and fruiting.
Poor yield. Stunted plants. Roots appear to have knots or beads on them. Problem most common on sandy soils.	**Nematodes**	See pages 169–182.
Fluffy white mycelium grows over leaves. stems, or pods; general rotting of affected parts.	**White mold.** Caused by *Sclerotinia sclerotiorum* fungus.	See page 148.
Mottled, distorted leaves. Stunted plants. Leaves may be thickened and brittle, and easily broken from plants. Poor yields.	**Mosaic viruses.** This group of diseases is spread from plant to plant by aphids and leafhoppers.	See pages 156–160. Pull up plants showing symptoms; they will be unproductive and may allow disease to spread to nearby healthy plants. Keep broadleaf weeds down, because they may serve as reservoir of viruses and facilitate spread to beans.
Tiny, white grubs inside seed within pod; circular exit holes may be visible in seed where adult weevils emerged.	**Bean weevil**, *Acanthoscelides obtectus.* Adult weevil is ⅛ inch long, light brown and mottled. Pest of beans in the field and in storage.	Be sure seeds are free of weevils before planting. Remove and destroy bean plants immediately after harvest.
Holes in pod; seeds hollowed and eaten.	**Lycaenid pod borers** (grublike caterpillars which become tiny butterflies as adults) or the **corn earworm**.	See page 46. (Management of caterpillars, generally).
Thin, white powdery growth on leaves and pods. Yellow spots on upper leaf surface.	**Powdery mildew**	Some resistant varieties. See page 140.
Chewing damage on flower or pods.	**Corn earworm**	See page 73.

For problems on seedlings please see page 213.

BEANS

CARROTS

WHAT THE PROBLEM LOOKS LIKE	PROBABLE CAUSE	COMMENTS
Carrots fail to emerge.	**Soil crusting, high soil temperatures or seedling pests.**	Emergence will be best in light soils such as sandy loam. Maintain uniform soil moisture at seed depth until seedlings emerge. Protect soil surface from rain or sprinklers. Avoid planting during high temperatures and keep soil as cool as possible when it is warm. Be sure seeds are not planted too deep. See page 213 for seedling pests.
Roots hairy, forked, misshapen.	**Root knot nematodes** are the chief cause. However, **overwatering, hard soil, rocks,** or **roots** in contact with **fertilizer** pellets or fresh **manure** can also cause poorly shaped carrots.	For nematodes, see page 169–182. Do not expect all carrots to be perfectly shaped, especially if grown in heavy soils. Remove rocks from soil.
Carrots twist around one another.	**Planted too close** together.	Thin carrots 1 to 2 inches apart when plants are small.
Carrots rot or have enlarged white "eyes."	**Overwatering**	Water less often. Do not plant carrots in heavy soils.
Roots with surface tunnels filled with rusty mush, presence of stiff white maggots (no aboveground symptoms).	**Carrot rust fly.** A small fly that lays its eggs in the crowns of carrots.	Peel off damaged areas before using. Harvest carrots as soon as possible and in blocks rather than selectively. Flies may enter field through holes left by harvested carrots. Do not store carrots in ground through winter. Control weed hosts (all *Umbelliferae*). Protective covers and other recommendations for cabbage maggot (page 126) may help.
White growth on leaves.	**Powdery mildew**	If extensive, may require fungicide; sulfur may be used. See page 140.
Thin or spindly growth.	**Weed competition**	Control weeds. See page 183–208.
Roots rotted, white fungus growth on soil surface and clinging to root. Small oval, honey-colored to brown sclerotia in fungal growth.	**Southern blight** or **white mold.** Caused by *Sclerotium rolfsii.*	Avoid planting in infested soil. Ammonium carbonate or nitrate fertilizers may discourage. See page 148.

For problems on seedlings please see page 213.

CARROTS

COLE CROPS
(CABBAGE, BROCCOLI, CAULIFLOWER, BRUSSELS SPROUTS)

WHAT THE PROBLEM LOOKS LIKE	PROBABLE CAUSE	COMMENTS
Irregular holes in leaves. Small or seedling plants may be destroyed.	**Caterpillars:** cabbage looper, imported cabbageworm, army-worms. Also **snails** or **slugs.**	See pages 67–78 and 132.
Small holes in leaves, growing points of young plants may be chewed causing stunted growth. Loose woven cocoons about ⅓ inch long on leaves.	**Diamondback moth caterpillar.** A small, green-yellow caterpillar with black hairs that is only ⅓ inch long when mature. The body is slender, and pointed at both ends, with a distinctive "V" formed by two prolegs at the rear end.	*Bacillus thuringiensis* very effective. Older plants not usually seriously damaged. Destroy culls and mustard-type weeds several weeks before planting.
Deformed or curled leaves. Colonies of small soft-bodied, usually gray green, insects on leaves. Sticky honeydew may be present.	**Aphids**	See pages 96–102. Hard to control once in heads. Choose planting times to avoid outbreaks.
Plants with a wilted appearance, or with distorted leaves turning brown.	**Harlequin bug.** Attractive shield-shaped insects usually black with bright red, yellow or orange markings. Adult bugs are ⅜ inch long. Injury is caused by bugs sucking plant tissue.	Handpick the bugs and their egg masses; eggs look like miniature white barrels with black hoops. Eliminate old, nonproductive cole crops, and wild radish and mustard, because they provide breeding sources. See page 88.
Feeding injury (engraving) on the surface of roots, or tunneling through the roots, of young plants. Plants fail to grow properly and may wilt and die. Feeding tunnels in germinating seedlings, which fail to produce plants.	**Cabbage maggot** (root maggot).	See page 126. All control measures are preventive. Nothing practical can be done when maggots occur on the growing crop.
Head of cauliflower looks cracked instead of smooth. Leaves may grow through head.	**Overmature, hot** or **dry weather,** or excess **nitrogen fertilizer.**	Grow varieties adapted to your area. Plant so crop develops during mild or cool weather. Fertilize properly. Do not let soil dry out.
Yellow or brown head on cauliflower instead of white head.	**Sunburn**	As soon as white head has 3 inch diameter, tie outer leaves around head with twine. Harvest 4 to 7 days later.
Plants stunted, wilted. Leaves yellowish-colored. Small glistening white specks on roots.	**Cyst nematode**	Do not plant cole crops year after year on same site in garden. See page 171.
Plants wilt. Swollen, misshapen roots. In later stages roots rot and plant dies.	**Clubroot.** Caused by the fungus *Plasmodiophora brassicae.*	Most common in acid soils. Add lime annually to affected soils below pH 7.2. Rotate out of crucifer crops for at least 2 years. Solarize soil.
Leaves with irregular yellowish areas on upper leaf surface, greyish "powder" on underside of yellow areas.	**Downy mildew,** especially on broccoli, cauliflower.	Tolerate. Improve air circulation. Some varieties of broccoli are tolerant. See page 143.
Heads suddenly split.	**Watering.** Sudden, heavy watering after prolonged dry period results in too fast growth, and heads crack.	Do not allow soil to get too dry; if it does, apply water slowly at first. When heads are firm, give plants one-quarter twist to break some roots, slowing growth.
Cabbage leaves turn yellow, older ones first, often one-sided. Entire plant may yellow and collapse.	**Fusarium yellows.** (cabbage mainly)	See page 160.

For problems on seedlings please see page 213.

CUCURBIT CROPS
(SQUASH, CUCUMBER, PUMPKIN, CANTALOUPE, WATERMELON)

CUCURBIT CROPS

WHAT THE PROBLEM LOOKS LIKE	PROBABLE CAUSE	COMMENTS
Deformed or curled leaves. Small soft-bodied insects on undersides of leaves. Sticky honeydew or black sooty mold sometimes present.	**Aphids**	See page 96–102.
Fine stippling on leaves. Yellowing or browning of leaves. Undersides of leaves appear silver-gray and have fine webbing with small yellow, orange or red dots about size of finely ground pepper.	**Spider mites**	See page 116.
Leaves turn yellow. Honeydew or sooty mold present. Clouds of tiny white insects fly up when plant is disturbed.	**Whiteflies**	See page 114. Get rid of dry, heavily infested plants in garden as quickly as possible. Remove lower, heavily infested leaves of plants which are not completely infested.
Coarse white stippling on upper surface of leaves. Severe attack is followed by browning of leaves.	**Leafhoppers**	See page 94.
Holes in leaves, scarring of runners, young fruit, and crown of plant, and wilting.	**Cucumber beetles**	See page 85.
Leaves with small specks which turn yellow, then brown; vines wilt from point of attack to end of vine.	**Squash bug**	Trap adults beneath boards in spring—turn over boards in morning and kill bugs. Handpick the adult bugs, egg masses, and young bugs on plants. See page 92.
Angular, necrotic areas on leaves.	**Angular leafspot.** Caused by a bacterium that spreads in water.	Avoid wetting foliage with irrigation water.
White powdery spots on leaves and stems. Spots may enlarge and completely cover leaf. Defoliation may occur and yields are reduced.	**Powdery mildew.** Spores of powdery mildew fungus are spread by wind and air currents. Disease is less severe in hot, dry weather.	A few resistant varieties available. Dusting sulfur effective, but may burn some varieties. See page 140.
Leaves have irregularly shaped light and dark green spots. Plants stunted, leaves small.	**Mosaic virus.** Several types, mostly transmitted by aphids.	Remove infected plants as soon as they are detected. Not practical to try to control by using insecticides to control aphids. Aluminum foil as a soil mulch may reduce or delay infections. See page 156. Fruit produced is edible even if deformed. Yield is often reduced.

CUCURBIT CROPS, *continued*

WHAT THE PROBLEM LOOKS LIKE	PROBABLE CAUSE	COMMENTS
Blossoms fall off.	**First-appearing flowers** of these crops are male which naturally drop. If later fruiting flowers drop in large numbers, it is probably due to **insufficient pollination.**	Ignore male flowers which drop. Frequent bee visits are needed for pollination. If bees do not visit your plants, hand pollination is possible using an artist's paint brush. Look for bees between 7 and 8 a.m.
Swellings or beads on roots. Poor yields. Mostly on sandy soils.	**Nematodes.** Nearly microscopic worms which attack root tissue.	Consider rotation, fallow or soil solarization; encourage vigorous root system. See pages 169–182.
Plants wilt and die beginning with older crown leaves. Light brown streaks inside lower stem, runners and root show when split lengthwise.	**Verticillium wilt.** A disease caused by Verticillium fungus.	Avoid ground previously planted with potatoes, peppers, eggplants, tomatoes, or cucurbits. Some resistance in zucchini and burpless cucumbers. See page 160.
Cantaloupe runners turn yellow and wilt; entire plant collapses. A one-sided brown lesion may form on affected runner, extending one to two feet.	**Fusarium wilt** of cantaloupe. Caused by *Fusarium* fungus.	Use resistant varieties or rotate out of cantaloupe for 5 years, see page 160.
Plants wilt suddenly, roots rotted.	**Sudden wilt.** Caused by species of *Pythium* fungus.	Avoid water stress after fruit set. Avoid wetting the soil to the crown. Plant on raised beds.

For problems on seedlings please see page 213.

CUCURBIT CROPS

LETTUCE

WHAT THE PROBLEM LOOKS LIKE	PROBABLE CAUSE	COMMENTS
Curled or distorted leaves. Stunted plants. Green, pink or black aphids on undersides of leaves or on stems or roots.	**Aphids**	See page 96–102.
Seedling plants severely damaged; caterpillars feeding in crown of seedling stunt growth or kill plant.	**Armyworms, corn earworms**	See page 67 and 73.
Leaves with ragged holes. Leaves devoured. Holes bored into heads of lettuce.	**Loopers**	See page 70.
Leaves skeletonized or almost totally destroyed.	**Armyworms**	See page 67.
Small holes in leaves, or leaves heavily skeletonized.	**Vegetable weevil**	Because adults do not fly, infestation of new areas takes place slowly and damage is usually spotty. This insect also attacks other vegetables, including carrots, radish, and turnip. See page 87.
Inner leaves of head lettuce black on edges.	**Hot weather**	Do not plant head lettuce for harvest in warmest months.
Upper leaf surface has yellow to light green areas delineated by leaf veins. Under surface has soft downy white growth of spores.	**Downy mildew**	Most prevalent under cool, moist conditions. Damaged outer leaves can usually be removed at harvest. Remove sources of the disease—old lettuce plants and wild lettuce—before planting. See page 143.
Lower leaves and entire plant wilts rapidly; hard black sclerotia and white cottony growth under lower leaves and crown of plant.	**Sclerotinia drop**, also called **lettuce drop.** Caused by a fungus.	See page 148.
Leaf veins lose green color and appear enlarged. Leaves appear puckered, ruffled and stand upright.	**Big vein.** Caused by an unidentified viruslike agent, which is transmitted by a soil fungus.	Associated with fine-textured, poorly drained soils; avoid these where big vein has been a problem. Do not over irrigate. Symptoms are not as severe if air temperatures are above 60° F, so a change in planting time can help. Some varieties more tolerant.
Edges of internal leaves are brown and often rotting; this symptom may not be visible from outside of head.	**Tipburn.** A physiological disorder related to calcium deficiency, aggravated by high soil fertility and high temperatures.	Hold nitrogen levels as low as possible, avoid excess potash. Avoid water stress; be sure calcium is adequate.
Leaves temporarily wilt and may turn dark green or gray green. Cross section of taproot shows that center has turned yellow, red or brown. Smaller roots may be dead.	**Ammonia injury.** Occurs when fertilizers break down to form free ammonia in amounts toxic to lettuce.	Most common with aqua ammonia and urea fertilizers. Chicken manure can also be a problem. Cool, waterlogged, compacted soil exacerbates situation. Keep ammonium forms of nitrogen out of seed row. Avoid overfertilizing.
Leaves silvery, upper leaf surface separated from rest of leaf with water droplets.	**Frost injury**	A sudden temperature drop below freezing will injure lettuce.

WHAT THE PROBLEM LOOKS LIKE	PROBABLE CAUSE	COMMENTS
Plants yellow and stunted. Leaves faintly mottled.	**Viruses**	Buy and plant virus-free seed if available. No practical control after symptoms on plants occur. Aluminum foil mulches may help prevent infection by aphid and leafhopper transmitted viruses. Plants showing symptoms near harvest are edible. Plants which show symptoms early may produce no or only small heads. See page 156.
Silvered areas on undersides of leaves of head lettuce. Yellowed lower leaves with tiny brown spots.	**Smog**	Do not grow head lettuce in smoggy months.
Plants begin to grow tall and send up flower stocks.	**High temperatures**. Prolonged hot weather causes seed stalks to form.	Grow lettuce only during cool months.
Torn areas on leaves.	**Birds**	See Publication 21385.

For problems on seedlings please see page 213.

ONIONS AND GARLIC

WHAT THE PROBLEM LOOKS LIKE	PROBABLE CAUSE	COMMENTS
Tiny bulbs. Roots look white and normal.	**Wrong variety** or planted at **wrong time**, or **weed competition**.	Plant right variety at proper time. Onions and garlic do not compete with weeds; must be controlled throughout crop cycle.
Onions develop seedstalks.	Period of **cold weather** after plants were about 6 to 10 weeks old.	Plant right variety at proper time. If onion sets are planted, use only small sets.
Roots rotted, pink. Yield drastically reduced.	**Pink root**. A disease caused by a soil fungus.	Grow a pink root resistant variety. Rotation helpful in reducing severity of disease.
Tunnels and cavities in bulb and underground stem. Plant may die or become wilted and yellow.	**Onion maggot** (root maggot)	See page 126. All control measures are preventive. Nothing practical can be done when maggots occur on the growing crop. Destroy cull onions immediately after harvest.
Leaves turn silvery.	**Onion thrips**. Tiny, slender insects usually hide in the angles of leaves.	Most common during dry, warm weather. See also page 125.
Yellow areas on leaves, soft purple growth of spores during wet or humid weather. Leaves and stalks later bend, wilt and die.	**Downy mildew**. Caused by a fungus that affects plants only in the onion family.	Destruction of old onion crop. Well drained soil, freely circulating air, and allowing plants to dry between irrigations help prevent the disease. See page 143.
Plants collapse; leaves and/or bulbs with white fuzzy growth specked with black bodies; bulbs with a soft watery rot.	**White rot**. Caused by a soil fungus, *Sclerotium* spp., northern California only.	Destroy diseased plants. To prevent spread in soil, do not compost. Do not replant onions in that area again, fungus survives in soil for years. See page 148.
Seedlings pale, thickened, deformed. Older plants stunted, limp and leaftips dying. Bulbs swollen at base, spongy, sometimes split.	**Stem and bulb nematode**	Use certified seed. Do not plant onions in area where onions, garlic, leeks or chives grew in previous years. Parsley and celery are also hosts. Remove and destroy infested culls. See page 172.

For problems on seedlings please see page 213.

ONIONS AND GARLIC

PEAS (and SUGAR PEAS)

WHAT THE PROBLEM LOOKS LIKE	PROBABLE CAUSE	COMMENTS
Surface scarring of pods; pods may be deformed.	**Thrips.** Tiny, very active insects that feed on flowers and pods.	Primarily a problem in sugar peas. Control weeds before they dry up and drive thrips into peas. Dusting with sulfur or abrasive powders may control. See page 125.
Leaves and stems covered with sticky honeydew and black, sooty mold.	**Aphids**	See pages 96–102.
Holes in leaves, black-spotted, greenish yellow beetles present.	**Cucumber beetles**	See page 85.
Leaves skeletonized, may be groups of tiny caterpillars feeding together.	**Armyworms**	See page 67.
Very fine whitish to yellowish stippling on upper leaf surface. Fine webbing on undersurface.	**Spider mites**	See page 116.
Winding white trails mined in leaves, stems or pods.	**Leafminers**	See page 124.
Semi-circular notches on margins of leaves, young plants sometimes appear chewed off at ground level when infestations are severe.	**Pea leaf weevil adults**	Once plants have grown past the 6-leaf stage, treatment normally not necessary as plants can grow away from threat of serious damage.
Leaves with white-purple cottony growth on undersides only. Tops of leaves have yellow blotches. Pods may have dark spots. Affected parts may be water-soaked.	**Downy mildew.** Caused by a soil or seedborne fungus.	Favored by low temperatures and moist conditions, especially fog. See page 143.
Leaves with white powdery growth on top sides, leaves may curl and dry out.	**Powdery mildew.** A fungus disease.	Favored by warm, dry days and cool damp nights. Dust with sulfur. Compost or bury infected residues to destroy over-wintering fungus. See page 140.
New growth distorted, curled, and mottled; pods distorted; plants eventually may die.	**Virus disease.** Usually spread by aphids.	Plant resistant varieties if available. Remove and destroy infested plants as soon as noticed. See page 156.
Yellowing of lower leaves, stunted growth. Cross section of lower part of stem may show reddish orange discoloration.	**Fusarium wilt.** Caused by a soilborne fungus.	Pull up and destroy infected plants as soon as you notice them. Do not replant peas in the soil for 5 to 10 years. See page 160.
Plants stunted, vines appear off-color and may dry up, roots rotted or absent; tends to occur in patches along rows or in fields.	**Root rot complex** often associated with low or wet spots.	Avoid wet soil or low areas where water may collect; raised beds improve drainage. Rotate crops.
Pods partially or entirely removed.	**Birds**	See Publication 21385.
Fat white grubs with brown heads inside peas. Circular holes on pods.	**Pea weevil**	Obtain clean seed. Eliminate infested debris and overwintering adults. Larvae bore into pods and peas. Adults eat blossoms.

For problems on seedlings please see page 213.

PEAS

PEPPERS AND EGGPLANTS

WHAT THE PROBLEM LOOKS LIKE	PROBABLE CAUSE	COMMENTS
Buds or fruits turn yellow. Buds or young pods may drop from plant. Pods remaining may become misshapen and develop yellow or red blotches. Pods marred by holes.	**Pepper weevil.** Adults are dark, robust snout beetles ⅛ inch long, with beak longer than head and thorax.. Larvae are less than ¼ inch long, white, legless, and found inside fruit. Pupae, also in fruit, are white to light brown.	Destroying pepper plants as soon as harvest is over should reduce problem the next year. Destroy nightshade plants, an alternative host, wherever they occur.
Curled and distorted leaves. Stunted plants.	**Aphids.** Small green to black soft-bodied insects which cluster on under-sides of leaves.	See pages 96–102.
Leaves mottled and distorted. Reduced yield.	**Mosaic viruses**	Grow tobacco-mosaic-resistant pepper varieties. See page 156.
(Peppers, primarily) Dark-colored dieback from growing tip. Sometimes pods have orange or yellow rings.	**Spotted wilt virus.** Disease is spread by thrips.	Control nearby weeds, which are reservoir host of the virus and thrips.
Plants do not grow, blossoms drop off, peppers do not form, peppers do not develop.	**Climate too cool**	Wait for warmer weather.
Plants wilt and die, brown streaks inside root and lower stem—shows when stem split lengthwise.	**Verticillium wilt.** Caused by a soilborne fungus.	Peppers fairly tolerant. Avoid ground previously planted with potatoes, tomatoes, or cucurbits. See page 160.
Leaves roll downward, generally on all plants, with no yellowing or new leaves and no stunting.	**Physiological leaf roll.** Very common symptom, not caused by a pathogen.	No action needed.
Small holes in leaves, more damage to lower leaves than to top ones.	**Flea beetle.** Tiny black beetles that jump.	See page 82.
White frothy foam on stems, mostly eggplants.	**Spittle bugs.** Small green insects can be found beneath the foam.	Tolerate, usually do not cause significant damage. Handpick.
Peppers with worm; small hole where worm entered.	**Corn earworm** **Omnivorous leafroller**	See pages 73 and 60.
Leaves wilt and eventually yellow and turn brown; tiny white flies flutter around when disturbed.	**Whiteflies**	See page 114.
Fruit of normal color but small or flattened in shape. Few or no seeds inside.	**Poor** or **incomplete pollination.**	Low light or low temperature. Plant in full sunlight. Tap flowers during midday to aid pollination.

For problems on seedlings please see page 213.

POTATOES

WHAT THE PROBLEM LOOKS LIKE	PROBABLE CAUSE	COMMENTS
Tunneling in tubers. Tuber eyes may be pink with excrement tangled in strands of silk around holes. Tunneling may be seen in stalks and leaves. Shoots wilting and dying.	**Potato tuberworm**	See page 76.
Curled and distorted leaves. Stunted plants.	**Aphids**	See pages 96–102.
Leaves curled upward. Older leaves turn yellow, then brown. Margin of young leaves may be purple. Nodes and petioles may be enlarged, twisted. Foliage may be rosetted; aerial tubers may be visible. Tubers small, may be produced in chains. Entire plants may be brown, stiff and upright.	**Potato psyllid.** Adults are about the size of aphids; always winged and light grey to dark brown. Immature forms are flat, disklike, and yellowish with marginal fringe. Found on undersides of leaves.	Dust with sulfur. Serious damage primarily to young plants. Once tubers are formed, psyllids can be tolerated.
White stippling on upper leaf surface. Leaves with margins or tips yellow or brown. Plants stunted.	**Leafhoppers**	See page 94.
Leaves full of small holes. Tubers with bumps and shallow winding trails on surface.	**Flea beetles.** Brown, black, or striped jumping beetles about $1/16$ inch long. Larvae are tiny, white grubs that tunnel in tuber surface.	Control usually not necessary. Peel away damage on tubers. See page 82.
Bumps or pimples on tubers. Swellings on roots. Brown spots in outer part of tuber flesh.	**Nematodes**	See pages 169–181.
Poor crop in spite of good growing practices. Leaves may have rough or crinkled appearance.	**Viruses**	Plant only certified seed potatoes in commercial plantings. See page 156.
Leaves turn greyish brown with rings of grey white downy spores when humidity is high. Later leaves and stems die. Tubers have brown purple scars on surface and often rot in storage.	**Late blight.** Caused by a fungus that affects potatoes, tomatoes, and other potato family plants.	A problem mostly in coastal and Bakersfield areas. Destroy and remove volunteer potatoes before planting. Use resistant varieties such as Kennebec or Nooksack. Plant certified tubers. Keep tubers covered with soil "hills." Bordeaux mixture may be used as a protectant. Cut vines 1 inch below soil surface and remove 10-14 days before harvest. Avoid harvesting under wet conditions. Eliminate all tubers and plants after harvest.
Plant leaves rolled, often with loss of dark green color and slowed growth. Brown speckling in stem end of tubers (called "net necrosis").	Leafroll and other viruses. Aphid-transmitted diseases.	Use certified seed. Avoid saving seed potatoes from gardens. Severe reduction in yield. Use resistant potato varieties. Russett Burbank most susceptible to net necrosis on tubers. Brown areas in tubers can be cut away before eating, if desired.
Plants defoliated. Yellow striped beetles and bright yellow larvae present.	**Colorado potato beetle** (Not present in California or Nevada)	Apply *Bacillus thuringiensis san diego* variety on larvae. Handpick. Destroy infested plant material.

POTATOES, *continued*

POTATOES

WHAT THE PROBLEM LOOKS LIKE	PROBABLE CAUSE	COMMENTS
Tubers knobby-shaped or with cavities.	**Alternate wet and dry conditions.**	Keep soil moisture uniform. Potatoes do not usually perform well in heavy soils.
Plants fail to emerge from ground after planting seed pieces.	**Potatoes bought at markets** often have been treated to prevent sprouting. **Soil organisms** sometimes rot seed pieces.	Plant only certified seed potatoes. Cut them when sprouts just begin to form and plant the seed pieces immediately. Plant when soil temperature is above 45° F.
Vines progressively decline and die earlier than normal; brown streaks inside lower stem show when stem split lengthwise.	**Verticillium wilt.** Caused by a soilborne fungus.	Avoid ground planted with tomatoes, peppers, eggplant, or cucurbits. Rotate to new ground. See page 160.
Tubers with brown streaks with what appears to be a root inside.	**Nutsedge rhizomes** penetrate the tuber.	Potato fields must be kept free of nutsedge. See page 207.
Scabby spots or pits on surface of tubers.	**Scab.** Disease caused by a bacterium in soil.	Disease is largely cosmetic; affected tubers are edible. Many varieties are resistant. If scab occurs, change varieties in the next year. Disease is favored by neutral to basic soil. Sulfur may be worked into soil to make it slightly acid and thus reduce disease.

For problems on seedlings please see page 213.

RADISHES

WHAT THE PROBLEM LOOKS LIKE	PROBABLE CAUSE	COMMENTS
Small plants may wilt and die. Grooves on surface of roots; winding tunnels through roots. Fleshy part may become streaked with brown from tunneling.	**Cabbage maggot** (root maggot).	All control measures are preventive. Nothing practical can be done when maggots occur on the growing crop. See page 126.
Deformed foliage with whitish or yellowish spotting of leaves. Plants may have wilted appearance.	**Harlequin bug**. Attractive shield-shaped insects usually black with bright red, yellow, or orange markings. Adult bugs are ⅜ inch long. Injury is caused by bugs sucking fluids from plant tissue.	Handpick the bugs and their egg masses. Eliminate old nonproductive cole crops, and wild radish and mustard because they provide breeding sources. See page 89.
Small holes in leaves.	**Flea beetles**. Small dark beetles about 1/16 inch long that jump like fleas. Also **cucumber beetles**.	Ignore problem if harvest approaching. See page 82, or page 85 for cucumber beetles.
Leaves with light green areas and a violet colored downy growth on underside. Roots with black network inside.	**Downy mildew**. A fungus disease.	See page 143.

For problems on seedlings please see page 213.

SPINACH

WHAT THE PROBLEM LOOKS LIKE	PROBABLE CAUSE	COMMENTS
Plants begin to grow tall and send up flower stalks.	**Bolting.** Caused by long daylight periods from late spring to early fall.	Plant spinach in fall or early spring. Choose varieties carefully.
Leaves partly or entirely consumed, presence of light green caterpillars.	**Loopers**	See page 70.
Leaves become generally yellow (faded).	**Aphids**	Hosing may not be practical on low foliage. Wash leaves before eating. See page 96–102.
Leaves with light green to yellow tan blotches; pull back skin of blotch to find maggots or their frass in the mine.	**Leafminers**	Pick off and destroy infested leaves. See page 124.
Yellow to pale green areas on leaves. Fluffy gray spores develop on undersurface of leaves after rain or heavy dew.	**Downy mildew**. Mostly a problem when weather is wet and humid or under frequent sprinkling.	Resistant varieties. See page 143.

For problems on seedlings please see page 213.

SWEET CORN

WHAT THE PROBLEM LOOKS LIKE	PROBABLE CAUSE	COMMENTS
Worms up to 1¾-inches long eating down through kernels of ears. Prior to tasseling, worms found in whorl of plant, feeding on developing tassel.	**Corn earworm.** Range in color from green to almost black with lengthwise stripes of various colors.	See also page 73. With a medicine dropper, apply insecticidal oil to the silk just inside the tip of each ear. 3 to 7 days after silks first appear. Use 20 drops per ear. or Break off and discard wormy end of ear. Insecticides will not control worms that are inside the ear. Preventive treatments to silks are intended to kill worms before they enter ears.
Holes in leaves	**Armyworms, corn earworm,** various **beetles,** or **grasshoppers.**	Ignore, or handpick insects. Loss of small amount of leaf tissue will not reduce yields. See also pages 67, 73, 82–87 and 130.
Incomplete kernel development or shriveled kernels.	**Poor pollination** caused by failure to plant enough corn at one time; insufficient **soil moisture**—especially from silking to harvest; **hot weather** or **high winds** 2 to 3 weeks before harvest; not **fertilized** as directed.	Plant at least 3 to 4 rows at least 8 feet long each time planting is made. Water and fertilize as directed. Grow varieties adapted to area.
Mottled leaves, poor or slow growth. Some death of leaf tissue along margin.	**Mosaic virus**	No control. All varieties can become infected but certain ones are more tolerant and produce well even when infected.
Ears, tassels, leaves with gray gnarled growths that become powdery.	**Common smut.** A fungus disease.	Remove and destroy growths as soon as noticed; keep black powder in galls from getting into soil. Plant early; more common problem in later harvests. Resistant varieties available. See page 149.
Brown spots on leaves with powdery rust-colored spots. Leaves may discolor.	**Rust.** A disease caused by a fungus.	Favored by cool temperatures and high humidity or overhead sprinklers. Resistant varieties available. Fungicides available.
Ears only partly filled, shortened silks, presence of earwigs on silks.	**Earwigs.** Feed on silk and prevent pollination and development of kernels.	Traps; check daily for earwigs and destroy. See also page 128.
Sticky or shiny leaves. Clusters of aphids. Smaller plants may be stunted.	**Aphids**	See pages 96–102.
Pink discoloration of kernels, may have mycelia.	**Fusarium ear rot**	Fungus gains entry through damage caused by thrips and other insects. Some varieties less susceptable. Early plantings usually escape injury.
Roots and stalk rot. Pith turns pink.	**Fusarium stalk rot**	Stressed plants more susceptable. Provide plants with good cultural care.

For problems on seedlings please see page 213.

TOMATOES

WHAT THE PROBLEM LOOKS LIKE	PROBABLE CAUSE	COMMENTS
Worms up to 1¾-inches long in immature or ripe tomatoes.	**Tomato fruitworm**	See page 73.
Worms up to ⅜-inch long tunneling in fruit.	**Potato tuberworm**	Do not plant tomatoes in a garden where potatoes were grown the year before. Destroy volunteer potato plants. See page 76.
Worms never longer than ¼ inch, tunneling in core and fleshy parts that radiate from core of fruit. Leaves may be mined and folded together.	**Tomato pinworm**	Pinworm is most common in southern California and central to southern end of San Joaquin Valley and occurs earlier in season that fruitworm. See page 75.
Leaves eaten, only stems remain. Fruit with small to large gouged-out areas. Very large caterpillars may be present.	**Hornworms.** Distinctive horn on rear end.	See page 72.
Fruit surface eaten away or fruit hollowed out.	**Snails** feed on surface of fruit. **Slugs** hollow out fruit.	Stake tomatoes to get fruit off the ground and away from slugs and snails. See page 132.
Creamy to yellowish cloudy spots lacking definite margin on ripe tomatoes. Tissue beneath the spots is spongy.	**Stink bugs.** Green to gray shield-shaped bugs ¼ inch long.	Stink bugs overwinter beneath boards in refuse piles, weedy areas and the like. Remove these from the garden area. See page 89.
Leaves almost totally eaten off of young plants. Small dark weevils on plants.	**Vegetable weevil**	Weevil attacks many vegetables but does not fly so spreads slowly through garden. Handpicking adults off plants at night can be effective if population low. See page 87.
Lower leaves and stems become a bronze or oily brown color. Discoloration gradually moves higher on plant. Lower leaves dry and drop from plant; in extreme cases plant may lose leaves.	**Tomato russet mite.** Very tiny mites not visible without magnification. With 20 power hand lens, mites appear as whitish yellow pear-shaped bodies which move slowly.	Do not grow tomatoes near petunias or any solanaceous plant such as potato— These are other hosts of the russet mite. Sulfur may be applied for control, excessive rates may injure plants.
Leaves yellowish and slightly curled. Some leaves and fruit with small shiny spots; others may appear blackened. Clouds of small white insects fly up when plant is disturbed. Tiny oval yellowish to greenish scalelike bodies fasten to undersides of leaves.	**Whiteflies**	See page 114.
Leaves curled downward. Some leaves of fruit with small shiny spots; others may be blackened. Undersides of some leaves, or stems, with colonies of small green to pinkish insects. Heavily infested plants may be stunted.	**Aphids**	See pages 96–102. Are not a problem unless honeydew or sooty mold becomes obvious.

TOMATOES (side tab)

WHAT THE PROBLEM LOOKS LIKE	PROBABLE CAUSE	COMMENTS
Seedlings or small transplants with small holes in leaves. In severe cases entire plants may be completely destroyed.	**Flea beetles.** Small beetles about 1/16 inch long that jump like fleas. May be black, brown, greenish or yellowish. A problem only during the first 2 to 3 weeks after plant emergence or transplanting.	Rarely damaging except on seedlings. See page 82.
Plants with poor vigor and reduced yields. Foliage yellows and turns brown from bottom up, may look wilted. Many beads or swellings on roots.	**Root knot nematode.** Nearly microscopic eelworms which attack feeder roots.	Plant varieties resistant to root knot nematodes. These are labeled VFN. See also page 169.
Blossoms fall off.	**Night temperatures** too low (below 55°F), **day temperatures** too high (above 90°F), **smog during blossoming period, excess nitrogen fertilizer;** too much **shade** from trees, house, etc. Some **varieties** are more susceptible than others.	Fertilize properly and do not plant too early. Hormone sprays can improve fruit set during low temperatures but will not help during high temperatures. Keep soil moderately moist. Plant tomatoes in full sun. Pounding the stakes of staked tomatoes or tapping on blossom stems three times per week in midday when flowers are open may help set fruit. Remember that early blossoms do not consistently set fruit. Some varieties are not adapted to California's hot summers and these often fail to set fruit.
Leaf veins turn purple and bronze. Leaves curl upward and feel thick, leathery, or brittle. Plant growth stops. If fruit is present, it ripens prematurely.	**Curly top virus.** Disease is spread by leafhoppers. However, some varieties characteristically roll leaves upward when they are perfectly healthy. Purple leaves can also indicate a phosphorus deficiency.	No practical control in home garden after plants are infected.
Leaves have irregular light and dark green color pattern and may be wrinkled or frilly. Terminal growth may be spindly with narrow wrinkled leaves.	**Mosaic viruses**	Tobacco mosaic virus (TMV) resistant varieties available. Do not handle plants more than necessary. Plant tomato seeds rather than transplants. Do not smoke and handle plants since TMV can be spread in tobacco. No cure for viruses in edible fruit but yield, size, and quality are reduced. See page 156.
Leaves roll upwards, leathery, plant may look wilted after wet spring conditions.	**Tomato leafroll.** A physiological disorder.	Symptoms will disappear when temperatures warm and soil dries out.

WHAT THE PROBLEM LOOKS LIKE	PROBABLE CAUSE	COMMENTS
Plants turn yellow starting with one side or branch and gradually spreading to the plant which eventually dies. Main stem when cut off at base is dark reddish brown instead of normal ivory color.	**Fusarium wilt.** Disease is caused by a soil fungus that infects tomatoes only.	Grow varieties labeled VF. These have resistance to most (but not all) races of *Fusarium*. See page 160.
Older leaves begin to yellow and eventually die. Yellowing begins between main veins of leaves. Internal stem is very slightly tan colored, usually in small patches.	**Verticillium wilt.** Disease is caused by a soil fungus that infects many different plants. Favored by cool soil and temperatures, but symptoms are most severe when plants are stressed for water in hot weather and when they have a fruit load.	Grow varieties labeled VF. Avoid ground previously planted with tomatoes, potatoes, peppers, eggplant or cucurbits. See page 160.
Plants grow slowly and wilt. Roots have water-soaked areas that turn brown, and dry up.	**Phytophthora root rot.** Caused by a soil fungus.	Most common in heavier claytype soils Irrigate affected plants carefully to maintain them. Do not saturate soil for extended periods and water more frequently for short periods. See page 162.
Plants wilt; white, cottony growth on stem near soil line.	**Southern blight.** Caused by *Sclerotium rolfsii*.	Rotation to corn or other nonhost crops for 2 to 4 years. See page 148.
Water-soaked brown areas on leaves and stems. Grayish white fungus grows on undersides of leaves and leaves die. Fruit discolored but firm.	**Late blight.** Caused by a fungus and favored by high humidity and temperatures around 68°F.	Avoid sprinkler irrigation. Destroy all tomato and potato debris after harvest.
Irregular yellow blotches on leaves; blotches turn brown and die but leaves usually don't drop, unless disease is very severe. No symptoms on stems or fruit.	**Powdery mildew.** Caused by a fungus.	Disease usually occurs late in summer or fall and does not cause significant losses unless very severe, so no control normally needed. If young plants are attacked, sulfur dust will control the disease. Avoid water stress. See page 140.
Fruit turns light brown and leathery on side exposed to the sun.	**Sunscald.** Caused by over-exposure of fruit to sun.	Maintain plant vigor to produce adequate leaf cover.
Fruits are brown-black on bottom (blossom) end. Affects both green and ripe fruit.	**Blossom end rot.** A physiological disease (not caused by microorganism).	Disease involves calcium nutrition and water balance in plant, is aggravated by high soil salt content or low soil moisture and is more common on sandier soils. Maintain even soil moisture. Some varieties are more affected than others.
Fruit distorted and disfigured. Blossom ends deeply scarred or lumpy.	**Catfacing.** A physiological disorder.	Caused by cool and cloudy conditions at bloom. Fruit is edible. Resistant varieties may be available.

TOMATOES, *continued*

WHAT THE PROBLEM LOOKS LIKE	PROBABLE CAUSE	COMMENTS
Lower leaves yellow with tiny brown specks. Leaves die. Blossoms drop. Poor plant growth.	**Smog**	Some varieties more susceptible than others.
Fruit with large cracks in concentric circles around stem.	Usually follows **rainfall**, especially after dry spell (sudden, rapid growth).	Remove ripe fruit immediately after a rain to prevent cracking.
Fruit with black mold along growth cracks (see fruit cracking above).	**Fruit rot.** Develops on damaged, cracked tissue.	Prevent fruit cracking (see above). Handle fruit carefully.
Fruit with large cracks radiating from stem.	Occurs during periods of **high temperatures** (above 90° F) and **high sunlight.**	Keep soil evenly moist. Maintain good leaf cover. In very hot regions, choose planting time to avoid fruit maturity when temperatures will be consistently above 90° F.

For problems on seedlings please see page 213.

TURNIPS

WHAT THE PROBLEM LOOKS LIKE	PROBABLE CAUSE	COMMENTS
Distortion, stunting, curling and wilting of plants. Soft-bodied insects in colonies on undersides of leaves.	**Aphids**	See page 96–102.
Holes in leaves. Buds or roots may show chewing injury. Stems of plants may be cut off at soil level, much like cutworm damage.	**Vegetable weevil**	Because adults do not fly, infestation of new areas takes place slowly and damage is usually spotty. See page 87.
Irregular holes in leaves. Small or seedling plants may be destroyed.	**Caterpillars:** cabbage looper, imported cabbage-worm, armyworms.	See pages 67–77.
Deformed leaves with whitish or yellowish spotting. Plants may have wilted appearance.	**Harlequin bug.** Attractive shield-shaped insects usually black with bright red, yellow, or orange markings. Adult bugs are ⅜ inch long. Injury is caused by bugs sucking fluids from plant tissue.	Handpick the bugs and their egg masses. Eliminate old, nonproductive cole crops, and wild radish and mustard because they provide breeding sources. See page 89.
Feeding injury (engraving) on the surface of roots, or tunneling through the roots, of young plants. Plants fail to grow properly and may wilt and die. Feeding tunnels in in germinating seedlings, which fail to produce plants.	**Cabbage maggot** (root maggot).	All control measures are preventive. Nothing practical can be done when maggots occur on the growing crop. See also page 126.

For problems on seedlings please see page 213.

ALMONDS

WHAT THE PROBLEM LOOKS LIKE	PROBABLE CAUSE	COMMENTS
Nuts infested with worms at harvest. Nut meat eaten, covered with webbing and brownish fecal material.	**Navel orangeworm**	See page 57.
Young shoots die back one to several inches from tip in spring. Small caterpillar with dark brown bands may be found inside each affected shoot. Surface of nut kernels may be scoured.	**Peach twig borer**	Can be tolerated in mature home orchards since nut damage is rarely of any consequence. See page 65.
Young shoots die back one to several inches from tip. Small pinkish caterpillar may be found in affected shoot. On rare occasions, nut meats are eaten.	**Oriental fruit moth**	Most home orchardists ignore this pest since occasional damage to nut is too minor to be of concern. See page 66.
Leaves tied together with webbing and eaten in spring. Young fruit may be eaten.	**Leafrollers**	Damage insignificant—does not warrant control. See pages 59 and 60.
Blossoms and associated twigs and leaves shrivel and die. Gummy ooze at base of flower. During humid weather, shriveled petals bear tiny, grayish brown masses (fungus spores).	**Brown rot.** Fungus survives mostly on diseased twigs. Spores spread by air currents and rain splash.	Nonpareil, Price and Solano cultivars less susceptible. See page 144.
Small, round, purplish spots on leaves and fruits enlarge to about ⅛ inch. Centers of leaf spots turn brown and fall out, leaving leaves with many "shot holes." Young leaves may fall off in large numbers. Affected fruits are gummy and some fall off.	**Shot hole disease** (Coryneum blight). Fungus survives on infected buds and twigs during winter. Rain or overhead irrigation spreads spores to leaves and twigs. Wind may also carry water-splashed spores.	Symptoms worse on lower portions of tree, where foliage remains wet longer. See page 150.
Other Common Pests on Almonds (See discussions on indicated pages.)	San Jose Scale and other scales	Page 103.
	Spider mites and other mites	Page 116.
	Borers–Peachtree Borer	Page 78.
	American Plum Borer	
	Pacific Flatheaded Borer	Page 80.
	Shothole Borer	Page 81.
	Phytophthora Crown and Root Rot	Page 162.
	Oak Root Fungus	Page 165.
	Bacterial Canker	Page 152.
	Verticillium wilt	Page 160.
	Root knot and other nematodes	Page 169.
	Birds, squirrels and other vertebrates	See UC ANR Publication 21385

ALMONDS

APPLES

WHAT THE PROBLEM LOOKS LIKE	PROBABLE CAUSE	COMMENTS
Fruit worm, with masses of damp, brown granular material (frass) at the core, often protruding from holes in the skin. Fruit may drop off prematurely.	**Codling moth**	See page 52.
Leaves and blossoms eaten and tied together with webbing. Young fruit may be deeply gouged and may fall to ground. Less severely damaged fruits reach maturity badly misshapen or with deep bronze-colored scars with a roughened surface.	**Leafrollers**, including fruittree leafroller, omnivorous leafroller and green fruitworms; in coastal areas, orange tortix and apple pandemis may also cause problems.	See pages 58–61 and 64.
Dark, bruise-like spots on mature fruit, may be depressed.	**Bitter pit,** a physiological disorder.	Avoid practices that encourage excessive vegetative growth like over-fertilizing, irrigating or pruning. Summer foliar sprays of calcium may help.
Leaves eaten in spring. Small gouges in newly set or young fruit; damaged areas later scab over.	**Western tussock moth**	See page 63.
Various irregular spots on upper sides of leaves. Corresponding areas on under surface have only a thin layer of tissue remaining, resembling blisters.	**Tentiform leafminer.** Small flattened caterpillar about ⅛ inch long with whitish body and brownish wedge-shaped head. Found inside "blisters" in leaf tissue.	Four generations per year. Last one spends winter as pupa in leaves on ground. Moth emerges in early February, lays eggs as leaves emerge. Seldom causes enough damage to warrant control. Parasites effectively control this pest by late summer.
White, cottony masses on woody parts of trees, often near pruning wounds. Warty growths on limbs and roots. Clear, sticky honeydew and black sooty mold on foliage and fruit. Overall tree decline if root-infestation is heavy for many years.	**Woolly apple aphid.** Reddish insect less than ⅛ inch long, covered with white cottony material. Very sluggish movements. See page 97.	Aphids spend winter on roots and branches, with little visible cottony material. During summer and fall successive generations migrate simultaneously from roots to branches and vice versa. Bands of sticky substances around trunks and branches can help prevent migration. Dormant oil sprays will kill colonies on branches. Not easily washed off with soapy water solution. No chemical control for root colonies. Tiny parasitic wasps are important natural control agents. Aphids often become numerous when these wasps are killed by insecticides applied during growing season.
New leaves severely distorted and curled, shoots become twisted. Large amounts of clear, sticky honeydew drip onto foliage and fruit. Many young apples become puckered and fail to grow. Clusters of small, sedentary purplish insects within curled leaves and on young fruit stems.	**Rosy apple aphid.** Pear-shaped insect less than ⅛ inch long with two tiny "pipes" protruding from back end. Young are dark green, becoming reddish to purplish and covered with a powdery bloom when mature. Winged individuals are brownish green.	Spring pest only. Overwinters as egg on bark of apple tree. Young appear with first new leaves. Migrates to plantain and ribgrass in June. New growth will cover up damage by summer. A dormant oil and insecticide spray will kill overwintering eggs. See pages 96–102 for other controls for aphids.

WHAT THE PROBLEM LOOKS LIKE	PROBABLE CAUSE	COMMENTS
New growth stunted; large amounts of clear, sticky honeydew and black sooty mold on foliage and fruit. Colonies of small sedentary, yellowish green insects on new shoots.	**Apple aphid (green apple aphid)**. Pear-shaped insects less than ⅛ inch long with two tiny "pipes" protruding from back end. Young are dark green, becoming bright yellowish green when mature. Winged individuals are yellowish brown.	Most serious where climate remains cool and moist. Remains on apple all year. Overwinters as egg on bark of tree. Young aphids appear at bud burst. Dormant oil and insecticide sprays kill overwintering eggs. However, infestations can still be severe due to movement of winged aphids from other trees. Shoot damage primarily a concern on young trees. See page 96–102 for other controls for aphids.
Young blossoms, shoots, and fruit wilt and collapse, turning brown to black.	**Fire blight**. Bacterium survives in cankers, spread by splashing rain and insects. Favored by warm, humid weather during bloom.	More common on pear (especially Bartlett) than apple. Also affects quince, loquat, crabapple, and pyracantha. Blossoms and succulent shoots are most susceptible. Cut diseased branches back about 12 inches into healthy wood, removing all affected tissue. Sterilize pruning tools in household disinfectants before making each cut.
Dark, olive green to black spots on foliage and fruits. Fruit spots later become scablike.	**Apple scab**. Fungus survives in dead leaves on ground. Spores are released during spring rains, landing on and infecting leaves and fruits. Later rains may cause secondary infections.	More common in heavy rainfall areas. Disk under or remove leaves on ground during winter. Sulfur or synthetic fungicides can be used as preventive treatments applied to young growth in spring. See page 154.
Leaves and shoots powdery white. Young growth stunted and distorted.	**Powdery mildew**. Fungus survives winter in infected buds. Spores spread by wind.	Prune out infected buds and diseased twigs. Use sulfur or synthetic fungicides as preventive treatment in spring. Warm days and cool, moist nights favor this disease. See page 140.
Other Common Pests on Apples (See discussions on indicated pages.)	Bacterial Blast and Canker Birds and squirrels Borers Orange tortrix, green fruit-worms, omnivorous leaf-roller, and other caterpillars Phytophthora crown and root rot San Jose scale Spider mites and other mites Stink bugs and other bugs Sunburn	See page 152. See UC ANR Publication 21385. See page 78. See page 52–67. See page 162. See page 106. See page 116. See page 89.

WHAT THE PROBLEM LOOKS LIKE	PROBABLE CAUSE	COMMENTS
Young shoots die back one to several inches from tip in spring. Small worm with dark brown bands may be found inside each affected shoot. Ripening fruit infested with worms feeding near stem end or along the seam.	**Peach twig borer**	See page 65.
Young shoots die back one to several inches from tip. Small pinkish caterpillar may be found in shoot. Ripening fruit with holes on sides or stem end and caterpillar inside.	**Oriental fruit moth**	See page 66.
In spring, leaves or leaves and fruit webbed together and eaten. Young fruits may suffer feeding injuries, which heal over into sunken brown scars at harvest.	**Leafrollers, tussock moth and green fruitworms**	See pages 59–64.
Fruits may contain whitish worm and/or drop off prematurely; hole(s) in skin; mass of damp, brown granular material (frass) at the pit, often protruding through holes.	**Codling moth.** Worm up to ¾ inch long with whitish to pinkish body and amber head with dark markings.	Rarely a pest of apricots outside of San Francisco Bay Area. Nonchemical control measures should suffice. See page 52.
In spring, foliage infested with pale green, slow-moving insects which produce a white mealy substance. Leaves and fruit covered with clear, sticky honeydew which turns black because of sooty mold growth.	**Mealy plum aphid.** Pale green pear-shaped insect less than ⅛ inch long with two tiny "pipes" protruding from back end.	Overwinters as egg on tree. Leaves tree for weeds in July, returns in early winter to lay eggs. Dormant oil spray kills overwintering eggs. Often only one or two limbs affected. Control measures seldom necessary. New growth will cover up damage by summer. See page 96–102.
Irregular holes eaten in leaves with no apparent webbing. Ripening fruit with shallow irregular gouges in surface or with holes extending to pit.	**Earwig.** Reddish brown beetlelike insect, up to ⅘ inch long, with short wing covers and pair of "pinchers" or "forceps" on back end. Sometimes found inside fruit, next to pit.	See page 128.
Small holes in ripening and ripe fruit. Greenish beetles with black spots or stripes present.	**Cucumber beetles**	See page 85.
Blossoms and associated twigs and leaves shrivel and die. Gummy ooze at base of flower. During humid weather, shriveled petals bear tiny grayish-brown masses (fungus spores).	**Brown rot.** Fungus survives on diseased twigs and old, mummified fruits. Spores spread by air currents, rain splash and insects.	See page 144.
Dark brown, firm, circular spots spread rapidly over fruit as it ripens. Fruit becomes rotten, will stick on tree until following season.	**Brown rot**	See page 144.

WHAT THE PROBLEM LOOKS LIKE	PROBABLE CAUSE	COMMENTS
Sudden limb death during summer heat. Rough, dark cankers at pruning wounds. Gum may ooze from edges of cankers.	**Eutypa dieback (Cytosporina canker).** Fungus survives on old, diseased wood. Spores spread by rain and infect trees through pruning wounds.	Cankers can spread 6 inches per year; fungus may reach trunk if affected branches not removed. Prune trees in summer when risk of rain is minimal. Remove infected limbs, cutting at least 6 inches below discolored area.
Small, round, purplish spots on leaves and fruits enlarge to about ⅛ inch. Centers of leaf spots turn brown and fall out, leaving leaves with many "shot holes." Affected fruits become scabby as they mature.	**Shot hole disease.** Fungus survives on infected buds and twigs during winter. Rain or overhead irrigation spreads to leaves and twigs. Wind may also carry water-splashed spores.	See page 150.
Other Common Pests on Apricots (See discussions on indicated pages.)	Bacterial canker Birds and squirrels Borers Green fruit beetle Oak root fungus Orange tortrix, fall webworm, redhumped caterpillar and other caterpillars Phytophthora crown and root rot San Jose scale and other scales Spider mites and other mites Stinkbug and other bugs	See page 152. See UC ANR Publication 21385. See page 78. See page 87. See page 165. See pages 58–66. See page 162. See page 106. See page 116. See page 88.

APRICOTS

AVOCADO

WHAT THE PROBLEM LOOKS LIKE	PROBABLE CAUSE	COMMENTS
Leaves skeletonized and sometimes webbed Fruit may be scarred and webbed to leaves.	**Amorbia.** Green caterpillar about ¾ to 1 inch long when full grown. A black stripe on the side of the head and the segment behind distinguishes it from orange tortrix. Wiggles when disturbed. Light green eggs laid in flat masses of 5 to 50 near midrib on upper leaf surface.	Remove leafrollers and egg masses. Many natural enemies. Can be controlled with *Bacillus thuringiensis* (Bt) or releases of *Trichogramma platneri*. See pages 49–50.
Leaves skeletonized. No webbing. Scars on fruit surface.	**Omnivorous looper.** Typical looper with prominent yellow, green, pinkish and/or black stripes. Grows to 2 to 2½ inches long.	Many natural enemies. Bt gives some control, but can be more effectively controlled with releases of *Trichogramma platneri*. See pages 49–50.
Leaves smaller than normal, usually pale or yellow green. New growth absent or stunted. Feeder roots blackened and dead.	**Avocado root rot.** Caused by the fungus *Phytophthora cinnamoni*	Associated with excess soil moisture due to overirrigation or poor drainage. Careful irrigation prolongs life of diseased trees and slows spread. Dry barriers of two rows of nonirrigated trees can prevent spread in an orchard. Trees planted on mounds less likely to be affected. See also page 162.
Purple discoloration along leaf veins. Leaves drop. Tiny yellow spotted mites.	**Six spotted mite**	See page 116. Apply insecticidal oils and water if necessary.
Leaves spotted. Colonies of tiny yellow mites with webbing.	**Persea mite**	See page 116. Apply insecticidal oils and water if necessary.
Leaves turn brown. Dark spider mites.	**Avocado brown mite**	See page 116. Apply insecticidal oils and water if necessary.
Brown discoloration on fruit and leaves with tiny black varnish spots.	**Greenhouse thrips**	See page 125. Apply insecticidal oils if necessary. Release *Thripobius* parasitic wasp.
Other Common Pests of Avocados (See designated pages for more information.)	Orange tortrix Verticillium wilt Phytophthora root and crown rot	See page 58. See page 160. See page 162.

AVOCADO

BLACKBERRIES, RASPBERRIES, AND OTHER CANEBERRIES

WHAT THE PROBLEM LOOKS LIKE	PROBABLE CAUSE	COMMENTS
Ripening blackberries do not darken in color—they become brighter red, hardened and sour and remain on old canes through winter. All or only part of berry may be affected.	**Redberry mite**. Extremely small, elongate, whitish creature, invisible to naked eye.	Many generations throughout growing season. Both mature and immature mites remain inactive during winter, hiding under bud scales, migrate to flowers in spring, and attack developing fruit. Apply lime sulfur in spring when new shoots are ¾ inch long. This treatment also helps control powdery mildew. Himalaya and Evergreen varieties most susceptible.
Leaves stippled with yellow; eventually become totally yellow, then dry and turn brown while still attached to canes. Undersides of leaves often covered with extemely fine strands of silk webbing.	**Spider mites**	Spider mites commonly become a problem during August-September and are encouraged by dusty conditions. Most serious on water-stressed plants. See page 116.
Tips of young shoots wilt during spring; cane may suffer dieback by summer. Cutting open affected portion of cane reveals thick, white worm or tunnel containing brownish granular material.	**Raspberry horntail**. S-shaped, segmented worm up to one inch long with white body and dark brown head. Has three pairs of legs near head end and short spine on tail end. Adults are wood wasps with long cylindrical bodies with a spine at the tip.	One generation per year. Adult emerges through hole cut in side of old cane during April or May and lays eggs inside new shoots, causing pronounced swelling. Remove and destroy infested canes. If this insect has been a problem in past years, an insecticide spray in April or May may be justified.
New shoots wilt in spring. White legless grub found burrowing toward base of cane.	**Raspberry cane borer**. Adult is slender shiny black beetle with bright yellow or orange pronotum.	Adults lay eggs in stems of new shoots in June, girdling stems above egg punctures. Grubs grow within canes, taking two years to develop. Prune off and destroy infected shoots and grubs as soon as wilting noticed.
Plants lack vigor, portions stunted and weakened. Lateral growth wilts in spring and entire cane may later die. Cutting open lower canes or crown area reveals worms tunneling through plant tissue.	**Crown borers**. Worms up to 1 inch long; whitish body and brown head. Adults are clear winged moths with black and yellow bands on bodies.	Raspberry crown borer requires 2 years to complete one generation. Adults emerge from crown area in late summer and lay eggs on leaves and stems. Larvae penetrate bark and remain there through winter. Feeding occurs inside canes and crown area during next two growing seasons. Strawberry crown moth has one generation per year. Adults emerge from crown in summer and lay eggs on lower leaves. Larvae tunnel into crown area or lower canes and feed until following summer. Keep plants properly irrigated and vigorous, since borers are attracted to stressed plants. Prune out and destroy infested shoots and canes.
Narrow, white, winding "trails" appear on canes in late spring or in summer. Trails may coalesce, giving canes a silvery appearance.	**Bushberry cane miner**. One-fourth inch long, flattened, light-colored caterpillar.	One generation per year. Worms mine under epidermis, then pupate. Adult moths appear in summer and lay eggs on canes. No control necessary. Damage to canes is superficial and often only noticed after insect is gone.

WHAT THE PROBLEM LOOKS LIKE	PROBABLE CAUSE	COMMENTS
Young fruiting stems wilt, often fall off in spring after apparently being chewed partially or all the way through.	**Cutworms**. Fat, smooth-skinned caterpillars, up to 1½ inches long, with mottled brown or grayish markings.	See page 77.
Buds and tender leaf growth eaten in spring. Foliage, especially near ends of canes, webbed together and eaten during summer. Worms sometimes found inside berries at harvest time.	**Leafrollers or orange tortrix**	See pages 58, 59 and 60.
Small holes appear in leaves and increase in size in May and June until only the veins remain, the leaves becoming "skeletonized." Greenish worms (often in large groups) on undersides of leaves.	**Raspberry sawfly**. Worm up to ⅔ inch long with smooth, pale green body with dark brown stripe down back. Adults are small, thick bodied wasps.	One generation per year. Adults emerge from soil in April or May and insert eggs in leaf tissue. Larvae feed until June then drop to ground and pupate in soil. Does not warrent control unless plants threatened with defoliation.
Canes become covered with white encrusta-tions or tiny grayish to brownish bumps.	**Scale insects**. Rose scales are ½₅ inch long with pure white scale covering which may be elongate or round. San Jose scales are slightly larger with round, grayish scale coverings. Oystershell scales are ¹⁄₁₀ inch long with brownish coverings shaped like miniature oyster shells.	Rose and San Jose scales have several generations per year; crawlers (young) appear first around early May. Oystershell scales have one generation per year with crawlers appearing in May or June. Pruning out old canes after harvest will decrease scale population. Apply dormant oil spray during winter. See also page 103–107.
Small, green sedentary insects found along stems, sometimes clustering on new growth.	**Raspberry aphid**. Small, green, pearshaped insect up to ¹⁄₁₆ inch long. May or may not have wings.	Damage not serious. Although virus diseases of caneberries are spread by aphids, control measures will not prevent spread of virus diseases. See page 96–102.
Tiny white spots appear on leaves in spring. When numerous, they may coalesce, resulting in bleached looking foliage. Undersides of leaves inhabited by small whitish insects which crawl or jump quickly.	**Rose leafhopper**. Small, narrow, whitish insect up to ⅛ inch long. Young are wingless and have red eyes.	Damage rarely serious enough to justify treatment. See page 94.
Berries deformed or scarred. Tiny slender insects on berries and flowers.	**Flower thrips**	Thrips feed on pollen and deform berries. They may be controlled naturally by Argentine ants and other predators. See also page 125.
Whitish gray powdery covering on leaves, young canes and buds; new shoots become twisted, affected older leaves curl upward.	**Powdery mildew**. Fungus survives on diseased plant tissue. Spores spread by wind.	Sulfur dust or sprays can be used as preventive measure. Reapply weekly to protect new growth and to renew deposits removed by rain or irrigation. Lime sulfur should be used if the disease is already evident. **Caution**: Plant damage may result if sulfur is applied when temperatures exceed 80° F. See page 140.

CANEBERRIES

WHAT THE PROBLEM LOOKS LIKE	PROBABLE CAUSE	COMMENTS
Leaves turn yellow, wither and fall, beginning at base of canes and progressing upward. Fruiting canes may take on a bluish-black cast and die during summer as fruits are maturing. Small groups of plants may be affected.	**Verticillium wilt**. Fungus survives in soil, building up on other host plants.	No cure. Remove and destroy infected plants. Avoid planting cane fruits in soils formerly planted to other hosts of the fungus. Soil solarization may be attempted before planting if Verticillium is known to be a problem. Ollalie variety is resistant to most strains of Verticillium. See page 160.
Entire plant weakened or killed at any time, often a group of plants affected. White fungal growth between bark and wood near ground level, but wood remains firm. Black "shoestrings" on roots.	**Armillaria root rot** (oak root fungus).	See page 165.
Small, red-bordered spots with whitish centers on leaves and canes. Plant reduced in vigor and may lose some leaves prematurely, leading to sunburn of canes.	**Leaf and cane spot**. Fungus survives on infected canes and leaves. Spores dispersed by splashing water.	Serious only on blackberries and boysenberries. Avoid overhead irrigation. After harvest and before fall rains, prune out and destroy old wood and apply a Bordeaux or a fixed copper fungicide. Spray again in spring when new laterals are leafing out, and again when flowers begin to open.
Leaves with wine red discoloration around the leaf veins. Fruit is dry, shrivelled and split.	**Downy mildew.** Fungus overwinters in roots and shoots and grows into new growth in spring.	Remove affected plants or spray with systemic fungicide. Spray adjacent plants. See page 143.
Small, yellow, blisterlike pustules appear in spring, first on fruiting canes and then on leaves. Canes dry out and crack, preventing proper ripening of fruit.	**Yellow rust**. Fungus overwinters on fruiting canes. Spores released from infected canes, spread by wind during spring and summer.	Mostly on boysenberries and blackberries. Avoid overhead irrigation. Prune out and destroy diseased canes before fall rains and apply a fixed copper fungicide. Spray with a fixed copper fungicide in spring when new laterals are leafing out, and again when flowers begin to open.
Orange, blisterlike pustules cover undersides of leaves in spring. Diseased shoots seem to recover by midsummer, but developing canes are smaller than normal and bear no fruit the following year.	**Orange rust**. Fungus is systemic, remains in host plant. Spores released from pustules on leaves, spread by wind in spring.	Red raspberries not affected. Fungicides are of little value. Remove and destroy infected plants, including roots.
Wartlike growths on roots, crown area or canes. Severely infected plants may be stunted.	**Crown gall** (cane gall). Bacteria survive in soil, spread by splashing water, pruning or cultivating tools and machinery.	Plants infected through pruning wounds, growth cracks, injured roots, or freezing injuries. Cut out infected canes during hot, dry weather (disinfect pruning tools before using on healthy plants). Remove and destroy severely infected plants. Lower soil pH with applications of sulfur. Be careful not to injure plants while planting, trellising, cultivating or harvesting. Obtain quality plants from nursery. Biological control agents (Galltrol) are available; see page 167.

CANEBERRIES, *continued*

WHAT THE PROBLEM LOOKS LIKE	PROBABLE CAUSE	COMMENTS
Weak, spindly canes. Leaves cup downward and redden prematurely in fall. Plants become unproductive in 2 to 3 years.	**Dwarf virus**. Transmitted by aphids when they feed on infected plant and then on healthy one.	Loganberry and wild blackberry most susceptible. Boysenberries and Young-berries unaffected. No cure. Remove infected plants immediately. Obtain virus-free plants from nursery.
Canes and laterals wilt and die back at tips in early spring as first leaves are unfolding. Associated with delayed leafing out.	**Dieback**. A physiological disorder.	Cause not yet determined, but may be associated with freezing injury, winter drought or insufficient chilling. Late fall and winter irrigations may reduce dieback.
Other Common Pests on Blackberries and Raspberries (See discussion on indicated pages.)	Gray mold (Botrytis fruit rot)	See page 146.

CHERRIES

WHAT THE PROBLEM LOOKS LIKE	PROBABLE CAUSE	COMMENTS
In spring, leaves severely curled by clusters of small, sedentary, shiny black insects.	**Black cherry aphid**. Shiny black, pear-shaped insect less than ⅛ inch long with two tiny "pipes" protruding from back end. Some individuals have clear wings, others wingless.	Overwinters as egg on bark of cherry tree, departs in mid-summer for various weeds, and returns in autumn to lay eggs. Spring pest only. New growth will cover up damage in summer. Dormant oil spray will reduce spring population. See page 96–102 for other controls.
Leaves tied together with webbing and eaten in spring.	**Leafrollers**	Damage insignificant—usually does not warrant control in cherries. See also page 59–60.
Leaves with brownish patches resulting from top layer having been eaten. Later, leaf tissue eaten completely through, leaving fine network of veins.	**Pearslug** (cherryslug). Olive green to blackish, slimy insect up to ½ inch long with head end widest part of body. Adult is a sawfly.	Two generations per year. Adults emerge from soil in early spring and lay eggs in leaf tissue. Mature larvae enter soil to pupate. Second generation appearing in summer does most damage. Pick pear-slugs off by hand and dislodge from foliage with strong stream of water. Insecticidal soap may be effective. Road dust or ash applied to foliage has been effective in killing pearslugs. Such dusts should be washed off after several days to discourage spider mites.
Dark brown, firm, circular spots spread rapidly over fruit as it ripens. Fruit becomes rotten, and sometimes sticks on tree through summer.	**Brown rot**. Fungus survives on diseased twigs and old, rotten fruits. Spores spread by air currents, rain splash and insects.	Most common in high rainfall areas. If rain occurs while fruit is growing, you can apply sulfur dust or wettable sulfur several weeks before harvest to prevent fruit rot. See page 144.

CHERRIES, *continued*

WHAT THE PROBLEM LOOKS LIKE	PROBABLE CAUSE	COMMENTS
Cherries fail to ripen and are conical, tasteless and tan colored. Leaves in midsummer are yellow. Only one or a few limbs may be affected. Trees on Mahaleb rootstock may die within weeks, as if girdled.	**Buckskin.** A disease caused a mycoplasmalike organism. Spread by leafhoppers.	Avoid planting cherries near privet hedges. Remove infected limbs or trees.
Masses of amber-colored gum on branches, cankers on wood. Infection does not go to roots, suckers may grow at base of tree.	**Bacterial canker** also known as **gummosis** or **soursap.**	See page 152. Mostly a problem on young trees.
Wilting of leaves, gummy exudate may occur near base of tree.	**Crown and root rot.** Caused by species of *Phytophthora* fungus.	Avoid saturating soil for long periods. Plant trees on raised mounds in wet areas, keep weeds and water away from trunk and soil line well below graft union. Avoid Mahaleb rootstock where drainage is poor. See also page 162.
Tree weakened. White fan-like plaques of fungus between bark and wood of crown and trunk. Mushrooms may appear at tree base in wet weather.	**Oak root fungus** (**Armillaria root rot**).	See page 165.
Leaves misshapen, deformed and mottled, light colored, with deep indentations; fruit pointed and small.	**Cherry crinkle.** A genetic disorder. Most common in Bing and Black Tartarian varieties.	Disease not infectious. Cut out affected limbs.
Long, narrow leaves and fruit with creases that are much deeper than normal.	**Deep suture.** A genetic disorder. Most common in Bing and Black Tartarian varieties.	Disease not infectious. Cut out affected limbs.
Other Common Pests on Cherries (See discussions on indicated pages.)	Bacterial canker Birds Borers Botrytis fruit rot (gray mold) Green fruitworm Green fruitbeetle San Jose and other scales Spider mites and other mites Stink bug and other bugs Western tussock moth, redhumped caterpillar, and other caterpillars	See page 152. See UC ANR Publication 21385 See page 78. See page 146. See page 64. See page 87. See page 103. See page 116. See page 88. See pages 62–65.

CHERRIES

CITUS

WHAT THE PROBLEM LOOKS LIKE	PROBABLE CAUSE	COMMENTS
Round, red brown scales on fruit leaves and twigs. Leaves may yellow and drop and twig dieback may occur.	**California red scale**	See page 108.
Fruit and leaves covered with honeydew and sooty mold. Tree vigor may be reduced. Scales on leaves or twigs.	**Various soft scales** including citricola, brown soft, or black scale. Look for scales on leaves or twigs, they are rarely on fruit.	Natural enemies usually provide effective control. *Metaphycus* parasites can be purchased and released for additional control. Keep ants out of trees; they protect scales from natural enemies. Oil sprays also effective; time treatments to target new brood right after hatching. See pages 110–111.
Leaves curled, honeydew and sooty mold.	**Aphids**	Control needed only for heavy infestations on young trees. See pages 96–102.
Fruit and leaves covered with honeydew and sooty mold. Tiny whiteflies fly out when branches disturbed.	**Whiteflies**	Natural enemies usually control; however ants must be eliminated. Dust control important. Chemicals not effective. See page 114.
Fruit and leaves covered with honeydew and sooty mold. Cottony secretion on scales on twigs.	**Cottony cushion scale.** Newly hatched nymphs are red and are found on leaves and twigs. Older scales are on twigs and covered with a cottony secretion. Eggs are in a fluted white egg sac about ½ inch long.	Becomes a pest when its natural enemies including vedalia beetle and a parasitic fly, are destroyed by insecticides. Reestablish natural enemies and avoid use of insecticides. See page 112.
Fruit and leaves covered with honeydew and sooty mold. Mealybugs present.	**Mealybugs.** Soft oval, distinctly segmented insects covered with a mealy white wax. Adults about ⅛–¼ inch long.	Natural enemies usually control. A A predator, the mealybug destroyer is available commercially for release. Control ants. See page 113.
A ring or partial ring of scarred tissue around stem on fruit skin. Young leaves may be deformed and scarred.	**Citrus thrips.** A tiny yellow insect about 1 mm. long. Very active.	Damage is primarily esthetic; ignore if fruit is for home use. See page 125.
Fruit is scarred, but scarring does not form a ring around stem.	**Wind abrasion**	Ignore. Create a windbreak or plant in a nonwindy area.
Surface feeding or holes in blossoms, leaves or very young fruit.	**Citrus cutworm.** Brown to green, smooth-skinned caterpillar with a prominent white stripe on each side. Curls up when disturbed.	A problem primarily in the San Joaquin Valley. Damage occurs in spring. Natural enemies often effective. *Bacillus thuringiensis* effective.
New leaves have holes and are webbed and rolled together.	**Leafrollers.** Orange tortrix. Amorbia.	See page 58.
Leaves and green fruit have a pale yellow stippling.	**Citrus red mite.** A tiny red mite, barely visible without a hand lens, found mostly on young leaves. No webbing.	Natural control should usually be sufficient in unsprayed backyard trees. An oil spray made between August and September will control problems in most areas. See page 119.

CITRUS, *continued*

WHAT THE PROBLEM LOOKS LIKE	PROBABLE CAUSE	COMMENTS
Oddly misshapened flowers and fruit. Primarily lemons in coastal areas.	**Citrus bud mite**. A very small, barely visible elongated yellow mite.	Can be controlled with a petroleum oil spray during May and June or September through November. See page 121.
Holes in leaves and fruit, slimy trails.	**Snails**	See page 132.
Leaves turning yellow and dropping. No mites. May be an abnormal number of blossoms.	**Overwatering**	Decrease irrigations. Once a week is too much. Avoid planting ferns, annual flowers or plants that need lots of water around trees.
Leaves turn pale green to yellow, especially in winter and spring. No mites.	**Nitrogen deficiency**	Check to see that fertilizer requirements are met. Symptoms sometimes appear in in spring when soil temperatures are cold and trees are unable to take up nutrients despite adequate amounts in soil.
Leaves turn yellow and drop. Beads of sap found on trunk, trunk cracking.	**Brown rot gummosis**. Caused by a fungus that infects the trunk and may spread to crown and woody roots.	Keep trunk dry; do not allow sprinkler water to hit trunk. Scrape away all diseased bark and a buffer strip of healthy light brown to greenish bark around margins. Allow to dry. Repeat if infection recurs.
Leaves turn yellow, root bark slides off easily when pinched.	**Phytophthora root rot**. Caused by same fungus that causes gummosis. Survives in soil for a long time.	If damage not severe, careful irrigation to avoid waterlogging with shorter more frequent irrigations may help. If severe, remove tree, use tolerant rootstock or fumigate if replanting. See also page 162. Trifoliate orange, alemow, and sour orange rootstocks are highly tolerant.
Ripe fruit turning light brown and soft.	**Brown rot fungus**. Fungus spores on ground are splashed onto fruit on lower branches.	Occurs primarily on fruit near ground during wet weather. Remove diseased fruit. Do not store infected fruit with clean fruit. A preventive Bordeaux treatment applied before first fall rains can be applied to tree skirts up to 4 feet and ground beneath trees. See page 146.
Older fruit drop.	Sudden change in **temperature** or low **moisture** or **fertility**.	Check to see that fertilizer and water are adequate.

FIGS

WHAT THE PROBLEM LOOKS LIKE	PROBABLE CAUSE	COMMENTS
Premature fruit drop.	**Cool weather, insufficient irrigation, weak trees.**	Choose varieties suited for your area that do not require cross pollination. Follow guidelines for irrigating and fertilizing.
Premature fruit drop, loss of leaves, stunted growth, small knots or galls on roots.	**Nematodes**	See page 169.
Yellow stippled leaves with webbing.	**Spider mites**	No control necessary. Irrigating will usually help. See page 116.
Ripe fruit sour and fermented.	**Yeasts** brought into figs by insects, especially fruit flies and small beetles, entering eyes.	Difficult to prevent. Destroy infested fruit. Choose varieties with small eyes, such as Mission.
Ripe fruit contains small (⅛ inch) black beetles sometimes with dull yellow bands on wings. White, short-legged grubs may be present.	**Dried fruit and sap beetle**. May also spread yeasts causing sour rots.	Eggs are laid on ripe and rotting fruit of all types. A generation is complete in 3 weeks in warm weather. Remove and destroy fallen fruit and cull piles. Beetles can be trapped in containers of fermenting fruit. Fig varieties with small eyes, such as Mission and Tina, are less affected. Selective pruning to allow more sunlight on figs dropped beneath trees or rapid removal of dropped fruit may reduce problems.
Ants or earwigs in ripening fruit.	**Ants or earwigs**	Apply sticky material around trunk of tree, prune away twigs and branches providing other routes for insects to reach ripening fruit. See pages 131 and 128.
Fruit has warty appearance. Oystershell--shaped scales on leaves, twigs or fruit.	**Fig scale**	Dormant season sprays with superior type oil. Parasites often effective. See Scale Insects page 103.
Surface of fruit scarred, large green beetles with yellow band around margins of wings present.	**Green fruit beetle**	See page 87.
Limbs girdled. Conspicuous sap and sawdust coming out of branches and trunk.	**Carpenterworm.** Damage is caused by boring larvae of a large moth.	Apply commercial preparations of the entomophagus nematode, *Neoplectana carpocapsae*, to burrows according to product label.
Bark and wood develop localized dead areas near pruning wounds or other injuries. Branches eventually girdled, leaves wither.	**Phomopsis canker.** Caused by a fungus that survives in cankers in trees or dead wood in surrounding areas.	Prune late in dormant season. Remove diseased branches from orchard and burn them. Kadota and Calimyrna varieties most susceptible.

FIGS

GRAPES

WHAT THE PROBLEM LOOKS LIKE	PROBABLE CAUSE	COMMENTS
Pale stippling on upper surface of leaves. Bleached areas may enlarge until leaf becomes mottled or dies, turns brown, and falls. Undersurface of leaf often shows small white cast skins of insects and tiny, dark, varnishlike spots. Older leaves affected first. Fruit covered with dark, sticky drops of excrement.	**Grape leafhopper**. Adult is narrow, ⅛ inch long, pale yellow with reddish and dark brown markings. Young are pale green with red eyes. Winged adults fly quickly when plants are disturbed; young run sideways. Damage leaves by sucking plant juices. –or– **Variegated leafhopper**. Similar to grape leafhopper but richly mottled with brown, green, red, white and yellow.	Not usually enough of a problem in home vineyards to justify chemical control measures but can be a serious problem in commercial vineyards. Adjacent plantings of prunes may improve biological control. See page 94.
Tiny yellowish stippling on upper leaf surface. Yellowing spreads along main veins and then throughout entire leaf. Fine webbing on undersides of leaves may increase until entire shoots are webbed over and leaves turn brown. Upper, sunnier parts of vine more affected.	**Spider mites**	See page 116. Zinfandel and Riesling varieties most susceptible.
Young leaves show bright pinkish or reddish swellings on upper surfaces. Corresponding areas on lower surfaces are concave and densely lined with a felty mass of plant hairs.	**Grape erineum mite**. Much smaller than spider mites. Can be seen only under magnification.	Control not necessary. Not an important pest. Early leaf distortion can be tolerated with no resulting loss in yield. See also page 121.
Shiny honeydew and sooty mold fungus, leading to sticky, black speckling of grapes. Clouds of tiny white insects fly up when plant is distrubed.	**Grape whitefly**. Adults are tiny mothlike insects 1⁄16 inch long with powdery white wings and bodies. Young are oval, flattened, and scale-like with lemon-yellow body circled by a narrow, white, waxy fringe.	Appears on grapes in the spring and has several generations. Usually more of a problem in foothill areas where buckthorn (Rhamnus) species grow wild. Removal or winter spraying of native or ornamental species of buckthorn with foliage spray oil will control pest because it overwinters on these shrubs. See page 114.
Large quantities of sticky honeydew run and drip onto fruit clusters, which later turn black from sooty mold. Some berries may crack. White, powdery insects ⅛ inch long are found in clusters on stems and trunk.	**Grape mealybug**. Wingless, oval, flattened insects ⅛ to 3⁄16 inch long. Appear to be dusted with a white, waxy secretion and have filaments around the margin of the body.	Remove loose bark in winter; young mealybugs and eggs are concealed in such places until spring. High temperatures in June kill much of the most damaging brood. Control of ants (which interfere with naturally occurring parasites and predators) is important. See also page 113.
Scarring and dwarfing of new shoots in early spring. Berries develop small dark scars surrounded by light "halos" or large irregular scars.	**Western flower thrips**. Tiny, slender insects 1⁄24 inch long, which are yellow to brown in color and resemble slivers of wood.	Damage, which occurs by the time grapes reach ¼ inch, is esthetic and not serious enough to justify treatment. See page 125.

WHAT THE PROBLEM LOOKS LIKE	PROBABLE CAUSE	COMMENTS
Brown, ⅛ inch bumps on stems of current season's growth, sometimes on leaves or bunches. Berries shiny from honeydew or blackened from sooty mold.	**European fruit lecanium.** Hemispherical, brown, beadlike insects ¼ inch long, clustered on stems of current season's growth or on 1- to 3-year-old wood. Occasionally on leaves or bunches.	Young scales become apparent in early spring and mature in late spring. Severe pruning reduces overwintering populations on young branches. Dormant spray with oil. Control of ants (which interfere with naturally occurring parasites and predators) is important. See also page 111.
Whitish ¹⁄₁₀-inch bumps, sometimes covering trunk, arms, or canes. Growth can be stunted.	**Grape scale.** Grayish white, round, flattened insects ¹⁄₁₀ inch long. Dry scale cover can be peeled off, revealing insect beneath.	Young scales become apparent in early spring and mature in late spring. Remove loose bark, under which young scales settle. Severe pruning and dormant oil sprays also recommended. See also page 103.
Lower leaves folded together in early season. Upper leaves affected later. Berries webbed together and eaten. Bunch rot often results.	**Omnivorous leafroller**	See page 60.
Pencil-size leaf rolls; leaf tissue eaten away. Berries may be eaten late in summer if worm populations are large and foliage sparse.	**Grape leaffolder.** Pale to bright green caterpillars up to 1 inch long, with yellowish heads. When disturbed they wiggle vigorously and fall to the ground without producing a silken thread.	Spends the winter as a pupa; first worms appear in late April. Three generations per year. Naturally occurring parasites work well. Removing suckers and thinning the leaves during the growing season removes many eggs. Handpick leafrolls. *Bacillus thuringiensis* effective.
Leaves and berry clusters webbed together. Berries partially eaten, stems girdled.	**Orange tortrix**	See page 58.
Undersurface of leaves eaten at first; later all leaf tissue between main veins is eaten, causing "skeleton" effect.	**Western grapeleaf skeletonizer.** Mature caterpillars are ½ inch long and yellow with two prominent purplish transverse bands and several black ones. Each body segment has four tufts of long, black stinging spines.	First caterpillars appear in early summer. Three generations per year. This pest is slowly spreading from southern California and will probably appear in all grape growing areas of the state. Treat with *Bacillus thuringiensis*. Good spray coverage on undersides of leaves is essential.
Buds eaten out, new shoots eaten part-way through.	**Cutworms**	Spring pest only. Vines often outgrow the damage, even when no control methods used. Working grasses and weeds into the soil in fall or early winter destroys many eggs and young larvae. See page 77.
Small, round holes in leaves become larger until whole leaves are eaten.	**Achemon sphinx moth.** Green worm with a long black horn on the back end. When fully grown, it is 4 inches long, has a dark button where the horn used to be, may vary in color, and has at least six diagonal white stripes down the sides.	Occurrence is cyclic, noticeable damage uncommon. First worms appear in late May. Second brood of worms in August does the most damage. Naturally-occurring parasites normally control this pest.

GRAPES

WHAT THE PROBLEM LOOKS LIKE	PROBABLE CAUSE	COMMENTS
Large amounts of leaf tissue eaten.	**Grasshoppers**	Large (2½ inch) green or tan grasshoppers appear in vineyard in spring. Smaller (1 inch) grasshoppers enter the vineyard in mid- to late summer when surrounding vegetation dries up. See page 130.
Young leaves and developing bunches eaten when new shoots are 12 to 14 inches.	**Hoplia beetle.** Beetle ¼ to ⅓ inch long, reddish brown on top, silvery and shiny underneath.	Spring pest; rarely does much damage. If beetles have been a problem in past seasons, disk weeds in early spring to destroy underground pupae.
New shoots (8 to 10 inches) wilt or break off during windy weather. Close inspection reveals a hole or gnawed area in crotch formed by shoot and spur. Tunnels throughout dead and living wood plugged with sawdustlike frass.	**Branch and twig borer.** Immature stage (grub) is white, fleshy and appears legless. It is blunt-ended with one end much thicker than the other.	Adult beetle emerges from inside vine in April. Grubs tunnel through wood from May until March or April. Prune out dying and dead parts of vine and remove or burn all prunings before March, since these are breeding sites for beetles. Keep vineyard free of prunings and wood cut from orchard trees and ornamentals.
Premature yellowing of leaves, stunted growth. Root tips swollen with yellow brown galls. Roots may be dead or dying.	**Grape phylloxera.** Look for groups of tiny yellow soft-bodied insects on roots, especially in summer and fall. They are oval and less than ⅒ inch long; a hand lens is required to see them.	No satisfactory controls available once vineyard is infested. Good irrigation and fertilizer practices help offset damage to roots. Use resistant rootstocks.
Reduced growth rate, vigor and fruit formation. Galls on roots.	**Nematodes**	See page 169. Resistant rootstocks available.
White, powdery, weblike growth on any green tissue. Young, rapidly growing tissue first affected; leaves become distorted, shriveled. Berries become scarred, crack open as they grow.	**Powdery mildew.** Caused by a fungus.	As a preventive, thoroughly dust all surfaces of vine with a fungicide beginning when shoots are 2 to 4 inches long. Repeat at 7 to 10 day (sulfur) intervals. Treating before disease occurs is most important. If disease already evident, apply wettable sulfur and wetting agent. Do not apply sulfur if temperatures exceed 90° F. The powdery mildew fungus also infects Boston ivy. Remove this inoculum source. See also page 140.
Shriveled and rotting grapes (often with a gray-brown fuzzy growth). Damage appears in late summer when fruit is ripe.	**Botrytis bunch rot.** Caused by a fungus.	Cool temperatures and higher humidity near harvest time encourage this disease. Trim leaves around bunches soon after bloom to improve air circulation. Pick ripened grapes as soon as possible. Bird damage aggravates rots. Fungicides also available. See page 146.
Hard, irregular gall(s) on trunk of vine at ground level and just beneath soil, continues to grow for years. Smaller galls may appear higher up trunk.	**Crown gall.** Caused by a bacterium.	No cure. Do not injure trunk of vine. Be sure that new plants brought into vineyard are not already infected. See page 167.

GRAPES

WHAT THE PROBLEM LOOKS LIKE	PROBABLE CAUSE	COMMENTS
In spring shoots on affected limbs are dramatically stunted with shortened inter-nodes. Leaves small, misshapened and scorched and tattered looking later in the season.	**Eutypa dieback.** Caused by a fungus. Often only part of the vine affected.	Remove all affected wood. Be sure to cut back to healthy wood with no staining in it. Cut in dry weather and apply fungicidal wound protectant.
Vines look water stressed. Leaves turn yellow, then red or brown around edges.	**Pierces' disease.** Caused by a bacterium that is spread by a type of leafhopper called a sharpshooter.	Disease does not spread within vineyard so vine removal not helpful in preventing spread. Remove vines as they become unproductive.
Small dark brown to black spots with light colored margins on leaves, shoots and fruit. During dormant season canes are bleached and may have areas with black spots on them.	**Phomopsis cane and leaf-spot.** Caused by a fungus.	More serious following a wet spring. However, not a serious concern in home vineyards. Prune out bleached canes during dormant season.
Other Common Pests on Grapes (See discussions on indicated pages.)	Birds Green fruit beetle Oak root fungus Phytophthora crown and root rot Saltmarsh caterpillar and armyworms Verticillium wilt	See UC ANR Publication 21385. See page 87. See page 165. See page 162. See page 72 and 67. See page 160.

PEACHES AND NECTARINES

WHAT THE PROBLEM LOOKS LIKE	PROBABLE CAUSE	COMMENTS
Young shoots die back one to several inches from tip in spring. Small worm with dark brown bands may be found inside each affected shoot. Ripening fruit infested with worms, particularly near stem end.	**Peach twig borer**	See page 65.
Young shoots die back one to several inches from tip. Small pinkish worm may be found in each affected shoot. Apparently sound fruit contains worms; wormy fruit may also have holes filled with hardened gum.	**Oriental fruit moth**	See page 66.
Leaves become curled, covered with tiny, sedentary greenish insects.	**Green peach aphid and other aphids.**	Chemical control usually not necessary. See page 96–102.
First spring leaves become yellow to reddish, thickened and curled, and fall off when temperatures increase. A second set of leaves grows, decreasing tree growth and and fruit production. Fruits may develop reddish, irregular, roughened areas on skin.	**Leaf curl.** Fungus survives the winter as spores on twigs and buds. Spread by wind and rain during spring.	Apply lime sulfur or copper fungicides as buds begin to swell, but before any color appears in them. Alternatively, use copper fungicide in fall (mid-November to mid-December) to control both leaf curl and shot hole disease. In high rainfall areas, spray at both times. See page 151.

WHAT THE PROBLEM LOOKS LIKE	PROBABLE CAUSE	COMMENTS
Blossoms and associated twigs and leaves shrivel and die. Gummy ooze at base of flower. During humid weather, shriveled petals bear tiny, greyish brown masses (fungus spores). Dark brown, firm, circular spots spread rapidly over fruit as it ripens. Fruit becomes rotten, will stick on tree until following season.	**Brown rot.** Fungus survives on diseased twigs and old, rotten fruits. Spores spread by air currents, rain splash and insects.	See page 144.
Small, round, purplish spots on leaves and fruits enlarge to about ⅛ inch. Centers of leaf spots turn brown and fall out, leaving leaves with many "shot holes." Many of the leaves may shed in spring. One-year-old wood may develop brownish or black, circular infections with associated gumming.	**Shot hole disease.** Fungus survives within infected buds and twigs during winter. Rain or overhead irrigation spreads spores to leaves and twigs. Wind may also carry splashed spores.	See page 150.
Fruit splits along suture before ripening, allowing entry of earwigs, dried fruit beetles, etc.	**Peach split pit**	Peach split pit is a physiological disorder of unknown cause, it may be associated with thinning too early or too much heat.
Leaves yellow, trees decline rapidly. Roots may have galls.	**Root knot** or **lesion nematodes**	Nemaguard rootstock is resistant to root knot. See also page 169–182.
Other Common Pests on Peaches (See discussions on indicated pages.)	Bacterial canker Birds Borers Dried fruit beetles Earwigs Green fruit beetles Leafrollers and other caterpillars Oak root fungus Peachtree borer Phytophthora crown and root rot San Jose Scale and other scales Spider mites, peach silver mite, and other mites Stink bugs and other bugs Western flower thrips	See page 152. See UC ANR Publication 21385. See page 78. See page 128. See page 87. See pages 59–66. See page 165. See page 78. See page 162. See page 103. See page 116. See page 88. See page 125.

PEARS

WHAT THE PROBLEM LOOKS LIKE	PROBABLE CAUSE	COMMENTS
Fruit wormy with masses of damp, brown, granular frass at the core, often protruding from holes in the skin. Fruit may drop off prematurely.	**Codling moth**	See page 52.
Leaves and blossoms eaten and tied together with webbing. Young fruit may be deeply gouged and may fall to ground. Less severely damaged fruits reach maturity badly misshapen or with deep bronze-colored scars with a roughened surface.	**Leafrollers** including fruittree leafroller and omnivorous leafroller; in coastal areas, orange tortrix and apple pandemis may also cause problems.	See pages 58–60.
Leaves with pinkish or brownish patches resulting from top surface having been eaten. Later, leaf tissue eaten completely through, leaving a fine network of veins.	**Pearslug.** Olive green to blackish, slimy insect up to ½ inch long with head end widest part of body.	Two generations per year. Adults emerge from soil in early spring and lay eggs in leaf tissue. Mature larvae enter soil to pupate. Second generation appearing in summer does most damage. Pick off by hand and dispose or dislodge from foliage with strong stream of water. Insecticidal soaps may be effective. Road dust or ash applied to foliage has been used to kill pearslugs. Such dusts should be washed off after several days to discourage spider mites.
Various irregular spots on upper sides of leaves. Corresponding areas on under surface have only a thin layer of tissue remaining, resembling blisters.	**Tentiform leafminer.** Small flattened caterpiller about ⅛ inch long with whitish body and brownish, wedge-shaped head. Found inside "blisters" in leaf tissue.	Four generations per year. Last one spends winter as pupa in leaves on ground. Moth emerges in early February, lays eggs as leaves emerge. Seldom causes enough damage to warrant control. Parasites effectively control this pest by late summer.
Fruit and foliage sticky, becoming black with sooty mold. Vegetative growth stunted or tree defoliated. Beads of clear, sticky honeydew enclosing tiny yellowish insects on leaves.	**Pear psylla.** Tiny insect up to ¹⁄₁₀ inch long. Yellow in immature stages to reddish brown as adult. Adult has dark spot on back, holds clear wings rooflike over body.	About five generations per year. Adult overwinters in sheltered places in the bark or under the ground. Eggs laid on or near new foliage. Dormant oil spray will kill many overwintering adults. Eggs and mature young (nymphs) are resistant to insecticides, but adults and smaller nymphs can be controlled during growing season with 2 to 3 oil sprays applied weekly. Pear decline disease transmitted by this insect.
Fruit spotted with clear, sticky honeydew which becomes black from growth of sooty mold. Whitish, cottony masses in calyx end of fruit and at bases of twigs and fruit clusters.	**Mealybugs.** Oval, purplish insects up to ³⁄₁₆ inch, usually covered with whitish, powdery material and possessing white filaments which protrude from edges of body. Very sluggish movements.	Usually two generations per year. Eggs or young mealybugs (crawlers) overwinter in cracks and under bark scales and are thus protected to some extent from dormant oil sprays. Parasites and predators, particularly green lacewings, are important in controlling this pest. Hot weather in summer also will kill young mealybugs. See page 113.

WHAT THE PROBLEM LOOKS LIKE	PROBABLE CAUSE	COMMENTS
Leaves distorted, new foliage stunted. Maturing fruit takes on a russetted appearance. Clusters of sedentary, greenish insects on new shoots.	**Aphids**	Only a minor problem in home orchards; distorted leaves soon covered up by normal foliage. See page 96–102.
Fruit becomes rough, brown and russetted, especially at stem and flower ends. Foliage takes on dry, rusty appearance.	**Pear rust mite**. Microscopic, whitish, wedge-shaped mites.	Many generations per year. Spend winter as nonfeeding adults behind leaf bud scales or any other protected 1- to 2-year-old wood. Feed until no more new foliage develops. Control measures seldom necessary; damage is mostly to fruit appearance. An oil spray with lime sulfur in October or November will considerably decrease population for following season. See page 121.
Leaves and flower buds develop reddish blisters which later turn brown or black. Fruit develops sunken, brown russetted areas and sometimes becomes deformed.	**Blister mites**. Microscopic, whitish, elongate creatures found in large numbers inside blisters.	Many generations. All stages of mite spend winter under bud scales. They move to flowers and leaves as they emerge, move back to bud scales in fall. Oil spray with lime sulfur in October or November kills many of the mites before they move deeper into buds. See page 121.
Circular holes cut along leaf margins and within during spring. Sometimes entire leaf consumed except for midrib.	**California pear sawfly**. Bright green worm up to ½ inch long with light brown head and black eyespots. Head end widest with body tapering to tail end.	One generation per year. Adults emerge from soil in early spring and lay eggs in leaf tissue. Mature larvae enter soil to pupate by June. Trees can tolerate large numbers of worms without suffering much damage.
Young blossoms and fruits wilt and collapse, turn brown to black. Sticky brown ooze appears on diseased shoots during humid weather. Cankers appear on limbs and secrete a dark ooze in spring. Entire branches or even trees can die in one season.	**Fire blight**. Bacterium survives in cankers, spread by splashing rain and insects. Favored by warm, humid weather during bloom.	Blossoms and succulent shoots are most susceptible. Cut diseased branches back about 9 inches into healthy wood, removing all affected tissue, beginning as soon as active cankers are seen in the spring. Sterilize pruning tools in household disinfectants before making cuts through succulent tissues. In hot, dry weather and cuts in woody tissue, disinfectant may be omitted. Apply dilute copper fungicides during bloom period as preventive sprays.
Other Common Pests on Pears (See discussions on indicated pages.)	Borers Oak root fungus Orange tortrix, green fruitworms, and other caterpillars Pear scab San Jose scale and other scales Spider mites and other mites Stink bugs and other bugs	See page 78. See page 165. See page 58–66. See page 154. See page 103. See page 116. See page 88.

WHAT THE PROBLEM LOOKS LIKE	PROBABLE CAUSE	COMMENTS
Young shoots die back one to several inches from tip in spring. Small caterpillar found inside each affected shoot. Ripening fruit infested with worms feeding near surface.	**Peach twig borer**	See page 65.
Leaves tied together with webbing and eaten during spring.	**Leafrollers** and **orange tortix**	See pages 58–60.
Leaves attached to fruits with webbing during summer. Small holes chewed through skin.	**Eyespotted bud moth.** Caterpillar up to ½ inch long, brown with shiny black head and black plate on rear end.	Mostly a coastal pest. One generation per year. Spends winter in cocoon in crotches of twigs and buds. Eggs laid on undersides of leaves in early summer, hatching by mid-July. Dormant oil sprays will kill overwintering worms on trees. "Nests" of webbed-together leaves are easily seen and can be pruned out.
Leaves with brownish patches resulting from top layer having been eaten. Later, leaf tissue eaten completely through, leaving a fine network of veins.	**Pearslug.** Olive green to blackish, slimy insect up to ½ inch long with head end widest part of body. Adult is a sawfly.	Two generations per year. Adults emerge from soil in early spring and lay eggs in leaf tissue. Mature larvae enter soil to pupate. Second generation appearing in summer does the most damage. Pick pearslugs off by hand and dispose of them or wash from foliage with a strong stream of water. Insecticidal soap may be effective. Road dust or ash has been used in killing pearslugs. Such dessicants should be washed off after a few days to discourage spider mites.
New growth covered with pale green, slow-moving insects which produce a white, mealy substance. Leaves curl, fruits may split; both covered with sticky honeydew and black sooty mold.	**Mealy plum aphid.** Pale green pear-shaped insects less than ⅛ inch long with two tiny "pipes" protruding from back end.	Overwinters as egg on tree. Leaves tree for weeds in July, returns in early winter to lay eggs. Dormant oil sprays kill over-wintering eggs. New growth will cover up damage by summer. Japanese hybrid plums unaffected. Also see pages 96–102.
Blossoms and associated twigs and leaves shrivel and die. Gummy ooze at base of flower. During humid weather, shriveled petals bear tiny, grayish brown masses (fungus spores).	**Brown rot.** Fungus survives on diseased twigs and old, rotten fruits. Spores spread by air currents, rain splash and insects.	See page 144.
Waxy coating on fruit partially lacking. Dried fruit may be russetted.	**Lacy scab,** also called **russet scab,** cause is unknown.	A soil or foliar application of nitrogen should be applied to prevent this.
Leaves burn and fall off, exposed upper branches sunburned and die back.	**Sunburn.** Generally occurs in years when a heavy load of fruit is set and potassium reserves are depleted.	A soil or foliar application of nitrogen should be applied to prevent this.

PLUMS AND PRUNES, *continued*

WHAT THE PROBLEM LOOKS LIKE	PROBABLE CAUSE	COMMENTS
Other Common Pests of Plums and Prunes (See discussions on indicated pages.)	Bacterial canker	See page 152.
	Birds	See UC ANR Publication 21385.
	Borers	See page 78.
	Codling moth	See page 52.
	Green fruitworms	See page 64.
	Oak root fungus	See page 165.
	Phytophthora crown and root rot	See page 162.
	Powdery mildew	See page 140.
	Redhumped caterpillar and other leaf-feeding caterpillars	See page 62.
	San Jose scale and other scales	See page 103.
	Shot hole	See page 150.
	Spider mites and other mites	See page 116

STRAWBERRIES

WHAT THE PROBLEM LOOKS LIKE	PROBABLE CAUSE	COMMENTS
New leaves become crinkled and stunted; center of plant eventually becomes compact mass. Distorted foliage takes on bronzed appearance, flowers wither and die.	**Cyclamen mite.** Extremely small pinkish-orange mite, not visible to naked eye.	Remove and destroy infested plants immediately. Obtain quality plants from reliable nursery. Replanting yearly with quality plants may reduce cyclamen mite problems. See page 123.
Dry, brownish areas appear on lower leaf surfaces in spring. Entire leaves become brown underneath, then die and dry up. Sparse new growth is yellowish and distorted; plants become stunted. Many tiny yellowish creatures and fine webbing on undersides of leaves.	**Spider mites**	See page 116. Commercial plantings can tolerate 10 mites per leaflet before flowering and 10 to 20 after flowering.
Leaves rolled up or webbed together with silk strands; fruit sometimes riddled with shallow holes. Small caterpillars may be found feeding within webbed foliage or inside ripening fruit.	**Leafrollers,** including garden tortrix, strawberry leafroller and omnivorous leaftier.	Fruit rarely affected. Damage to leaves usually not serious. Remove dead leaves and trash from base of plants in late winter. See also orange tortrix page 58.
Ripe fruit with large holes, no dried slime evident. Some leaves eaten away, and young plants occasionally cut off at base.	**Cutworms**	See page 77.
Small, deep holes in berries with a dried, shiny slime deposit on surrounding fruit surface.	**Snails and slugs**	See page 132.
Dark gray, oval, many-legged "bugs" found in holes in ripening fruit.	**Sowbugs**	Initial damage usually caused by snails or earwigs. See page 129.

STRAWBERRIES

PLUMS AND PRUNES

WHAT THE PROBLEM LOOKS LIKE	PROBABLE CAUSE	COMMENTS
Small, deep holes in ripening fruit. No dried slime evident.	**Earwigs**	See page 128.
All or part of plant may wilt and die in spring or fall. Cutting open crown reveals whitish caterpiller and/or hollowed areas with brown, granular fecal material and old silken cocoon.	**Strawberry crown moth**	One generation per year. Moths emerge from crowns in June or July and lay eggs on undersides of leaves. Larvae crawl to crown area and burrow in, noticeably injuring plants in fall or when they again become active the following spring. Keep plants well watered and vigorous. Pull up and destroy infested plants.
Rapid decline of plants in March or April. Leaves sometimes have scalloped appearance on edges. Pulling up plant reveals smaller roots eaten and larger roots or crown scraped and girdled.	**Otiorhynchus root weevils.** Curved, legless grubs, up to ⅜ inch long, with whitish body and brown head.	One generation per year. Depending on the species, adults emerge from April to June, feed (at night) on leaves for several weeks then lay eggs in soil near crown. Larvae move to roots and feed until winter. Feeding resumes in spring, when greatest damage is done before pupation. Remove and destroy infested plants. Replanting yearly will minimize root weevil problem. No effective chemical control. Predaceous nematodes may be effective.
Leaves and fruit covered with clear, sticky substance (honeydew), often turning black due to growth of sooty mold fungus. Small yellow to whitish insects clustered on undersides of young, developing leaves and along veins of mature leaves.	**Strawberry aphid and other aphids**	Not normally a serious problem; natural enemies often keep populations low. Plants can tolerate 10-15 aphids per leaflet. See page 96–102.
Berries deeply furrowed, gnarled or twisted (catfaced). Deformed areas often associated with large, hollow seeds which become straw-colored.	**Lygus bugs.** Greenish or brownish oval insects, about ¼ inch long; yellow "V" or triangle on middle of back. Catfacing also caused by poor pollination due to adverse weather condtions, lack of pollinating insects, or a fungus disease.	Not serious enough in gardens to warrant control. At least one lygus per plant can be tolerated in commercial plantings. Removing misshapen berries will conserve plant energy for production of next set of fruit. See page 91.
Plants wilt during warm weather, even when irrigated. General decline in vigor. Pulling plants up reveals white, cottony deposits on roots.	**Ground mealybug.** Tiny white, eggshaped insects, ⅟₁₆ inch long, with short antennae and segmented body. Usually found in cottony masses they secrete.	Many generations per year. Overwinter as adults or immatures, increasing their feeding in spring. Start producing eggs in summer. Remain beneath surface of soil throughout year. Replanting yearly will minimize problems. No effective chemical control. See page 113.
Whitish powdery covering on leaves, flower calyces and fruit. Leaves cup upward, turn reddish on undersides, and in severe instances appear burned on margins. Vigor and productivity of plant decrease.	**Powdery mildew.** Fungus survives on diseased plant tissues. Spores spread by wind.	Apply wettable sulfur at 7- to 14-day intervals as preventive treatment, or use lime sulfur when mildew first appears. Caution: strawberries are extremely sensitive to sulfur at high temperatures. Do not treat if temperatures expected to reach 80° F or higher. See page 140.

STRAWBERRIES

WHAT THE PROBLEM LOOKS LIKE	PROBABLE CAUSE	COMMENTS
Ripe fruit has soft, dull-colored spots. These may become gray with fungus growth.	**Gray mold rot.** Fungus survives on diseased plant tissues. Spores spread by wind. Favored by wet conditions and low temperatures.	Fungus spores may infect dead or dying leaves and stems or delicate tissues such as unbroken skin of berry. Pick and destroy infected berries immediately. Plastic sheeting or a dry mulch on ground beneath berries may reduce incidence of mold and will prevent rots caused by other fungi. See page 146.
Plant loses vigor, foliage loses shiny green luster. Young and old leaves wilt during warmest part of day. Roots blackened and/or rotted. Plant usually collapses early in second growing season.	**Phytophthora root rot.** Several fungi cause similar symptoms; all survive in the soil, and are favored by cool weather and poor drainage.	Obtain uninfested plants from nursery. Do not plant strawberries in exceptionally heavy soils. Plant on raised beds where excess water can be carried away. Where root rots have been a problem in the past, consider soil solarization. See page 162.
Outer leaves wilt and dry; plant becomes stunted and eventually collapses during summer of first year. New leaves may appear at center, or spindly new runners may form, but these also die. Roots not discolored or rotted.	**Verticillium wilt.** Fungus survives in soil for many years, building up on other host plants.	See page 160.
Plants lose vigor, young central leaflets have distinct yellow margins and may cup upward or downward. Irregular yellow spots appear on crinkled leaves and main veins occasionally show streaking.	**Virus diseases.** Several viruses are transmitted by strawberry aphid and leafhoppers when they feed on infected plants and then healthy ones.	No cure. Remove affected plants immediately, and plant only virus-free plants from nursery. See page 156.
Terminal tissues of young, folded leaves killed in late spring or early summer; when leaves mature, they are irregular shaped and puckered, appearing burned back from their tips to about half normal size.	**Tip burn.** Physiological disorder, most common in southern California and inland areas.	Sudden spell of high temperatures after prolonged, cool spring weather seems to induce this condition. Hot weather during summer and fall does not affect heat-conditioned plants. No action necessary. Plants will replace damaged leaves with healthy foliage.
Margins of leaves burned or bronzed. Growth may be inhibited. Roots underdeveloped.	**Salt injury.** Strawberries are very sensitive to salt.	Often associated with manures with a high salt content. Never allow manures to get in root zone. Also may be due to saline water and insufficient leaching of salts from soil.

STRAWBERRIES

WALNUTS

WHAT THE PROBLEM LOOKS LIKE	PROBABLE CAUSE	COMMENTS
Green nuts drop or dry up on tree. Little webbing in nut. Older nuts worm infested. Hull with masses of brown fecal material protruding from entry holes. Most caterpillars leave nuts before harvest.	**Codling moth**	Early season varieties most susceptible. See page 52.
Nuts worm-infested at harvest time. Nut eaten, covered with webbing and brownish fecal material. Shell not stained.	**Navel orangeworm**	More common in trees damaged by codling moth or blight. See page 57.
Leaves tied together with webbing and eaten in the spring.	**Leafrollers**	Damage insignificant; does not warrant control in walnuts. See also pages 59–60.
Small black spots on husk become large blackened areas which remain soft, unsunken and smooth. Areas damaged by walnut blight dry up. Hull difficult to remove from shell. Shell darkly stained. Nut meat not affected. Infested nuts tend to remain on tree.	**Walnut husk fly.** White to yellowish maggots up to ⅜ inch long, rounded at one end, slightly pointed at other. Found in blackened areas of hull, never inside shell. Adult fly is small and brown with yellow spot on back and three dark bands on each wing.	One generation per year. Most eggs laid in hull from late July to early August. Larvae pupates in ground; adult fly emerges the following summer. Most home orchardists ignore this pest since nut meat unaffected. To remove husks from shells, place in damp burlap bag for a few days. Dispose of infested husks in a tightly sealed bag. Payne, Ashley, Placentia and Erhardt varieties less susceptible. Pherocon AM traps are available for monitoring flies in commercial orchards to time insecticide treatments.
Leaves and nuts covered with clear, sticky honeydew which may turn black because of sooty mold growth. Some leaf drop may occur. Nuts become sunburned. Tiny yellow insects along veins on underside of leaf. Reduction in size and quality of nuts.	**Walnut aphid.** Tiny (¹⁄₁₆ inch) oval- to pear-shaped insect which moves sluggishly. Occurs on underside of leaf. Some individuals may have smoke-colored wings. During fall some have two dark bands across back.	Many generations per year. Overwinters as egg in rough places on twigs. Not much of a problem because an introduced parasitic wasp usually controls this pest. See pages 96–102.
Leaves and nuts covered with clear, sticky honeydew. Some leaf drop may occur. Small yellow insects lined up along main vein on upper side of leaf. Reduction in size and quality of nuts.	**Dusky-veined walnut aphid.** Small pear-shaped yellow insect which moves sluggishly. Occurs on upper side of leaf only. Some individuals may have wings.	Many generations per year. Overwinters as egg in cracks in bark. Often controlled by beneficial insects. See pages 96–102.
Hull with rough, sunken, hard, blackened areas. If young nuts are affected, they drop prematurely. Older nut kernels turn black and shrivel. (Also see walnut husk fly.)	**Walnut blight.** Bacteria survive in buds, twig lesions and old nuts. Spread by rain splash.	Worse in early-leafing varieties such as 'Payne' and 'Ashley.' Avoid getting foliage wet with sprinklers. Avoid irrigation altogether during bloom. Open up trees with pruning to get better air circulation. In commercial orchards, spray with fixed copper fungicide when 50 percent of female flowers have "feathers" exposed. In heavy rainfall areas, spray when 1 percent of female flowers are in feather stage and repeat when 10 to 20 percent reach this stage.

WALNUTS

WHAT THE PROBLEM LOOKS LIKE	PROBABLE CAUSE	COMMENTS
Poor terminal growth and yellowing; drooping leaves lead to premature defoliation, particularly on top. Tree becomes stunted or eventually dies. Small holes and cracks in bark at graft union. When bark is removed, horizontal black line is evident at union.	**Blackline**. Virus is transmitted by grafting and possibly during pollination.	No known cure. Affected trees should be removed to protect adjacent healthy trees. All English walnut varieties on black walnut rootstocks are believed to be susceptible.
Leaves yellow, tree declines.	**Overwatering** creates conditions that favor diseases such as Phytophthora. Black line (above) has similar symptoms.	Problem commonly occurs when walnut tree in lawn is overwatered. Allow soil to dry out, dig grass and soil away from around trunk, fill with gravel and reduce irrigations.
Other Common Pests of Walnuts (See discussions on indicated pages.)	Bacterial cankers Crown gall Oak root fungus Phytophthora crown and root rot Redhumped caterpillar Spider mites Squirrels Walnut scale, San Jose scale and other scales	See page 152. See page 167. See page 165. See page 162. See page 62. See page 116. See UC ANR Publication 21385. See page 103–111.

WALNUTS

Appendix

USING DEGREE-DAYS FOR PREDICTING GROWTH AND DEVELOPMENT OF CROPS AND INVERTEBRATE PESTS

Temperature is very important in controlling the rate of development of crop plants and many pests, especially insects. For this reason, it is misleading to describe development in terms of time alone. For instance, someone may tell you that it takes three weeks for a certain caterpillar species to develop from egg to adult. However, in truth, because invertebrates develop faster when temperatures are consistently warm than when temperatures are much cooler, it may take anywhere from two to four weeks depending on ambient temperatures. The same is true for the development of crop plants. To monitor crop development and predict pest behavior, professional pest managers often use a system that takes into account the accumulation of heat with passing time. This system is based on the

degree-day unit (abbreviated °D).

One degree-day is the amount of heat that accumulates during a 24 hour period when the average temperature is one degree above the developmental threshold of the organism under consideration. The simplest way to estimate the number of degree-days accumulating on a given day is to use the formula:

$$\frac{\text{daily high} + \text{daily low}}{2} - \frac{\text{developmental}}{\text{threshold}} = \frac{\text{Degree-}}{\text{days}}$$

For example, on a day in which the highest temperature reached was 90° F and the lowest was 70° F, a pest with a developmental threshold of 60° F would accumulate 20° D:

$$\frac{90° \text{ F} + 70° \text{ F}}{2} - 60° \text{ F} = 20° \text{ D}$$

This formula is not very accurate when the daily low is below the developmental threshold of the organism. Degree-day tables and computer programs are available for more accurate calculations and should be obtained by anyone desiring to use degree-day calculations to schedule pest management practices. Obtain a copy of Leaflet 21373, *Degree-days: The Calculation and Use of Heat Units in Pest Management*, from University of California Agriculture and Natural Resources Publications for more information on calculation methods.

The key factor in determining degree-day requirements is accurate information about lower (and, if possible, upper) developmental thresholds of the organism. The lower developmental threshold is the temperature at and below which development stops. Above the lower developmental threshold, the growth rate increases until temperatures reach the upper developmental threshold above which the developmental rate decreases (Figure A-1). Lower developmental thresholds and in some cases upper developmental thresholds have been established for certain plants and pests through carefully controlled laboratory and field experiments. The total amount of heat required for the organism to develop from one point to another in its life cycle can also be calculated experimentally and expressed in total degree-days. Some examples are included in Table A-1.

The degree-day requirement for a given species remains constant regardless of location or temperature fluctuations. However, each species has differing requirements so degree-day calculations and charts cannot be used to estimate the development of other species in the same environment.

To develop degree-day accumulations for your orchard or farm you must take daily readings of maximum and minimum temperatures with a maximum-minimum thermometer or other device in your field or orchard or get readings from a nearby weather station. Temperature readings right in your cultivated area, shielded from direct sun, away from paved surfaces, and 3 to 5 feet from the ground are best. Local variations in terrain, vegetation cover, and elevation can make a significant difference in readings so regional weather information may not reflect the situation in your field.

Accumulation of degree-days is initiated at a specific point in the organism's development known as a biofix. The biofix will be different for various pests, but could be the time when male moths are first caught in pheromone traps, or the time when egg laying begins, or planting date in the case of crops. Once the biofix point is passed, you must start adding up the degree-day accumulations for each day. The most common method is to use a

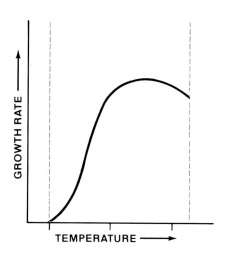

FIGURE A-1. Approximate relation between temperature and growth rate. As temperature increases above the developmental threshold (dashed line at left), growth rate increases to a maximum. Beyond the maximum, growth again slows as temperature approaches the upper limit for growth (dashed line at right).

degree-day table specially developed for that pest as shown in Figure A-3 to get degree-day totals for each day and then keep a record as shown in Figure A-2 of the degree-day accumulations over time. You can then compare your accumulations to the known degree-day requirements for that

MAX TEMPS		84	82	80	78	76	74	72	70	68	66	64	62	60	58
118	1	38	37	37	36	35	35	34	33	32	32	31	30	29	28
116	1	38	37	37	36	35	35	34	33	32	32	31	30	29	28
114	1	38	37	37	36	35	35	34	33	32	31	31	30	29	28
	1														
112	1	38	37	37	36	35	34	34	33	32	31	30	29	29	28
110	1	38	37	36	36	35	34	33	33	32	31	30	29	28	28
108	1	38	37	36	36	35	34	33	32	32	31	30	29	28	27
	1														
106	1	38	37	36	35	35	34	33	32	31	31	30	29	28	27
104	1	38	37	36	35	35	34	33	32	31	30	29	28	28	27
102	1	37	37	36	35	34	33	33	32	31	30	29	28	27	26
	1														
100	1	37	37	36	35	34	33	32	31	31	30	29	28	27	26
98	1	37	36	36	35	34	33	32	31	30	29	28	27	26	25
96	1	37	36	35	34	34	33	32	31	30	29	28	27	26	25

MINIMUM TEMPERATURES

FIGURE A-2. A portion of a degree-day table indicating the number of degree-days (28) that had accumulated on a day when the maximum temperature was 98 and the minimum temperature was 64.

species (e.g. Table A-1) to predict when new generations or other developmental stages of the organism may occur.

While degree-day calculations have been very useful in commercial agriculture for identifying best times for monitoring and pest control, they should never be used to substitute for careful field monitoring of the organism itself. Back up your degree-day predictions by checking field populations to make sure the pest is developing as expected. Occasionally, other factors such as overlapping pest generations, inaccurate weather information, natural enemies, pesticide sprays, or food quality may cause degree-day calculations to be somewhat imprecise.

Although it can heighten your understanding of pest biology, accumulation of degree-days is not necessary for management of pests in the home garden or very small farm. Larger scale growers should consult the various IPM manuals listed in the References for specific information on degree-day requirements and how to use degree-days to help in the management of pests on those crops.

DATE	DAILY LOW	DAILY HIGH	DEGREE-DAYS FOR DATE	TOTAL DEGREE-DAYS FOLLOWING OVIPOSITION
4/10	47	72	10.0	10.0
4/11	46	73	10.5	20.5
4/12	46	75	11.5	32.0
4/13	47	74	11.0	43.0
4/14	48	74	11.0	54.0
4/15	46	75	11.5	65.5
4/16	46	76	12.0	77.5

FIGURE A-3. Example of degree-day accumulations for peach twig borer to determine the optimum time for a May spray. Add together the total number of degree-days that accumulate after the April flight begins; spray after 400 to 500°D have accumulated.

TABLE A-1.

Developmental Thresholds (°F) and Degree-days Required to Develop Through One Generation for Selected Insect Pests.

INSECT	DEVELOPMENTAL THRESHOLD		DEGREE-DAYS PER GENERATION
	Lower	Upper	
California red scale	50	95	1030
Codling moth	52	94	1085
Corn earworm	57	94	968
Lygus bug	52	NA	1000
Oriental fruit moth	45	90	963
Peach twig borer	50	88	1060
San Jose scale	51	90	1050
Walnut aphid	41	NA	572

NA = not available

Source: Threshold and degree-day information from Zalom et al. 1983. *Degree-Days: The Calculation and Use of Heat Units in Pest Management*, UC ANR Leaflet 21373.

References

Chapter 2: Designing a Pest Management Program

Altieri, M. A., and L. L. Schmidt. 1986. Cover crops affect insect and spider populations in apple orchards. *California Agriculture.* 40 (1 and 2):15–17.

Altieri, M. A. 1981. Weeds may augment biological control of insects. *California Agriculture.* 35(5 and 6):22–24.

Atkins, E.L., D. Kellum, and K.W. Atkins. 1981. *Reducing Pesticide Hazards to Honeybees.* UC ANR Leaflet 2883.

Barclay, L.W., C.S. Koehler, C.S. Davis, and W.J. Moller. 1981. *Insect and Disease Management in the Home Orchard.* UC ANR Leaflet 21262.

Eaton, A. T. 1986. Pest Control for Organic Vegetable Growers. *Cooperative Extension Service,* University of New Hampshire, Durham, NH.

Fereres, E., D.W. Henderson, W.O. Pruitt, W.F. Richardson, and R.S. Ayers. 1981. *Basic Irrigation Scheduling.* UC ANR Leaflet 21199.

Fereres, E. ed. 1981. *Drip Irrigation Management.* UC ANR Leaflet 21259.

Flint, M. L., and R. van den Bosch. 1981. *Introduction to Integrated Pest Management.* Plenum Press, NY.

Hagen, K. S. 1976. Role of Nutrition in Insect Management. *Proceedings Tall Timbers Conference on Ecological Animal Control by Habitat Management* 6:221–261.

Hickman, G.W., and M.L. Flint. 1987. *Some Biological Control Agents Commercially Available in California.* UC ANR Leaflet 7115.

Johnson, Hunter, ed. 1987. *Proceedings Organic Farming Training Conference.* University of California, Riverside. 114 pp.

Johnson, Hunter. 1985. Fertilizing Vegetable Gardens. Mimeo. Vegetable Crops Extension. University of California, Riverside.

Koehler, C.S., and W.S. Moore. 1979. *Controlling Insects, Diseases and Related Problems in the Home Vegetable Garden.* UC ANR Leaflet 21086.

Koehler, Carl, W. Barclay and T. Kretchun. 1983. Companion plants. *California Agriculture* 37 (9 and 10): 14–15.

Marer, P.J. 1988. *The Safe and Effective Use of Pesticides.* UC ANR Publication 3324.

Perry, E. 1983. *Wood Ashes as a Garden Fertilizer.* UC ANR Leaflet 7138.

Pittenger, D.R. 1983. *Home Vegetable Gardening.* UC ANR Leaflet 2989.

Pullman, G.S., J.E. DeVay, C.L. Elmore, and W.H. Hart. 1984. *Soil Solarization, A Nonchemical Method for Controlling Diseases and Pests.* UC ANR Leaflet 21377.

Raabe, R. 1981. *The Rapid Composting Method.* UC ANR Leaflet 21251.

Rauschkab, R.S. 1980. *Soil and Water Management for Home Gardeners.* UC ANR Leaflet 2258.

Reisenaur, H.M. ed. 1983. *Soil and Plant Tissue Testing in California.* UC ANR Leaflet 1879.

Reuther, W., J.A. Beutel, A.W. Marsh, J.L. Meyer, F.K. Aljibury, R.S. Ayers, and H. Schulbach. 1981. *Irrigating Deciduous Orchards.* UC ANR Leaflet 21212.

Risch, S. J., D. Andow, and M. A. Altieri. 1983. Agroecosystem diversity and pest control: Data, tentative conclusions and new research directions. *Environmental Entomology* 12(3)625–629.

Stimmann, M.W., and Jack Litewka. 1979. *Using Pesticides Safely in the Home and Yard.* UC ANR Leaflet 21095.

van den Bosch, R., P. S. Messenger and A. P. Gutierrez. 1982. An Introduction to Biological Control. Plenum Press, NY.

van Horn, M. 1995. *Compost Production and Utilization: A Growers' Guide.* UC ANR Leaflet 21514.

Ware, G. W. 1980. *Complete Guide to Pest Control.* Thompson Publications, Fresno, CA.

Yuen, G. Y. and R. D. Raabe. 1984. Effects of small scale aerobic composting on survival of some fungal plant pathogens. *Plant Disease* 68(2):134–136.

Chapter 3: Common Insects, Mites, Other Arthropods, and Snails and Slugs

Anonymous. 1984. Earwigs: Friend or Foe? *Common Sense Pest Control* I (1):5–6.

Anonymous. 1986. The Hundred Years' War ... Gardeners vs. Slugs and Snails. *Sunset Magazine.* April 1986. 122–26.

Altieri, M., K. S. Hagen, J. Trujillo, and L. E. Caltagirone. 1982. Biological control of *Limax maximus* and *Helix aspersa* by indigenous predators in a daisy field in central coastal California. *Acta Ecologica* 3(4):387–90.

Barclay, L. W., C. S. Koehler, C. S. Davis, and W. J. Moller. 1981. *Insect and Disease Management in the Home Orchard.* UC ANR Leaflet 21262.

Bentley, W. J., and W. R. Bowen. 1980. *The Green Fruit Beetle—A Common Fruit Pest.* UC ANR Leaflet 21191.

Bethell, R. S. ed. 1978. *Pear Pest Management.* UC ANR Publication 4086.

Borror, D. J., and R. E. White (1970). *A Field Guide to the Insects of America North of Mexico.* Houghton-Mifflin Co., Boston.

Carroll, D. P and S. C. Hoyt. 1984. Augmentation of European earwigs for biological control of apple aphid in an apple orchard. *J. Econ. Entomology.* 77:738–40.

Daar, S., H. Olkowski, and W. Olkowski. 1989. Directory of producers of natural enemies of common pests. *IPM Practitioner,* 11(4):15–18.

Fisher, T. W., J. B. Bailey, and N. J. Sakovich. 1983. Skirt pruning, trunk treatment for snail control. *Citrograph* 68(12):292–97.

Flaherty, D. L., F. L. Jensen, A. N. Kasimatis, H. Kido, and W. J. Moller. ed. 1981. *Grape Pest Management,* UC ANR Publication 4105.

Gill, Raymond J. 1982. *Color-Photo and Host Keys to California Whiteflies.* Scale and whitefly Key #2. California Department of Food and Agriculture, Sacramento, CA.

Gill, Raymond J. 1982. *Color-Photo and Host Keys to the Mealybugs of California.* Scale and Whitefly Key #3. California Department of Food and Agriculture, Sacramento, CA.

Gill, Raymond J. 1982. *Color-Photo and Host Keys to the Soft Scales of California.* CDFA Scale and Whitefly Key #4. California Department of Food and Agriculture, Sacramento, CA.

Gill, Raymond J. 1982. *Color-Photo and Host Keys to the Armored Scales of California,* CDFA Scale and Whitefly Key #5. California Department of Food and Agriculture, Sacramento, CA.

Hagen, K. S. 1976. Role of Nutrition in Insect Management. *Proceedings Tall Timbers Conference on Ecological Animal Control by Habitat Management,* Number 6.

Hesketh, K. A., and W. S. Moore. 1979. *Snails and Slugs in the Home Garden.* UC ANR Leaflet 2530.

Hickman, G. and M.L. Flint. 1987. *Some Biological Control Agents Commercially Available in California.* UC ANR Leaflet 7115.

Hoy, M. A. 1984. *Managing Mites in Almonds, An Integrated Approach.* UC IPM Publication 1. UC IPM Project, University of California, Davis

Hunter, C. 1994. *Suppliers of Beneficial Organisms in North America.* Calif. Dept. Pesticide Regulation-PM 9403. Sacramento, CA.

Jepson, L. R., H. H. Keifer, and E. W. Baker. 1975. *Mites Injurious to Economic Plants.* University of California Press, Berkeley, CA.

Koehler, C. S. and W. S. Moore. 1979. *Controlling Insects, Diseases and Related Problems in the Home Vegetable Garden.* UC ANR Leaflet 21086.

Koehler, C. S., L. W. Barclay, T. M. Kretchun. 1983. Soaps as insecticides. *California Agriculture* 37(9–10):11–12.

Koehler, C., and W. Barclay. 1983. Snail Barriers. *California Agriculture* 37(9–10):15.

Kono, T., and C. S. Papp. 1977. *Handbook of Agricultural Pests: Aphids, Thrips, Mites, Snails, and Slugs.* California Department of Food and Agriculture. Sacramento, CA.

Krantz, G. W. 1975. *A Manual of Acarology.* Oregon State Univ. Book Stores, Inc. Corvallis, OR.

Levi, H. W., L. R. Levi (1968). *Spiders and Their Kin.* Golden Press, Western Publishing Company.

Metcalf, C. L., and R. L. Metcalf. 1993. *Destructive and Useful Insects.* McGraw-Hill, New York.

Metcalf, R. L. 1985. *Plant Kairomones and Insect Pest Control.* Illinois Natural History Survey Bulletin. Vol 33, Art. 3.

Moore, W. S., J. C. Profita, and C. S. Koehler. 1979. Soaps for home landscape control. *California Agriculture* 33:13–14.

Moore, W. S. and C. S. Koehler. 1980. *Ants and Their Control,* UC ANR Leaflet 2526.

Moore, W. S., and C. S. Koehler. 1981. *Aphids in the Home Garden and Landscape.* UC ANR Leaflet 21032.

Oatman, E. R., and G. R. Platner. 1971. Biological control of the tomato fruitworm, cabbage looper and hornworms on processing tomatoes in Southern California using mass releases of *Trichogramma pretiosum. J. Econ Ent,* Vol. 64 (2):501–506.

Oatman, E. R. and G. R. Platner. 1978. Effect of mass releases of *Trichogramma pretiosum* against lepidopterous pests on processing tomatoes in Southern California, with notes on host egg population trends. *J. Econ Ent,* 71 (6):896–900.

Oatman, E. R., and G R. Platner. 1985. Biological control of two avocado pests, *California Agriculture* 39 (11–12):21–23.

Oatman, E. R., M. E. Badgley, and G. R. Platner. 1985. Predators of the two-spotted spider mite on strawberries. *California Agriculture* 39 (1 and 2): 9–15.

Olkowski, Helga. 1987. Thrips on plants indoors. *Common Sense Pest Control* III (1):4–9.

Olkowski, Helga. 1986. Protecting roses and other plants from aphids. *Common Sense Pest Control* 11 (2):18–20.

Olkowski, Helga. 1985. Whitefly in the vegetable garden. *Common Sense Pest Control* I(3):22–23.

Olkowski, Helga. 1985. Sowbugs in the garden. *Common Sense Pest Control* (3):21.

Olkowski, H., and W. O. Olkowski. 1986. Controlling mites on plants indoors. *Common Sense Pest Control* II(2):13–18.

Olkowski, William. 1984. Readers' column on symphylans. *Common Sense Pest Control* 1(1): 22–23.

Olkowski, William, Sheila Daar, and Helga Olkowski. 1985. *Update: Codling Moth—Big Changes Ahead.* The IPM Practitioner Vol. VII (5):1–5.

Payne, C. C. 1982. Insect viruses as control agents. *Parasitology* 84:35–77.

Pickel, C. *Tree Bands for Control of Codling Moths.* Mimeo from Santa Cruz County Cooperative Extension Office.

Pickel, C., R. S. Bethell, and W. W. Coates. 1986. *Codling Moth Management Using Degree-Days.* UC IPM Publication 4, (Available from UC IPM Project, UC, Davis, CA 95616.)

Pinnock, D. E., R. J. Brand, J. E. Milstead, and N. F. Coe. 1974. Suppression of populations of *Aphis gossypii* and *A. spiraecola* by soap sprays. *J. Econ. Entomol.* 67(6):783–84.

Pinnock, Dudley, K. S. Hagen, D. V. Cassidy, R. J. Brand, J. E. Milstead, and R. L. Tasson. 1978. Integrated Pest Management in High way Landscapes. *California Agriculture* 32(2):33–34.

Powell, Jerry A., and Charles L. Hogue. 1979. *California Insects.* University of California Press, Berkeley.

Sakovich, N. J., J. B. Bailey, and T. W. Fisher. 1984. *Decollate Snails for Control of Brown Garden Snails in Southern California Citrus Groves.* UC ANR Leaflet 2184.

Skinner, G., and S. Finch. 1986. Reduction of cabbage root fly damage by protective discs. *Annals of Applied Biology* 108(1):1–10.

Swan, Lester A, and Charles Papp. 1972. *The Common Insects of North America.* Harper and Row, New York.

UC Statewide lPM Project. 1991. *Integrated Pest Management for Citrus.* UC ANR Publication 3303.

UC Statewide IPM Project. 1996. *Integrated Pest Management for Cotton.* UC ANR Publication 3305.

UC Statewide IPM Project. 1985. *Integrated Pest Management for Almonds.* UC ANR Publication 3308.

UC Statewide IPM Project. 1985. *Integrated Pest Management for Cole Crops and Lettuce.* UC ANR Publication 3307.

UC Statewide IPM Project. 1986. *Integrated Pest Management for Potatoes.* UC ANR Publication 3316.

UC Statewide IPM Project. 1987. *Integrated Pest Management for Walnuts,* Second Edition. UC ANR Publication 3270.

UC Statewide IPM Project. 1998. *Integrated Pest Management for Tomatoes,* Fourth Edition. UC ANR Publication 3274.

UC Statewide IPM Project. 1991. *Integrated Pest Management for Apples and Pears.* UC ANR Publication 3340

Westigard. P. H.. and L. J. Gut. 1985 Codling moth control in pears with modified programs using insect growth regulators. *J Econ Ent.* 79 (1): 47–49

Wyman, J. A.. N. C Toscano. K. Kido. H. Johnson, and K. S. Mayberry 1979. Effects of mulching on the spread of aphid-transmitted watermelon mosaic virus to summer squash *J. Econ Entomol.* 72: 139–43.

Zalom. Frank 1981. Effects of aluminum mulch on fecundity of apterous *Myzus persicae* on head lettuce in a field planting *Ent. Exp. & Appl.* 30:227–30.

Zalom. F. G., and W. S Cranshaw. 1981 Effects of aluminum foil mulch on parasitism and fecundity of apterous *Myzus persicae. Great Lakes Entomologist* 14 (4):171–76.

Zalom, F.G., P.B. Goodell, L.T. Wilson, W. W. Barnett, and W. J. Bentley. 1983. *Degree-Days: The Calculation and Use of Heat Units in Pest Management.* UC ANR Leaflet 21373.

Chapter 4: Diseases

Adams, P. B.. and W. A. Ayers. 1982. Mycoparasite for biocontrol of lettuce drop *Phylopathology* 72: 485–88.

Agrios. G. N. 1988. *Plant Pathology.* Academic Press, 3rd ed.

Barclay L W., C. S. Koehler, C. S. Davis. and W. J. Moller. 1981. *Insect and Disease Management in the Home Orchard.* UC ANR Leaflet 21262

Bernhardt, E., J. Dodson, and J. Watterson. 1988. *Cucurbit Diseases.* Petoseed Co., Inc., Saticoy, CA.

Chalfant, R. B., C. A. Jaworski, A. W. Johnson, and D. R. Sumner. 1977. Reflective film mulches, millet barriers. and pesticides: effects on watermelon mosaic virus, insects, nematodes, soil-borne fungi, and yield of yellow summer squash. *J. Amer. Soc. Hort. Sci.* 102:11–15.

Converse, R.H., ed. 1987. *Virus Diseases of Small Fruits.* USDA, Agriculture Handbook No. 631.

Daniell, J. W. 1973. Effect of time of pruning on growth and longevity of peach trees. *J. Amer. Soc. Hort. Sci.* 98:383–86.

Dixon, G. R. 1981. *Vegetable Crop Diseases.* AVI, Westport, CT.

Dowler, W. M., and D. H. Petersen. 1966. Induction of bacterial canker of peach in the field. *Phytopathology* 56:989–1000.

Duffus, J. R., and R. A. Flock. 1982. White-fly-transmitted disease complex of the desert southwest. *Calif. Agric.,* Nov.-Dec. 4–6.

English, H., J. E. DeVay, J. M. Ogawa, and B. F. Lownsbery. 1982. *Bacterial Canker and Blast of Deciduous Fruits.* UC ANR Leaflet 2155.

Erwin, D. C., S. Bartnicki-Garcia, P. H. Tsao, eds. 1983. *Phytophthora: Its Biology, Taxonomy, Ecology, and Pathology.* American Phytopathological Society. St. Paul, MN.

Frazier, N. W. ed. 1970. *Virus Diseases of Small Fruits and Grapevines.* Univ. of Calif. Div. of Agric. Sci., Berkeley, CA.

Gubler, W. D., J. J. Marois, A. M. Bledsoe, and L. G. Bettiga. 1987. Control of Botrytis bunch rot of grapes with canopy management. *Plant Dis.* 71:599-601.

Hagedorn, D. H. 1984 *Compendium of Pea Diseases.* Amer. Phytopath. Soc., St. Paul, MN.

Hall, H. L., W. D. Hamilton, and K. D. Gowans. 1975. *Reducing Root Rots in Plants.* UC ANR Publication 4004.

Hubbard, G. P., et al. 1983. Biocontrol of seed rots enhanced by application of iron. *Phytopathology* 73:655–59.

Kerr, A. 1972. Biological control of crown gall: seed inoculation. *J. Appl. Bact.* 35:493–97.

Kontaxis, D. G., and V. E. Rubatzky. 1983. Managing phytophthora root rot in cauliflower. *Calif. Agric.* Nov.-Dec., p. 12.

Kraft, J. M., and G. C. Papavizas. 1983. Use of host resistance, *Trichoderma,* and fungicides to control soilborne diseases and increase seed yields of peas. *Plant Disease* 67:1234–37.

Locke, J. C., J. J. Marois, and G. C. Papavizas. 1985. Biological control of fusarium wilt of greenhouse-grown chrysanthemums. *Plant Disease* 69:167–169.

Lumsden, R. D., et al. 1986. Suppression of lettuce drop caused by *Sclerotinia minor* with composed sewage sludge. *Plant Disease* 70:197–201.

Maas, J. L., ed. 1984. *Compendium of Strawberry Diseases.* Amer. Phytopathological Soc., St. Paul, MN.

Mace, M. E., A. A. Bell, and C. A. Beckman. 1981. *Fungal Wilt Diseases of Plants.* Academic, New York.

MacNab, A. A., A. F. Sherf, and J. K. Springer. 1983. *Identifying Diseases of Vegetables.* The Pennsylvania State University, University Park, PA.

Malajckuk, N. 1979. Biological suppression of *Phytophthora cinnamomi* in eucalyptus and avocado in Australia. In B. Schippers and W. GAms, ed. *Soil-Borne Plant Pathogens.* Academic Press, London.

McCain, A. H. 1978. Peach leaf curl control for home gardeners in the San Francisco Bay Area. *Calif. Plant Pathol.* 43:4–5.

McCain, A. H., R. D. Raabe, and S. Wilhelm. 1981. *Plants Resistant or Susceptible to Verticillium Wilt.* UC ANR Leaflet 2703.

Mills, W. D. and A. A. LaPlante. 1954. *Diseases and Insects in the Orchard.* Cornell Ext. Bull. 711:20–28.

Mircetich, S. M. and G. T. Browne. 1985. Phytophthora root and crown rot of apple trees. *Proc. Oregon Hort. Soc.*

Moller, W. J., A. H. McCain, and D. H. Chaney. 1979. *Leaf Curl Control in Peaches and Nectarines.* UC ANR Leaflet 2613.

Moller, W. J., D. Rouch, and R. Sanborn. 1979. *Coryneum Blight of Stonefruits in California.* UC ANR Leaflet 2863.

Moore, L. W. 1977. Prevention of crown gall on Prunus roots by bacterial antagonists. *Phytopathology* 67:139–44.

Nameth, S. T., F. F. Laemmlen, and J. A. Dodds. 1985. Viruses cause heavy melon losses in desert valleys. *Calif. Agric.,* July-Aug., 28–29.

Natwick, E. T., and A. Durazo III. 1985. Polyester covers protect vegetables from whiteflies and virus disease. *Calif. Agric.* July-Aug., 21–22.

New, P. B., and A. Kerr. 1972. Biological control of crown gall: field measurements and glasshouse experiments. *J. Appl. Bact.* 35:279–87.

Ogawa, J. M., H. English, W. J. Moller, B. T. Manji, D. Rough, and S. T. Koike. 1980. *Brown Rot of Stone Fruits.* UC ANR Leaflet 2206.

Ohr, H. D., G. A. Zentmyer, E. C. Pond, L. J. Klure. 1980. *Plants in California Susceptible to Phytophthora cinnamomi.* UC ANR Leaflet 21178.

Patterson, C. L., R. G. Grogan, and R. N. Campbell. j1986. Economically important diseases of lettuce. *Plant Disease* 70:982–987.

Paulus, A. O., F. Munoz, J. Nelson, W. L. Schrader, and H. W. Otto. 1985. Fungicides for control of powdery mildew of melons. *Calif. Agric.* July-Aug.

Pickel, C. and R. S. Bethell. 1985. *Apple Scab Management.* UC ANR Leaflet 21412.

Raabe, R. D. 1979. *Resistance or Susceptibility of Certain Plants to Armillaria Root Rot.* UC ANR Leaflet 2591.

Ross, N., M. N. Schroth, R. Sanborn, et al. 1970. Reducing loss from crown gall disease. *Calif. Agr. Exp. Sta. Bul.* 18454.

Sall, M. A., J. Wrysinski, F. J. Schnick. 1983. Temperature-based sulfur applications to control grape powdery mildew. *Calif. Agric.* July-Aug., 4–5.

Savage, S. D., and M. A. Sall. 1983. The influence of cultural practices on bunch rot of grapes. *Plant Disease* 67:771–74.

Savage, S. D., and M. S. Sall. 1982. Vineyard cultural practices may help reduce botrytis bunch rot. *Calif. Agric.* Mar.-April, 8–9.

Schroth, M. M., A. R. Weinhold, A. H. McCain, D. C. Hildebrand, and N. Ross. 1971. Biology and control of *Agrobacterium tumefaciens. Hilgardia* 40:537–52.

Sholberg, P. L., J. M. Ogawa, and B. T. Manji. 1981. Diseases of prune blossoms, fruits, and leaves. In D. E. Ramos, ed. *Prune Orchard Management.* UC ANR Publication 3269.

Shurtleff, M. C., ed. 1980. *Compendium of Corn Diseases,* 2nd ed. Amer. Phytopath. Soc., St. Paul, MN. pp. 60–68.

Siegler, E. A. 1938. Relations between crown gall and pH of the soil. *Phytopathology* 28:858–59.

Snyder, R. L., P. D. LaVine, M. A. Sall, J. E. Wrysinski, and F. J. Schick. 1983. *Grape mildew control in the central valley of California using the powdery mildew index.* UC ANR Leaflet 21342.

Spencer, D. M., ed. *The Powdery Mildews.* Academic Press, London, 1978.

Stephens, C. T. and T. C. Stebbins. 1985. Control of damping-off pathogens in soilless container media. *Plant Disease* 69:494–96.

Thomas, H. E., S. Wilhelm, and N. A. MacLean. 1953. Two root rots of fruit trees. Plant Diseases. *The Yearbook of Agriculture.* USDA.

Tjosvold, S. A. 1982. *Controlling Damping-off Diseases in the Garden.* UC ANR Leaflet 21299.

UC Statewide IPM Project. 1987. *Integrated Pest Management for Walnuts,* Second Edition. UC ANR Publication 3270.

UC Statewide IPM Project. 1984. *Integrated Pest Management for Citrus.* UC ANR Publication 3303.

UC Statewide IPM Project. 1989. *Integrated Pest Management for Tomatoes,* 3rd ed. UC ANR Publication 3274.

UC Statewide IPM Project. 1985. *Integrated Pest Management for Almonds.* UC ANR Publication 3308.

UC Statewide IPM Project. 1985. *Integrated Pest Management for Cole Crops and Lettuce.* UC ANR Publication 3307.

UC Statewide IPM Project. 1986. *Integrated Pest Management for Potatoes.* UC ANR Publication 3316.

UC Statewide IPM Project. 1984. *Integrated Pest Management for Cotton.* UC ANR Publication 3305.

UC Statewide IPM Project. (1991). *Integrated Pest Management for Apples and Pears,* UC ANR Publications.

Villapudua, J. R., and D. E. Munnecke. 1986. Solar heating and amendments control cabbage yellows. *Calif. Agric.* 40(5–6):11–13.

Wilson, E. E., and J. M. Ogawa. 1979. *Fungal, Bacterial, and Certain Nonparasitic Diseases of Fruit and Nut Crops in California.* UC ANR Publication 4090.

Yepsen, R.B., Jr. 1976. *Organic Plant Protection.* Rodale Press, Emmaus, PA.

Yuen, G. Y., and R. D. Raabe. 1984. Effects of small-scale aerobic composting on survival of some fungal plant pathogens. *Plant Disease* 68:134–136.

Zentmyer, G. A., and H. D. Ohr. 1981. *Avocado Root Rot.* UC ANR Leaflet 2440.

Chapter 5: Nematodes

Ferris, H, P G Goodell, and M. V. McKenry. 1980. *General Recommendations for Nematode Sampling.* UC ANR Publication 21234

McKenry, Michael V., and Philip A. Roberts. 1985. *Phytonematology Study Guide.* UC ANR Publication 4045.

Motsinger, Ralph and Gene Moody. 1979. *Marigolds for the Control of Nematodes.* BPI News, November 1979.

Rickard, David A. 1977. Nematode control based on the *Targetes* effect. N.C. D.A. *Nema Notes.* North Carolina Department of Agriculture. Note No. 1.

UC IPM Project. 1985. *Integrated Pest Management for Cole Crops and Lettuce.* UC ANR Publication 3307.

Chapter 6: Weeds

California Weed Conference. 1989. *Principles of Weed Control in California.* Thompson Publications, Fresno, CA.

Finch, C.U, and W.C Sharp. 1983. *Cover Crops in California Orchards and Vineyards.* Soil Conservation Service, USDA, Davis, CA.

Fischer, B.B, A.H. Lange, and June McCaskill. 1978. *Grower's Weed Identification Handbook.* UC ANR Publication 4030.

Hesketh, K.A., and C.E. Elmore. 1980. *Vegetable Planting Without Weeds,* UC ANR Publications 21153.

Robbins, W, M. Bellue, and W. Ball. 1970. *Weeds of California.* State of California Documents and Publications, P.O. Box 1015, North Highlands, CA 95660.

Ross, M. A. and C. A. Lembi. 1985. *Applied Weed Science.* Burgess Publishing Company, Minneapolis, MN.

Chapter 7: Crop Tables

Adams, E. B., A. L Antonelli, D. Bosley, R. S. Bosley, R. S. Byther, S. J. Collman, R. E. Hunter, and R. E. Thornton. *Home Gardens.* Cooperative Extension, College of Agriculture, Washington State University Publication EB 422, Pullman, Wash.

Alderman, D. C. 1975. *Growing Strawberries in Your Home Garden.* UC ANR Leaflet 2219.

Bailey, J. Blair, and M. P. Hoffman. 1980. *Amorbia: A California Avocado Insect Pest.* UC ANR Leaflet 21156.

Bailey, J. Blair and M. P. Hoffman. 1979. *Omnivorous Looper on Avocados in California.* UC ANR Leaflet 21101.

Barclay, L. W., and C. S. Koehler. 1981. *Managing Insects and Diseases in the Home Vineyard.* UC ANR Leaflet 21196.

Barclay, L. W., C. S. Koehler, and J. B. Bailey. 1982. *Insect and Disease Management for Home Berry Plantings.* UC ANR Leaflet 21320.

Barclay, Leslie W., Carlton S. Koehler, and C. S. Davis. 1981. *Insect and Disease Management in the Home Orchard.* UC ANR Leaflet 21262.

Bari, M. A, and H. K. Kaya. 1984. Evaluation of the Entomogenous Nematode, *Neoaplectana carpocapsae,* and the Bacterium *Bacillus thuringiensis* Var. Kurstaki for Suppression of the Artichoke Plume Moth. *J. Econ. Entomol.* 77: 225–229.

Bernhardt, E., J. Dodson, and J. Watterson. 1988. *Cucurbit Diseases.* Petoseed Co., Inc., Saticoy, CA

Bethell, R., ed. 1978. *Pear Pest Management.* UC ANR Publication 4086.

Beutel, J. A. 1981. *Figs in the Home Garden.* UC ANR Leaflet 2481.

Burton, V. E., S. Humphrey and A. W. Johnson. 1984. *Insect and Spider Mite Control Program for Beans.* UC ANR Leaflet 21386.

Davidson, R. H., and W. F. Lyon. 1979. *Insect Pests of Farm, Garden and Orchard.* John Wiley and Sons.

Flaherty, D. L., F. L. Jensen, A. N. Kasimatis, H. Kido, and W. J. Moller, ed. 1981. *Grape Pest Management.* UC ANR Publication 4105.

Haynes, K. F., M. C. Birch and J. A. Klun. 1981. Sex pheromone offers promise for control of artichoke plume moth. *Calif. Agriculture* 35 (1 and 2):13–14.

Kasmire, R. F., ed. 1981. *Muskmelon Production in California.* UC ANR Leaflet 2671.

Koehler, C. S., and W. S. Moore. 1979. *Controlling Insects, Diseases and Related Problems in the Home Vegetable Garden.* UC ANR Leaflet 21086.

Johnson, H. Jr., K. Mayberry, and J. Guerard. 1984. *Watermelon Production.* UC ANR Leaflet 2672.

LaRue, J. H., R. D. Copeland, and J. Pehrson. 1983. *Growing Avocados in the San Joaquin Valley.* UC ANR Leaflet 2904.

La Vine, Paul D. 1982. *Growing Boysenberries and Ollallie Blackberries.* UC ANR Leaflet 2441.

Maas,J. L., ed. 1984. *Compendium of Strawberry Diseases.* American Phytopathological Society, St. Paul, Minn.

MacNab, A. A., A. F. Sherf, and J. K. Springer. 1983. *Identifying Diseases of Vegetables.* The Pennsylvania State University, University Park, Pa.

Metcalf, C. L., W. P. Flint, and R. L. Metcalf. 1962. *Destructive and Useful Insects.* McGraw-Hill Book Co.

Micke, W., W. Schreader, and W. Moller. 1977. *Sweet Cherries for the Home Grounds.* UC ANR Leaflet 2951.

Micke, W., and D. Kester. 1978. *Almond Orchard Management.* UC ANR Publication 4092.

Oatman, E. R., M. E. Badgley, and G. R. Platner. 1985. Predators of the two-spotted spider mite on strawberry. *California Agriculture* 89(11–12):21–23.

Oatman, Earl R. and Gary R. Platner. 1985. Biological control of two avocado pests. *California Agriculture* 39 (11–12):21–23.

Perry, Ed, and Wes Asai. 1983. *Fertilizing Home Fruit Trees and Grapevines.* UC ANR Leaflet 21329.

Pickel, C. A., and R. S. Bethell. 1985. *Apple Scab Management.* UC ANR Publication 21412.

Platt, R. G. 1981. *Micronutrient Deficiencies of Citrus.* UC ANR Leaflet 2115.

Ramos, D. E., ed. 1981. *Prune Orchard Management.* UC ANR Publication 3269.

Ramos, D. E., ed. 1985. *Walnut Orchard Management* UC ANR Publication 21410.

Sakovich, N. 1980. *Citrus Pest and Disease Control in the Home Orchard.* UC ANR Leaflet 21166.

Sims, W. L., V. E. Rubatzky, R. H. Sciaroni, and W. H. Lange. 1977. *Growing Globe Artichokes in California.* UC ANR Leaflet 2675.

Takatori, F. H., F. D. Souther, J. I. Stillman and B. Benson. 1977. *Asparagus Production in California.* UC ANR Bulletin 1881. See also General References.

Trumble, J. T., E. R. Oatman, and V. Voth. 1983. Thresholds and Sampling for Aphids in Strawberries. *California Agriculture* 37(11&12):20–21.

Ulrich, Albert, M. A. E. Mostafa, and W. W. Allen. 1980. *Strawberry Deficiency Symptoms.* UC ANR Publication 1917.

University of California, Division of Agricultural Sciences, *Edible Pod Pea Production in California.* UC ANR Leaflet 21328.

UC Statewide IPM Project. 1985 *Integrated Pest Management for Almonds.* UC ANR Publication 3308.

University of California Statewide IPM Project. 1984. *Integrated Pest Management for Citrus.* UC ANR Publication 3303.

University of California Statewide IPM Project. 1991. *Integrated Pest Management for Cole Crops and Lettuce.* UC ANR Publication 3307. Berkeley, Calif.

UC Statewide IPM Project. 1986. *Integrated Pest Management for Potatoes.* UC ANR Publication 3316.

UC Statewide IPM Project. 1998. *Integrated Pest Management for Tomatoes* (Fourth Edition) UC ANR Publication 3274.

UC Statewide IPM Project. 1987. *Integrated Pest Management for Walnuts.* UC ANR Publication 3270.

Voss, Ronald, ed. 1979. *Onion Production in California.* UC ANR Publication 4097.

Welch, N. C., J. A. Beutel, R. Bringhurst, D. Gubler, H. Otto, C. Pickel, W. Schrader, D. Shaw, and V. Voth. 1989. *Strawberry Production in California.* UC ANR Leaflet 2959.

Wilhelm, Stephen. 1961. *Diseases of Strawberry.* California Agricultural Experiment Station Circular.

Wilson, E. E., and J. M. Ogawa. 1979. *Fungal, Bacterial and Certain Nonparasitic Diseases of Fruit and Nut Crops in California.* UC ANR Publication 4090.

Zentmyer, G. A., and H. D. Ohr. 1981. *Avocado Root Rot.* UC ANR Leaflet 2440.

University of California DANR Publications* of particular interest

Grape Pest Management, Second edition. UC ANR Pub. 3343.

Grower's Weed Identification Handbook. UC ANR Pub. 4030.

Integrated Pest Management for Walnuts. UC ANR Pub. 3270.

Integrated Pest Management for Tomatoes. UC ANR Pub. 3274.

Integrated Pest Management for Citrus. UC ANR Pub. 3303.

Integrated Pest Management for Cole Crops and Lettuce. UC DANR Pub. 3307.

Integrated Pest Management for Almonds. UC ANR Pub. 3308.

Integrated Pest Management for Potatoes. UC ANR Pub. 3316.

Integrated Pest Management for Apples and Pears. UC DANR Pub. 3340.

Integrated Pest Management for Strawberries. UC DANR Pub. 3351.

Integrated Pest Management for Stone Fruits. Forthcoming.

Managing Insects and Mites with Spray Oils. 1991. UC DANR Pub. 3347.

Natural Enemies Handbook. UC ANR Pub. 3386.

Pests of Landscape Trees and Shrubs. UC ANR Pub. 3359.

The Safe and Effective Use of Pesticides. UC ANR Pub. 3324.

Wildlife Pest Control around Gardens and Homes. UC DANR Pub. 21385.

* These publications may be ordered from ANR Publications, University of California, 6701 San Pablo Avenue, Oakland, CA 94608-1239 or call 1-800-994-8849.

See also:
Pest Notes series for home gardeners†
UC IPM Pest Management Guidelines for agricultural crops†

† Copies of these publications may be obtained at California county UC Cooperative Extension offices or on the World Wide Web at http://www.ipm.ucdavis.edu

Glossary

allelopathic. the ability of a plant species to produce substances that are toxic to certain other plants.

annual. a plant that normally completes its life cycle of seed germination, vegetative growth, reproduction, and death in a single year.

antagonists. organisms that release toxins or otherwise change conditions so that activity or growth of other organisms (especially pests) is reduced.

bacterium. a unicellar microscopic plant-like organism that does not produce chlorophyll. Most bacteria obtain their nitrogen and energy from organic matter; some bacteria cause plant or animal diseases (plural: bacteria).

beneficials. organisms that provide a benefit to crop production, applied especially to natural enemies of pests and to pollinators such as bees.

biological control. the action of parasites, predators, pathogens, or competitors in maintaining another organism's density at a lower average than would occur in their absence. Biological control may occur naturally in the field or be the result of manipulation or introduction of biological control agents by people.

Bordeaux mixture. a fungicide made of a mixture of hydrated lime and copper sulfate.

botanical. derived from plants or plant parts.

broad-spectrum pesticide. a pesticide that is toxic to many different species.

bulb. a short underground stem, with many fleshy scale-leaves filled with food, often remaining viable during times when leaves die back and the rest of the plant is dormant.

canker. a dead and discolored, often sunken, area (lesion) on stem, branch, or twig of a plant.

canopy. the leafy parts of vines or trees.

caterpillars. immature forms of butterflies and moths and sawflies.

certified seed or planting stock. seeds, tubers, or young plants certified by a recognized authority to be free of or to contain less than a minimum number of specified pests or pathogens.

cole crops. any of the group of crucifer family crops that are varieties of the species *Brassica oleracea*, including cabbage, broccoli, cauliflower, and brussels sprouts.

collar region. in grasses, the region where leaf blade and sheath meet.

companion planting. the practice of planting certain plant species—often herbs—in close association with crop plants to repel pests.

cotyledons. leaves formed within the seed and present on seedlings immediately after germination; seed leaves.

cover crops. cultivation of a second type of crop primarily to improve the production system for a primary crop; examples include grasses or legumes maintained in orchards or vineyards and legume or other crops grown during the winter season to improve soil condition, fertility, or pest control for a primary annual crop grown in the summer.

crown. point at or just below the soil surface where main stem (trunk) and roots join.

degree-day. a unit combining temperature and time, used in monitoring growth or development of organisms.

disk. a type of cultivator made up of many circular blades used for weed control and soil preparation.

dormant. to become inactive during winter or periods of cold weather.

dormant spray. treatment applied during the period when trees are inactive.

entomophagenous nematodes. nematodes that eat insects.

fallow. cultivated land that is allowed to lie dormant, with no crops growing on it, during a growing season.

frass. solid fecal material produced by insect larvae.

fruiting bodies. in fungi, reproductive structures containing spores.

fungus. multicellular lower plant lacking chlorophyll, such as a mold, mildew, rust, or smut. The fungus body normally consists of filamentous strands called mycelium and reproduces through dispersal of spores (plural: fungi).

fungicide. a pesticide used for control of fungi.

gall. localized swelling or outgrowth of plant tissue, often formed in response to the action of a pathogen or other pest.

girdled. to form a ring of dead or damaged tissue around the stem or root; usually girdling kills the plant.

graft union. place where the rootstock joins the scion or top part of a grafted tree or vine.

herbicide. a pesticide used for control of weeds.

honeydew. an excretion from insects, such as aphids, mealybugs, and soft scales, consisting of modified plant sap.

horticultural oils. highly refined petroleum (or seed derived) oils that are manufactured specifically to control pests on plants.

host. a plant or animal species that provides sustenance for another organism.

inorganic. containing no carbon; generally used to indicate materials (e.g. pesticides) that are of mineral origin.

instar. the period between molts in larvae of insects. Most larvae pass through several instars; these are usually given numbers such as first instar, second instar, etc.

integrated pest management (IPM). a pest management strategy that focuses on long-term prevention or suppression of pest problems through a combination of techniques such as encouraging biological control, use of resistant varieties, or adaption of alternate cultivating, pruning, or fertilizing practices or modification of habitat to make it incompatible with pest development. Pesticides are used only when careful field monitoring indicates they are needed according to pre-established guidelines or treatment thresholds.

larvae. immature forms of insects that develop through the process of complete metamorphosis including egg, several larval stages, pupa, and adult. In mites, the first stage immatures are also called larvae.

lesion. a localized area of diseased or discolored tissue.

ligule. in many grasses, a short membranous projection on the inner side of the leaf blade at the junction where the leaf blade and leaf sheath meet.

metamorphosis. the change in form that takes place as insects grow from immatures to adults.

microbial pesticides. pesticides that consist of bacteria, fungi, viruses, or other microorganisms used for control of weeds, invertebrates, or plant pathogens.

microorganism. an organism of microscopic size, such as a bacterium, virus, fungus, viroid, or mycoplasma.

mineral oils. synonymous with horticultural oils.

molt. in insects and other anthropods, the shedding of skin before entering another stage of growth.

monitoring. the process of carefully watching the activities, growth, and development of pest organisms or other factors on a regular basis over a period of time, often utilizing very specific procedures.

mulch. a layer of material placed on the soil surface to prevent weed growth, consume moisture or regulate heat. Plant-derived or synthetic materials may be used.

mycelium. the vegetative body of a fungus, consisting of a mass of slender filaments called hyphae (plural: mycelia).

nymphs. immature forms of insects that go through gradual metamorphosis with no larval stage. Also immature forms of mites after first larval stage.

organic. a material (e.g. pesticide) whose molecules contain carbon and hydrogen atoms. Also may refer to plants or animals which are grown without the use of synthetic fertilizers or pesticides.

parasite. an organism that derives its food from the body of another organism, the host, without killing the host directly; also an insect that spends its immature stages in the body of a host that dies just before the parasite emerges (this type is also called a parasitoid).

pathogen. a microorganism that causes a disease.

perennial. a plant that lives longer than two years—some may live indefinitely. Some perennial plants lose their leaves and become dormant during winter; others may die back and resprout from underground root structures each year.

pesticide. any substance or mixture of substances intended for preventing, destroying, repelling, or mitigating any insects, rodents, nematodes, fungi, or weeds, or any other forms of life declared to be pests; and any other substance or mixture of substances intended for use as a plant regulator, defoliant, or desiccant.

pheromone. a chemical produced by an animal to attract other animals of the same species.

photosynthesis. the process by which plants convert sunlight into energy.

predator. an animal that kills other animals and feeds on them.

protective coverings. any cloth, screen, plastic or other material placed over growing plants to prevent damage by pests or harsh weather.

pupa. the nonfeeding, resting stage between larva and adult of insect species that undergo complete metamorphosis.

resistant. able to tolerate conditions (such as pesticide sprays or pest damage) harmful to other strains of the same species.

resistant varieties. strains of a plant species able to resist or tolerate damage by a pest normally damaging to that plant species.

rhizome. a horizontal underground stem, especially one that forms roots at the nodes to produce new plants.

roguing. removal of individual diseased or undesirable plants

rootstock. the lower portion of a graft union, which develops into the root system.

rotation. the practice of purposefully alternating crop species grown on the same plot of land.

row covers. any fabric or protective covering placed over rows of plants to protect them from pest damage or harsh climate.

sclerotium. a firm, compact mass of mycelium that serves as a dormant stage for some fungi (plural: sclerotia).

secondary pest outbreak. an increase in a pest's population to economically damaging levels brought on when a pest control action applied for another pest disrupts natural control.

seed leaves. the first leaf (grasses) or two leaves (broadleaf plants) on a seedling; synonymous with cotyledons.

selective pesticides. pesticides that are toxic primarily to the target pest (and perhaps a few related species), leaving most other organisms, including beneficials, unharmed.

solarization. the practice of heating up soil to levels lethal to pests through the application of clear plastic to the soil surface for 4 to 6 weeks.

spore. a reproductive structure produced by some plants and microorganisms that is resistant to environmental influences.

stolon. an aboveground runner or rooting structure found in some species of plants.

synthetic organic pesticides. synthesized pesticides with carbon and hydrogen in their basic structure.

tuber. a much enlarged, fleshy underground stem.

vegetative growth. growth of stems, roots and leaves, not flowers and fruits.

vector. an organism, such as an insect, that can transmit a pathogen to plants or animals.

virus. a very small organism that can multiply only within living cells of other organisms and is capable of producing disease symptoms in some plants and animals.

Index